Low-Cost Sensors and Biological Signals

Low-Cost Sensors and Biological Signals

Editors

Frédéric Dierick
Fabien Buisseret
Stéphanie Eggermont

MDPI • Basel • Beijing • Wuhan • Barcelona • Belgrade • Manchester • Tokyo • Cluj • Tianjin

Editors
Frédéric Dierick
Centre National De Rééducation
Fonctionnelle et de Réadaptation
Luxembourg
Université catholique de
Louvain
Belgium
CeREF-Technique
Belgium

Fabien Buisseret
CeREF-Technique
Belgium
Université de Mons—UMONS,
Research Institute for Complex
Systems
Belgium

Stéphanie Eggermont
CeREF-Technique
Belgium

Editorial Office
MDPI
St. Alban-Anlage 66
4052 Basel, Switzerland

This is a reprint of articles from the Special Issue published online in the open access journal *Sensors* (ISSN 1424-8220) (available at: https://www.mdpi.com/journal/sensors/special_issues/LCSBS).

For citation purposes, cite each article independently as indicated on the article page online and as indicated below:

LastName, A.A.; LastName, B.B.; LastName, C.C. Article Title. *Journal Name* **Year**, *Volume Number*, Page Range.

ISBN 978-3-0365-0842-9 (Hbk)
ISBN 978-3-0365-0843-6 (PDF)

Contents

About the Editors

Frédéric Dierick holds a Ph.D. in Physiotherapy and Rehabilitation (Université Catholique de Louvain (UCLouvain), Belgium, 2006). From 2003 to 2019, he worked at the Haute Ecole Louvain in Hainaut (HELHa) as a teacher-researcher and developed the Forme et Fonctionnement Humain laboratory (Physiotherapy Department, Charleroi, Belgium). In 2020, he joined the CNRFR—Rehazenter (Luxembourg, Luxembourg) as a senior researcher and manager of national and international research projects. Currently he is also project manager at CeREF (HELHa's research center). His research is devoted to human motion analysis in healthy and pathological states, especially the development of multidisciplinary methods involving low-cost sensors, medical imaging, and nonlinear time series analysis.

Fabien Buisseret has a degree in Physics and a Ph.D. in Science (University of Mons (UMONS), Belgium). After pursuing research in hadronic physics as a post-doctoral fellow at UMONS, he became a teacher at Haute Ecole Louvain in Hainaut, a researcher at CeREF (HELHa's research center) and a scientific collaborator at UMONS. His research interests include human motion analysis through mechanical, nonlinear, and sensor-based approaches, as well as theoretical and hadronic physics. He also participates in the creation and management of interdisciplinary projects for CeREF.

Stéphanie Eggermont is a civil engineer (Faculté Polytechnique de Mons). She holds a Ph.D. in engineering sciences (Université catholique de Louvain (UCLouvain), Belgium, 2013) in the field of antennas involving metamaterials. Since September 2011, she has been professor at the Haute Ecole Louvain in Hainaut and head of research of the electricity/electronics department of CeREF (HELHa research center). She is involved in various research projects aiming, in particular, at the development of intelligent and connected electronic boards and platforms in medical or cultural applications.

Editorial

Low-Cost Sensors and Biological Signals

Frédéric Dierick [1,2,3,*], Fabien Buisseret [3,4] and Stéphanie Eggermont [3]

[1] Laboratoire d'Analyse du Mouvement et de la Posture, Centre National de Rééducation Fonctionnelle et de Réadaptation-Rehazenter, 2674 Luxembourg, Luxembourg
[2] Faculty of Motor Sciences, Université Catholique de Louvain, 1348 Louvain-la-Neuve, Belgium
[3] CeREF-Technique, Chaussée de Binche 159, 7000 Mons, Belgium; buisseretf@helha.be (F.B.); stephanie.eggermont@cerisic.be (S.E.)
[4] Service de Physique Nucléaire et Subnucléaire, UMONS, Research Institute for Complex Systems, Place du Parc 20, 7000 Mons, Belgium
[*] Correspondence: frederic.dierick@gmail.com

Citation: Dierick, F.; Buisseret, F.; Eggermont, S. Low-Cost Sensors and Biological Signals. *Sensors* **2021**, *21*, 1482. https://doi.org/10.3390/s21041482

Received: 2 February 2021
Accepted: 17 February 2021
Published: 20 February 2021

Publisher's Note: MDPI stays neutral with regard to jurisdictional claims in published maps and institutional affiliations.

Low-cost sensors, i.e., sensors typically cheaper than USD 100, are currently available, allowing the measurement of a wide range of physiological signals. These signals contain valuable information that can be used to increase the understanding of any physiological function of clinical interest. Hence, low-cost sensors are expected to play a key role in the future of clinical practice and medical diagnosis. In particular, they may facilitate the collection of big data and allow broader diffusion of evidence-based medicine, which is essential to improving medical practice. Low-cost sensors may also be of interest in virtual or augmented reality applications, including rehabilitation. Their use is associated with several challenges: First, sensors should be accurate enough to unambiguously compute relevant indicators from biosignals, in particular in patients with medical conditions. Second, the designed sensors should be as non-intrusive and ready-to-use as possible with fast calibration procedures. Third, they require user-friendly and cross-platform interfaces that provide secure data storage and easy data analysis and visualization. We invited authors to submit their latest results in the field, either research articles or reviews; 12 papers were accepted for publication in this Special Issue of *Sensors*, entitled "Low-Cost Sensors and Biological Signals." They are summarized in the next paragraphs.

Low-cost sensors allow for the full monitoring of human motion. In particular, inertial measurement units (IMUs) and magnetic angular rate and gravity sensors (MARGs) are compact devices able to measure the 3D acceleration and angular speed of a given anatomical landmark with an accuracy comparable to gold-standard material only available in research centers [1]. As shown in this work through the study of a clinical test assessing neck mobility, the precision reached is sufficiently high for daily use in clinical practice. More generally, physiotherapy is a field that can benefit from such motion sensors. Cappelle et al. [2] present a low-complexity wireless motion sensor based on IMUs designed to be physiotherapist-friendly. The small size and low weight as well as the wireless data transmission are needed to reduce the impact on patient motion and to allow for easy positioning on a patient's body.

Regarding daily use, calibration has to be as fast as possible compared to the typical time a clinician spends with a patient. Accurate calibration will allow the computation of angular position from acceleration and angular velocity. Angular amplitudes are one of the most commonly used indices to assess joint mobility. Calibration procedures are presented [3] for IMUs, leading to an accuracy of less than 3.4° on lower limb amplitude measurements. These results are coherent with those of Hage et al. [1], although Hage et al. focused on the neck rather than lower limbs. In real-life situations, some perturbations cannot be avoided, which jeopardize calibration efforts, e.g., magnetic disturbances for MARGs. It may be necessary to add extra information to compensate for the perturbations. An example is given in Wöhle and Gebhard [4], who show that eye-tracking data can be used to improve the accuracy on MARG head-orientation measurements.

Once human motion is measured, it can be used as an input signal to interact with a virtual environment or with more classical videogames. An example is provided in Foreman and Engsberg [5], who show that Microsoft Kinect® is a reliable tool for assessing trunk motion. The coupling between low-cost motion sensors and serious videogames opens the possibility to innovative methods in rehabilitation. A review [6] shows that the use of videogames and motion-capture systems in rehabilitation contributes to the recovery of the patient, mostly in post-stroke rehabilitation. Sensors may be relevant not only in rehabilitation but also in helping patients to improve their motor abilities and to recover autonomy. Krasovsky et al. [7] focus on adults and children with motor impairments such as stroke or cerebral palsy. They show that a spoon instrumented with an IMU allows for a clinically feasible assessment of self-feeding.

Kinematics is obviously not the only method to assess physical activity. Two other types of biosignals are discussed [8,9]. In Tahir et al. [8], a systematic design and characterization procedure for different pressure sensors is proposed for building low-cost smart insoles for detecting vertical ground reaction force in gait analysis. In Wójcikowski and Pankiewicz [9], a new algorithm for the measurement of the human heart rate using photoplethysmography is presented. The algorithm is less demanding in computing power than many others, which is an important advantage regarding the autonomy of wearable devices.

Low-cost sensors may not only be useful in characterizing an individual's state: they can also offer ways to classify individuals in different groups. Gabis et al. [10] show that accelerometers provide enough information to discriminate between typically developed children and children with autism spectrum disorder via a simple motor task (star jump). Li et al. [11] report that the kinematic patterns measured by IMUs are significantly different between the Baduanjin teacher, senior students, and junior students: changes in kinematics are, in this case, related to one participant's experience.

A direction in which low-cost sensors may be applied is in affective technologies: biosignals measured by sensors (temperature, skin humidity, etc.) reflect the emotional state of an individual. Such signals may be communicated to a wearable device worn by another person to enhance the methods of communicating with each other [12].

Institutional Review Board Statement: Not applicable.

Informed Consent Statement: Not applicable.

Data Availability Statement: Not applicable.

Acknowledgments: The Guest Editors thank all the authors, reviewers, and members of MDPI's editorial team whose work has led to the publication of this Special Issue. Financial support from the European Regional Development Fund (Interreg FWVl NOMADe) is acknowledged.

Conflicts of Interest: The authors declare no conflict of interest.

References

1. Hage, R.; Detrembleur, C.; Dierick, F.; Pitance, L.; Jojczyk, L.; Estievenart, W.; Buisseret, F. DYSKIMOT: An Ultra-Low-Cost Inertial Sensor to Assess Head's Rotational Kinematics in Adults during the Didren-Laser Test. *Sensors* **2020**, *20*, 833. [CrossRef] [PubMed]
2. Cappelle, J.; Monteyne, L.; Van Mulders, J.; Goossens, S.; Vergauwen, M.; Van der Perre, L. Low-Complexity Design and Validation of Wireless Motion Sensor Node to Support Physiotherapy. *Sensors* **2020**, *20*, 6362. [CrossRef]
3. Lebleu, J.; Gosseye, T.; Detrembleur, C.; Mahaudens, P.; Cartiaux, O.; Penta, M. Lower Limb Kinematics Using Inertial Sensors during Locomotion: Accuracy and Reproducibility of Joint Angle Calculations with Different Sensor-to-Segment Calibrations. *Sensors* **2020**, *20*, 715. [CrossRef] [PubMed]
4. Wöhle, L.; Gebhard, M. SteadEye-Head—Improving MARG-Sensor Based Head Orientation Measurements Through Eye Tracking Data. *Sensors* **2020**, *20*, 2759. [CrossRef]
5. Foreman, M.H.; Engsberg, J.R. The Validity and Reliability of the Microsoft Kinect for Measuring Trunk Compensation during Reaching. *Sensors* **2020**, *20*, 7073. [CrossRef] [PubMed]
6. Alarcón-Aldana, A.C.; Callejas-Cuervo, M.; Bo, A.P.L. Upper Limb Physical Rehabilitation Using Serious Videogames and Motion Capture Systems: A Systematic Review. *Sensors* **2020**, *20*, 5989. [CrossRef] [PubMed]

7. Krasovsky, T.; Weiss, P.L.; Zuckerman, O.; Bar, A.; Keren-Capelovitch, T.; Friedman, J. DataSpoon: Validation of an Instrumented Spoon for Assessment of Self-Feeding. *Sensors* **2020**, *20*, 2114. [CrossRef] [PubMed]
8. Tahir, A.M.; Chowdhury, M.E.H.; Khandakar, A.; Al-Hamouz, S.; Abdalla, M.; Awadallah, S.; Reaz, M.B.I.; Al-Emadi, N. A Systematic Approach to the Design and Characterization of a Smart Insole for Detecting Vertical Ground Reaction Force (vGRF) in Gait Analysis. *Sensors* **2020**, *20*, 957. [CrossRef] [PubMed]
9. Wójcikowski, M.; Pankiewicz, B. Photoplethysmographic Time-Domain Heart Rate Measurement Algorithm for Resource-Constrained Wearable Devices and its Implementation. *Sensors* **2020**, *20*, 1783. [CrossRef] [PubMed]
10. Gabis, L.V.; Shefer, S.; Portnoy, S. Variability of Coordination in Typically Developing Children Versus Children with Autism Spectrum Disorder with and without Rhythmic Signal. *Sensors* **2020**, *20*, 2769. [CrossRef]
11. Li, H.; Khoo, S.; Yap, H.J. Differences in Motion Accuracy of Baduanjin between Novice and Senior Students on Inertial Sensor Measurement Systems. *Sensors* **2020**, *20*, 6258. [CrossRef]
12. Alfaras, M.; Primett, W.; Umair, M.; Windlin, C.; Karpashevich, P.; Chalabianloo, N.; Bowie, D.; Sas, C.; Sanches, P.; Höök, K.; et al. Biosensing and Actuation—Platforms Coupling Body Input-Output Modalities for Affective Technologies. *Sensors* **2020**, *20*, 5968. [CrossRef] [PubMed]

Article

DYSKIMOT: An Ultra-Low-Cost Inertial Sensor to Assess Head's Rotational Kinematics in Adults during the Didren-Laser Test

Renaud Hage [1,2,*], Christine Detrembleur [1], Frédéric Dierick [2,3], Laurent Pitance [1], Laurent Jojczyk [2], Wesley Estievenart [2] and Fabien Buisseret [2,4]

[1]   Laboratoire NMSK, Institut de Recherche Expérimentale et Clinique, Université Catholique de Louvain, 1200 Brussels, Belgium; christine.detrembleur@uclouvain.be (C.D.); laurent.pitance@uclouvain.be (L.P.)
[2]   CeREF, Chaussée de Binche 159, 7000 Mons, Belgium; frederic.dierick@gmail.com (F.D.); jojczykl@helha.be (L.J.); estievenartw@helha.be (W.E.); buisseretf@helha.be (F.B.)
[3]   Centre National de Rééducation Fonctionnelle et de Réadaptation—Rehazenter, Laboratoire d'Analyse du Mouvement et de la Posture (LAMP), 2674 Luxembourg, Luxembourg
[4]   Service de Physique Nucléaire et Subnucléaire, UMONS, Research Institute for Complex Systems, 20 Place du Parc, 7000 Mons, Belgium
*   Correspondence: renaud.hage@uclouvain.be

Received: 27 November 2019; Accepted: 3 February 2020; Published: 4 February 2020

Abstract: Various noninvasive measurement devices can be used to assess cervical motion. The size, complexity, and cost of gold-standard systems make them not suited to clinical practice, and actually difficult to use outside a dedicated laboratory. Nowadays, ultra-low-cost inertial measurement units are available, but without any packaging or a user-friendly interface. The so-called DYSKIMOT is a home-designed, small-sized, motion sensor based on the latter technology, aiming at being used by clinicians in "real-life situations". DYSKIMOT was compared with a gold-standard optoelectronic system (Elite). Our goal was to evaluate the DYSKIMOT accuracy in assessing fast head rotations kinematics. Kinematics was simultaneously recorded by systems during the execution of the DidRen Laser test and performed by 15 participants and nine patients. Kinematic variables were computed from the position, speed and acceleration time series. Two-way ANOVA, Passing–Bablok regressions, and dynamic time warping analysis showed good to excellent agreement between Elite and DYSKIMOT, both at the qualitative level of the time series shape and at the quantitative level of peculiar kinematical events' measured values. In conclusion, DYSKIMOT sensor is as relevant as a gold-standard system to assess kinematical features during fast head rotations in participants and patients, demonstrating its usefulness in both clinical practice and research environments.

Keywords: inertial sensor; kinematics; head rotation; ecological research

---

1. Introduction

Neck pain is a common neuromusculoskeletal symptom with a prevalence ranging from 22% to 70%, increasing with age and affecting most often women around 50 years old [1]. It is the fourth leading cause of years lived with disability in 188 countries during the period 1990–2013 [2]. Therefore, the correct identification of the source of neck pain is paramount. However, probably due to imperfect diagnosis, the majority of patients with neck pain are still nowadays called "non-specific" [3].

According to the Bayesian inference, a medical diagnosis indicates that one disorder (e.g., muscular, discogenic, lack of sensorimotor control deficits) more than another is probably the cause of a patient's symptoms, and thus, investigations are needed to reinforce or refute the hypothetical diagnosis [4]. In accordance with the literature [2,5–7], diagnoses and therapeutic interventions for neck pain should be informed using quantitative (strength and range of motion) and qualitative (sensorimotor appraisal) assessment of neck rotation. Quantitative devices have been reported to be superior to visual estimation to assess the cervical range of motion [8], the most popular method used by clinicians being goniometry [9]. Although very easy to use, goniometry has a margin of error of about 5° [9]. Moreover, maybe more important than movement amplitude [5,10], the evaluation of sensorimotor function has demonstrated its importance in developing a better understanding of the pathophysiological mechanisms associated with cervical pain [6] both in cases of specific neck pain such as traumatic neck pain [11], as well as for idiopathic neck pain [12]. Therefore, in an attempt to better define the clinical picture of patients by focusing on head movement [5,13,14] especially in axial rotation [15], clinicians show increased interest in quantitative devices that can accurately monitor movement.

Various noninvasive three-dimensional motion capture systems are used in the field of cervical research in order to evaluate kinematic variables going beyond simple range of motion such as speed, acceleration and deceleration using electrogoniometers [16], ultrasound waves [17], optical-based systems [18,19] and inertial sensors [20] and so on. Nevertheless, their dimension, complexity, and cost make such systems often difficult to use in clinical practice. The need for compact, user-friendly and low-cost measurement devices that can bring relevant information in everyday clinical practice is therefore obvious and goes beyond neck exploration, although we chose to focus on that topic in the present study.

Inertial measurement units sensors (IMUs) began to be applied to human movement before 2000 [21]. IMUs consist of accelerometers and gyroscopes which are organized in orthogonal triads in order to obtain three-dimensional kinematics [22]. Most often, IMUs are now supplemented by magnetometers and thermometers and are called MARG sensors (magnetic angular rate and gravity sensors). This technology has the advantage of not requiring external equipment such as cameras to acquire the orientation and position of the human segments, and it does not limit the subject's movement to the volume covered by the cameras. IMUs or MARGs thus seem to be the appropriate basic tool to design a device which could be easily used in a clinical and ecological environment [23,24]. Note that this technology suffers from high measurement noise and drift [25] that can mostly be cured by a Kalman filter [25]. Nowadays, the large-scale production MARGs sensors make them affordable compared to gold-standard systems but nevertheless the prices are several thousand euros (e.g., Vicon®, XSENS®). Other MARGs may indeed be bought at typical prices less than 50 €, still without any packaging nor user-friendly interface.

It is in this context that our team designed a small-sized, light, and ultra-low-cost inertial sensor called DYSKIMOT. After first laboratory tests, our goal was to evaluate the accuracy of DYSKIMOT compared to a gold-standard optoelectronic system when performing a clinical sensorimotor test developed by Hage et al. (i.e., the DidRen-Laser Test) in small groups of asymptomatic and symptomatic neck pain participants [26]. We selected different dynamic outcomes to evaluate our DYSKIMOT [27,28]: range of motion, peak speed, average speed, peak acceleration, and peak deceleration.

2. Materials and Methods

2.1. Participants

Fifteen cervical non-disabled participants (NDP) (3 females, 12 males) and 9 cervical disabled patients (DP) (4 females, 5 males) were recruited from students in University hospital and among researchers' patients to participate in this study, see Table 1. Inclusion criteria for NDP were the absence of neck pain episodes in the last 6 months and a neck disability index (NDI) [29] score of less than or equal to 8%. Inclusion criteria for the DP were a numeric pain rating scale (NPRS) equal to or greater than 3/10 [30] and an NDI > 8%. Exclusion criteria were for NDP and DP: impaired cognition, blindness, deafness, dizziness, or vestibular disorders diagnosed by a physician. Participants and patients did not exhibit any neuromusculoskeletal or neurologic disorder that could influence the performance of head rotation in the horizontal plane. The participants signed informed consent and gave permission to publish their case details. The study was approved by the local ethics committee

(Comité d'Ethique Hospitalo-Facultaire Saint-Luc-UCL (IRB 00001530)) and conducted in accordance with the declaration of Helsinki.

Table 1. Characteristics of the participants in NDP and DP groups. Data are given either under mean ± SD or median [Q1–Q3] form.

	NDP (n = 15)	DP (n = 9)
Age (years)	24 ± 3	31 ± 14
Gender (men/women)	12/3	5/4
BMI (kg/m^2)	22.2 ± 2.7	21.8 ± 2.3
NDI (%)	0 [0–0]	14 [10–16]
NPRS (/10)	0 [0–0]	3 [0–0] [1]

[1] SD = Standard Deviation, BMI = Body Mass Index, Q1 = First Quartile, Q3 = Third Quartile, NDI = Neck Disability Index. NPRS = Numeric Pain Rating Scale, NDP = Non-Disabled Participants. DP = disabled participants.

2.2. The DidRen Laser Test

The DidRen Laser Test [26] was used to assess neck mobility through standardized axial rotations of the head in NDP and DP.

After watching an explanatory video, participants sat on a chair with backrest, without armrests, placed at 90 cm from a vertical panel equipped with 3 targets (LEDs) arranged horizontally and located 52 cm apart (Figure 1). Participants wore an adjustable helmet with a laser beam attached on the top was worn by the participant (Figure 1). The experimenter (RH) adjusted the helmet so that the laser hit the central target while the participant was in a neutral position before the test began. The instructions were the same for all participants: "You must reach the targets as fast as you can and perform the head movement without moving your shoulders". The targets were then turned on in a predefined sequence and the participant's task was to rotate his/her head so that the laser beam hit the target as quickly as possible. When the laser beam was stabilized by the participant on a target for at least 0.5 s, the target LED lit up and an audible signal was emitted. A complete test was composed of 5 cycles of cervical axial rotation to the right and left sides respectively.

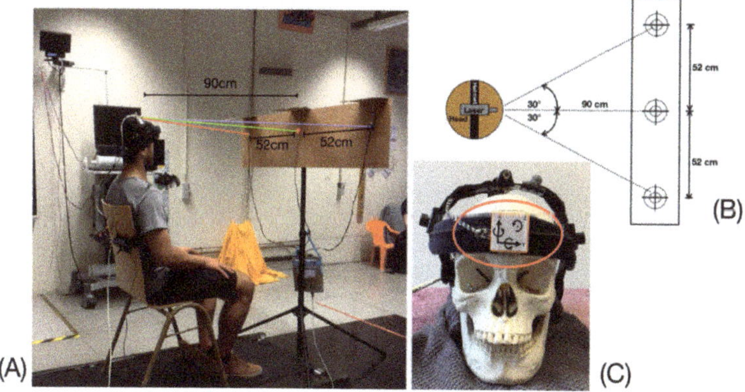

Figure 1. (**A**) DidRen Laser Test installation device. (**B**) Schematic view from above. The passage from one target to another induces an axial rotation of the head of 30° either to the left or to the right sides of the bodyline. (**C**) The Helmet worn by the participant with the Laser on the top. The DYSKIMOT sensor can be seen (red circle) at the front of the helmet.

A first test was carried out to familiarize the participant and a second for data recording and analyses [26].

2.3. Motion Sensors

2.3.1. Elite System (BTS)

An optoelectronic system composed of 8 infra-red cameras (ELITE, BTS, Milan, Italy) (Figure 2A) with sampling frequency of $\overline{f}$ = 200 Hz test carried out the three-dimensional recording of the markers on the helmet during the DidRen Laser. A kinematic model composed of 3 markers on a helmet and fixed during all experimentations representing the head was used and adapted from [31] (Figure 2B,C). Helmet markers were positioned such that one was just aside the top of the head (Top H) and positioned next to the laser, and two on each side of the head (R.H and L.H) (Figure 2C). Real time detection of head rotation markers was executed around a coordinate system such that the axis of rotation for head axial rotations was X (inferior-superior axis). The Y-axis was aligned with participant's mediolateral axis at the beginning of the test and the Z-axis was aligned with the antero-posterior axis. This is illustrated in Figure 2B. The system was previously calibrated within the infra-red camera's field of view [31] and the instantaneous X, Y, and Z coordinates of the three markers were recorded, leading to $\vec{X}_{Top\ H}$, $\vec{X}_{L.H}$, and $\vec{X}_{R.H}$. The vector $\vec{u} = \vec{X}_{Top\ H} - \frac{\vec{X}_{L.H} + \vec{X}_{RH}}{2}$ gives the orientation of antero-posterior axis (coinciding with that of the laser beam).

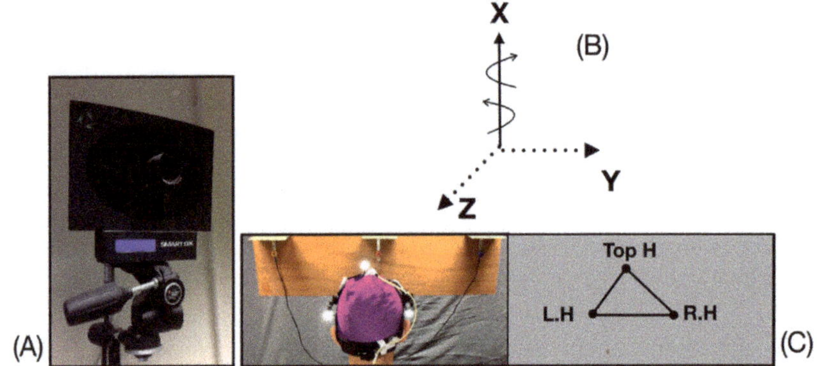

Figure 2. (**A**) Infra-red cameras (ELITE, BTS, Milan, Italy). (**B**) Head axis of rotation is denoted as X. (**C**) Placement of the reflective markers on the head.

The angular displacement time series of the head, $\overline{\theta}_i$, has been computed from the coordinates of the markers as described in details in [32]: $\overline{\theta}_i = cos^{-1}\left(\frac{\vec{u}_i \cdot \vec{u}_0}{\|\vec{u}_i\|\|\vec{u}_0\|}\right)$. The index i denotes the vector at time $i\ \overline{\Delta t}$, $\overline{\Delta t} = 1/\overline{f}$. The angular velocity was then computed as $\overline{\omega}_i = \frac{\overline{\theta}_{i+n} - \overline{\theta}_{i-n}}{2.n.\overline{\Delta t}}$ with n = 5 and, similarly, the angular acceleration has been computed as $\overline{\alpha}_i = \frac{\overline{\omega}_{i+n} - \overline{\omega}_{i-n}}{2.n.\overline{\Delta t}}$. The choice n = 5 guaranteed an optimal smoothness of the curves both for Elite and DYSKIMOT time series (assessed by visual inspection).

2.3.2. DYSKIMOT

The DYSKIMOT sensor is a MARG sensor based on the Micro-Electro-Mechanical Systems (MEMS) IMU LSM9DS1 (SparkFun, 14 €), with a mass of 10.44 gr and size of 3 × 3 cm (Figure 3A,C). It is composed of 3-axis accelerometer, gyrometer and magnetometer, plus a temperature sensor (Figure 3B). These internal components respectively measure acceleration (in [g], ±16 [g]), angular velocity (in [°/s], ±2000 [°/s]) and magnetic field (in [gauss], ±16 [gauss]).The apparatus can operate between −40 °C and +85 °C. The sensitivity depends on the sensor and on the selected range; detailed information is given in the datasheet (https://www.st.com/en/mems-and-sensors/lsm9ds1.html). For example, the gyrometer sensitivity is 8.75 10^{-3} °/s /LSB at the range ± 245 °/s, i.e., the range we use in the present study. Communication with other electronic components is made via serial peripheral interface bus (SPI) or inter-integrated circuit (I2C) protocol. The data recorded at a sampling frequency f = 100 Hz are transmitted to a PC via an Arduino Uno Rev 3 (23 €) and a USB cable (RS232 serial link). That sampling frequency was actually the maximal reachable with the devices used. The Arduino contains the data recovery program, using the SparkFun library provided for this sensor, and transfers them to a home-made acquisition software.

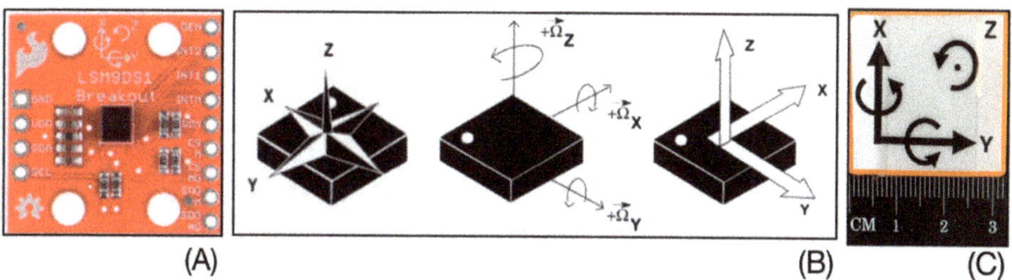

Figure 3. (**A**) Micro Electro-Mechanical Systems (MEMS). (**B**) 3-axis accelerometer, a 3-axis gyrometer, a 3-axis magnetometer and a temperature sensor. (**C**) Dimension of the DYSKIMOT.

The DYSKIMOT sensor was placed in front of the helmet (Figure 1C) with the X-axis in the vertical direction (inferior-superior axis). The Y-axis was aligned with participant's mediolateral axis at the beginning of the test and the Z-axis was aligned with the antero-posterior axis. This choice has two advantages. From a clinical point of view it is the most reliable position to record cervical axial rotation as shown in [33]. From an algorithmic point a view, the sensor orientation is such that the relevant information about the DidRen Test is fully contained in the X-component of angular velocity measured by the gyroscope. The latter time series was denoted ω_i. A trapezoidal integration gave the head's rotation angles θ_i, where the constant of integration was chosen such that the angle was zero at the beginning of the test. The derivative $\alpha_i = \frac{\omega_{i+n} - \omega_{i-n}}{2.n.\Delta t}$ with $n = 5$ and $\Delta t = 1/f$ provided the head's angular acceleration. Angles computed from the gyroscope showed a linear drift. Since the DidRen Laser Test consists of quasi-periodic rotations of 30° around a neutral position, a straightforward way of removing the drift is to subtract the least square regression line from the time series θ_i. Notice that Elite (DYSKIMOT) time series are written with (without) a bar.

Before using the DYSKIMOT in this study, a test was performed using a sensor attached to a servo motor (see Figure 4) to mimic the sequence of the cervical axial rotation during the DidRen test, i.e., angles from 30° to the left and right by going back through the 0 angle (Figure 4). The servo motor with an Arduino Uno Rev 3, was programmed to perform the sequence repeatedly. The result can be seen in Figure 5. The sensor was kept static during the first 20 s of the test. The linear drift is clearly observable on the raw angular data and the parameters of this line are computed by a least squares regression. Then comes the activation of the actuator and the beginning of the sequence (around 25 s) started. The fitted linear drift was eventually subtracted from the raw angular data. Such a procedure is satisfactory for time series displaying the typical behavior of the DidRen Laser Test (Figure 5). Such a procedure may actually work in all cases, including non-periodic tests. The regression line parameters may even be stored provided they do not change over time or with temperature. We checked that the drift stays linear at larger time scales (30 min).

2.4. Data Analysis

Signals from DYSKIMOT and Elite were synchronized by an external digital trigger (National Instrument, Austin TX, United States). Since the frequencies of both sensors were different (100 Hz vs 200 Hz), the accuracy of the synchronisation of the time series $\bar{\theta}$, $\bar{\omega}$, $\bar{\alpha}$ (Elite) and θ, ω, α (DYSKIMOT) is in the order of 5 ms.

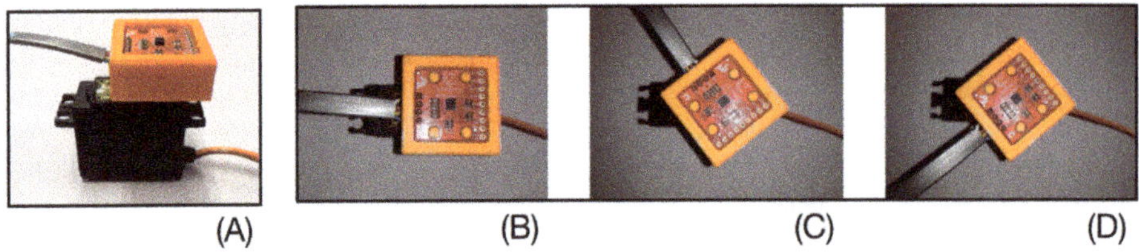

Figure 4. (**A**) Servo motor + housing adapted to its axis to fix the MARG. (**B**) Angle = 0°, (**C**) Angle = +30°, (**D**) Angle = −30°.

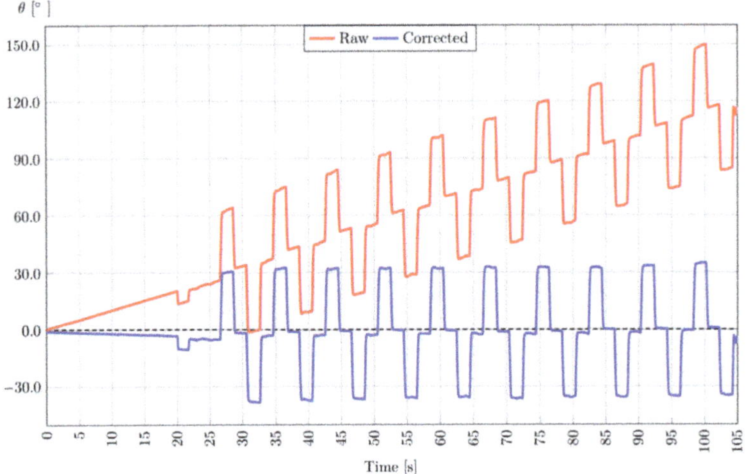

Figure 5. Example of linear drift due to integration of DYSKIMOT's raw angular velocity (red line) and correction of the drift of the test angle Z with the servo motor (blue line). The corrected angle (blue line) is obtained by subtraction of the regression line to the raw angle. This plot has been obtained by fixing the DYSKIMOT sensor on a servo motor (MG995, Tower Pro) performing successive and opposite rotations of amplitude 30°.

Then the following parameters were calculated during each cycle and averaged on the 5 cycles achieved by each participant, see Figures 6 and 7 for a graphical illustration of our computational procedure: (1) angle (range of motion, in °); (2) peak angular velocity (maximum angular velocity reached, in °s^{-1}); average angular velocity (in °s^{-1}); (3) peak angular acceleration (maximum angular acceleration reached, in °s^{-2}); (4) peak angular deceleration (minimum angular acceleration reached, in °s^{-2}). The beginning of all cycles has been manually marked by one of the authors (RH) within a homemade software that performed the averages over the 5 cycles for each trial. The peak value of a given time series X_i has been computed to be $\max(X_i)$ unless the maximal value was judged to be an artefact by visual inspection of the curves. Then, the value below this maximum was retained.

Although our goal was to measure the agreement between Elite and DYSKIMOT sensors for ND and NDP participants, the computed parameters were of clinical interest, as neck velocity during fast rotation can discriminate between nonspecific neck pain and healthy control [13,14].

A Passing–Bablok regression [34], which allows to compare the DYSKIMOT vs Elite data, was performed on the individual values of the parameters for DP and NDP simultaneously so that the agreement between both sensors could be appraised and summarized by a "calibration line".

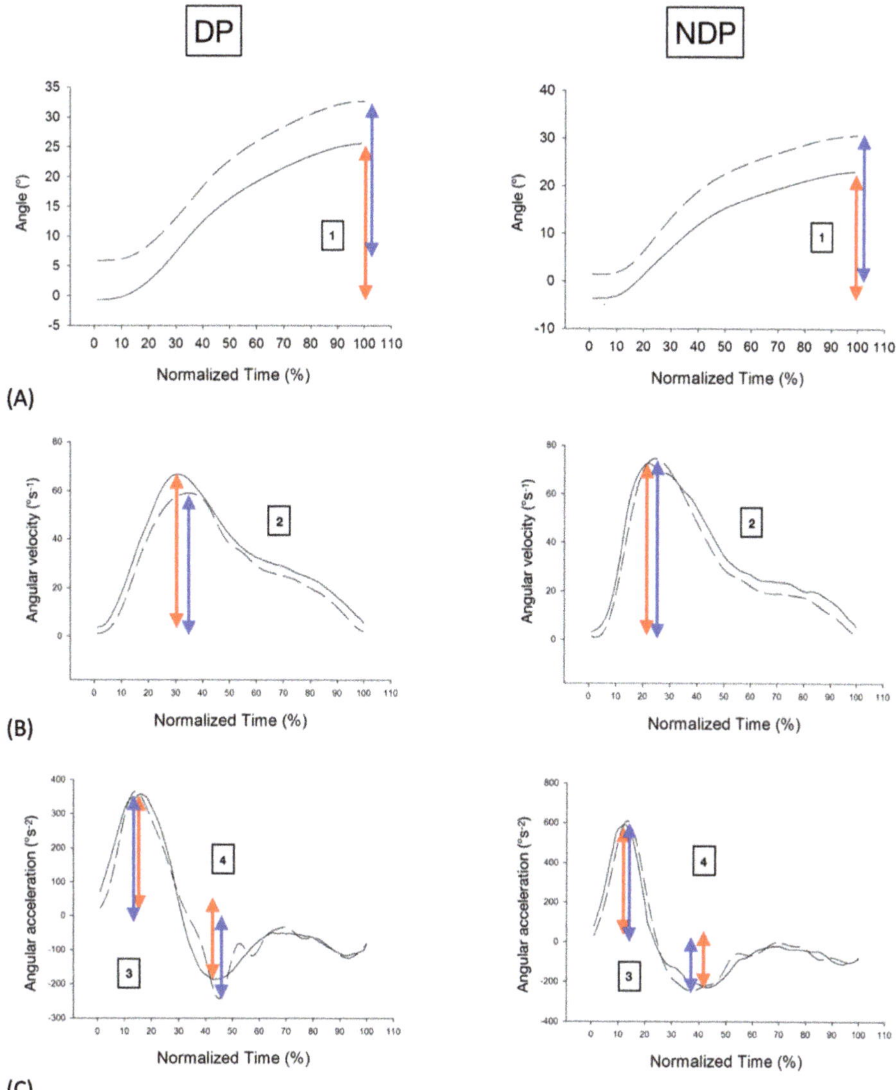

Figure 6. Typical plots of variables analyzed during one right rotation in a DP (34 years, Male, NDI = 22, NRPS = 5) and an NDP (25 years, Male, NDI = 0, NPRS = 0): (**A**) Angle; (**B**) Angular velocity; (**C**) Angular acceleration. Elite curves (dotted lines) can be compared to DYSKIMOT ones (solid lines). Computed parameters are illustrated by blue (Elite) or red (DYSKIMOT) arrows. (1) angle; (2) peak angular velocity; (3) peak angular acceleration; (4) peak angular deceleration.

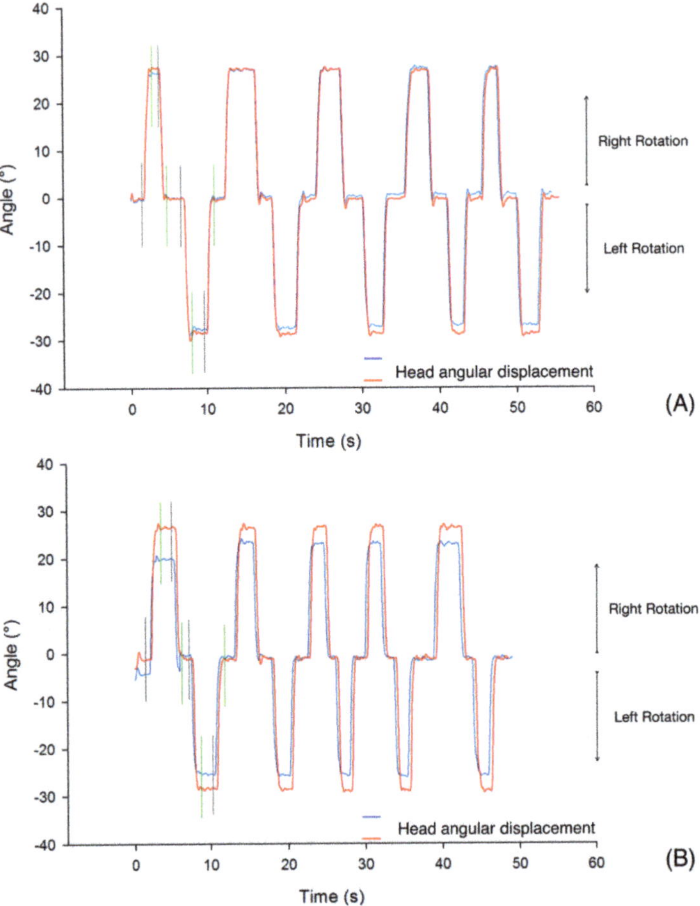

Figure 7. Typical traces of the head motion during the 5 cycles of the DidRen Laser Test showing comparison of Elite (red line) and DYSKIMOT (Blue line) angle discrepancies. In (**A**), the best angle agreement between Elite and DYSKIMOT (difference = 0.6°, mean angle during 5 cycles = 25.7°) in an DP (34 years, male, NDI=22, NPRS = 5), and in (**B**) the worst agreement (difference = 4.0°, mean angle during 5 cycles = 27.5°) in a NDP (22 years, male, NDI = 0, NPRS = 0). Cursors indicating the beginning (grey) and end (green) of one axial rotation movement are shown.

A two-way ANOVA was then used to assess potential differences between the two systems (System factor: Elite or DYSKIMOT) and between the groups (Status factor: NDP or DP) for the parameters mentioned above. When the ANOVA indicated significant interaction, a post hoc Holm-Sidak analysis with pairwise multiple comparisons was carried out. Significance was fixed at $p < 0.05$ and all statistical procedures were performed with SigmaPlot 13 (Systat Software, Inc).

Finally, a dynamic time warping (DTW) analysis (without windowing) was carried out on the z-normalized data and the Euclidian DTW distances between the time series angle $\left(\theta, \overline{\theta}\right)$, angular velocity $(\omega, \overline{\omega})$, angular acceleration $(\alpha, \overline{\alpha})$ for DYSKIMOT and Elite were calculated for all the participants and then averaged. The z-normalization consisted in replacing a time series X by $\frac{X - E(X)}{SD(X)}$, E and SD denoting the average and standard deviation respectively.

The Passing–Bablok regressions and the DTW were performed by using R v3.4.2 and the packages mcr and dtw.

It is worth saying that the accuracy of synchronization is not a matter of concern: the parameters have been independently computed from Elite and DYSKIMOT time series, and no locality constraint has been added in the DTW procedure through a window parameter. Synchronization was mainly a facilitating tool for graphical exploration of the data.

3. Results

A total of twenty-four participants were recruited with their demographics charateristics detailed in Table 1.

Results of the two-way ANOVA are shown in Table 2. A statistically significant difference was observed in the angle ($p < 0.001$) and average angular velocity ($p < 0.022$) for the System factor, i.e., Elite and DYSKIMOT lead to different means. Neither the status factor nor the interaction effects lead to statistically significant differences.

Table 2. Results of the two-way ANOVA performed on the parameters. p values are given for the differences between Elite and DYSKIMOT (System), between DP and NDP (Status) and for the interaction effect System x Status. p values lower than 0.05 are given in bold font.

	ANOVA	**Difference of the Means (Dyskimot-Elite or DP-NDP)**	p
Angle (°)	System	1.76	**<0.001**
	Status	−0.398	0.157
	System x Status		0.094
Average angular velocity ($°s^{-1}$)	System	−5.50	**0.022**
	Status	1.72	0.462
	System x Status		0.655
Peak angular velocity ($°s^{-1}$)	System	5.74	0.498
	Status	6.22	0.630
	System x Status		0.708
Peak angular acceleration ($°s^{-2}$)	System	89.9	0.282
	Status	59.8	0.473
	System x Status		0.880
Peak angular deceleration ($°s^{-2}$)	System	−81.3	0.261
	Status	−46.6	0.517
	System x Status		0.955

The results of Passing–Bablok regressions are shown in Table 3. The slopes were close to 1, with the best agreement observed for the peak angular velocity, and Pearson's coefficients range from 0.431 to 0.922, i.e., there is a moderate to excellent linear correlation between DYSKIMOT and Elite results. The agreement between both systems can be graphically appraised in Figure 8. Angles show the poorest linear correlation, resulting in a large uncertainty in the best fit (large 95% confidence interval). The other parameters show better linear correlation, and the best fit is known with better accuracy (smaller 95% confidence intervals).

Table 3. Results of Passing–Bablok regressions performed on the computed parameters. Slope and Offset are given with their 95% confidence intervals (between brackets).

	Slope	**Offset**	**r**
Angle (°)	0.908 [−2.09, 1.86]	4.34 [−20.6, 82.5]	0.431
Average angular velocity ($°s^{-1}$)	0.922 [0.713, 1.32]	−0.518 [−18.0, 7.57]	0.694
Peak angular velocity ($°s^{-1}$)	1.01 [0.942, 1.11]	2.80 [−5.27, 10.2]	0.906
Peak angular acceleration ($°s^{-2}$)	1.10 [0.939, 1.20]	7.71 [−50.4, 125]	0.922
Peak angular deceleration ($°s^{-2}$)	1.04 [0.906, 1.13]	43.1 [−0.623, 131]	0.918

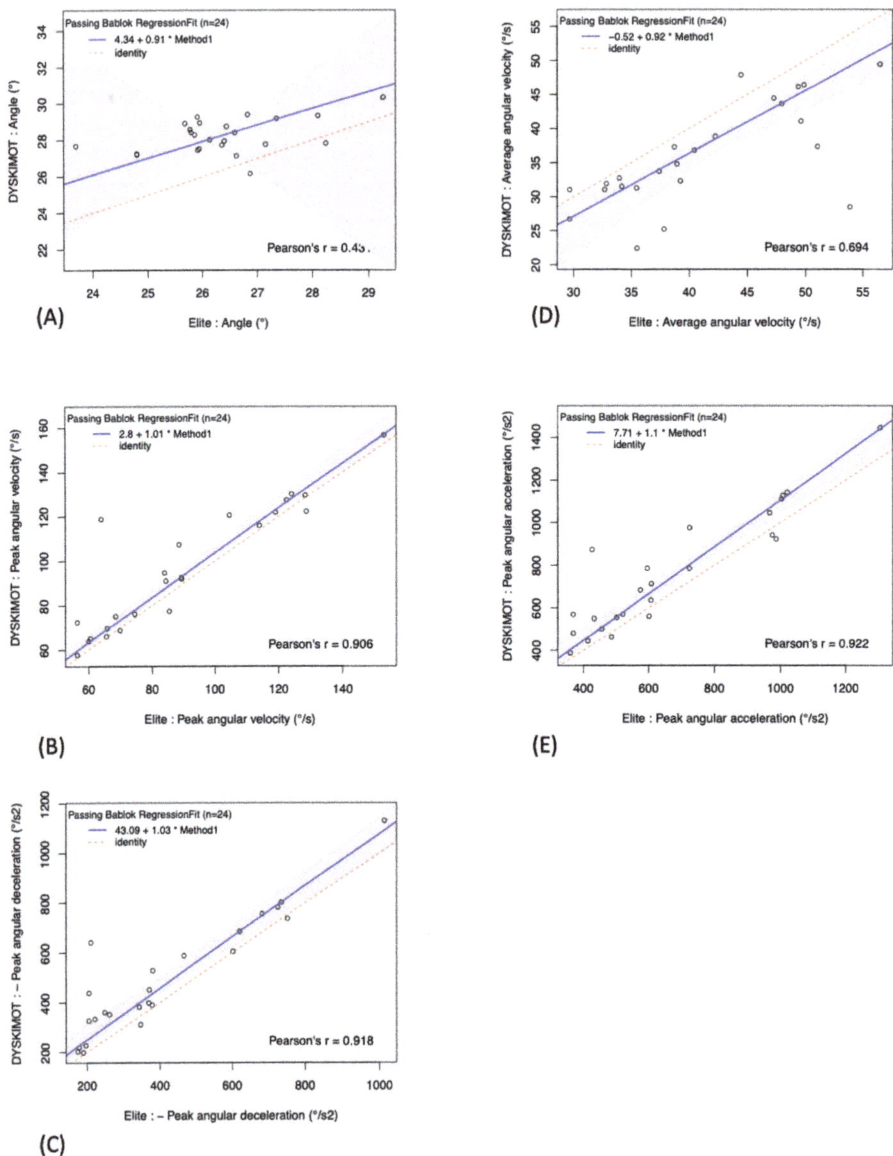

Figure 8. Passing-Bablok regressions showing the individual parameters computed from the DYSKIMOT and Elite data (points): (**A**) Angle, (**B**) Peak angular velocity, (**C**) Peak angular deceleration, (**D**) Average angular velocity, (**E**) Peak angular acceleration. The regression line (solid line) is given and compared to the identity line (dotted line). The 95% confidence interval for the linear fit is also displayed (colored band).

The DTW distance (d) allows for an estimation of the closeness of DYSKIMOT and Elite systems for the whole time series, not only for selected valued. We obtained 5.16 ± 2.68, 8.82 ± 5.80, and 14.40 ± 7.14 for angle, angular velocity, and angular acceleration time series respectively. Typical traces are shown in Figure 9.

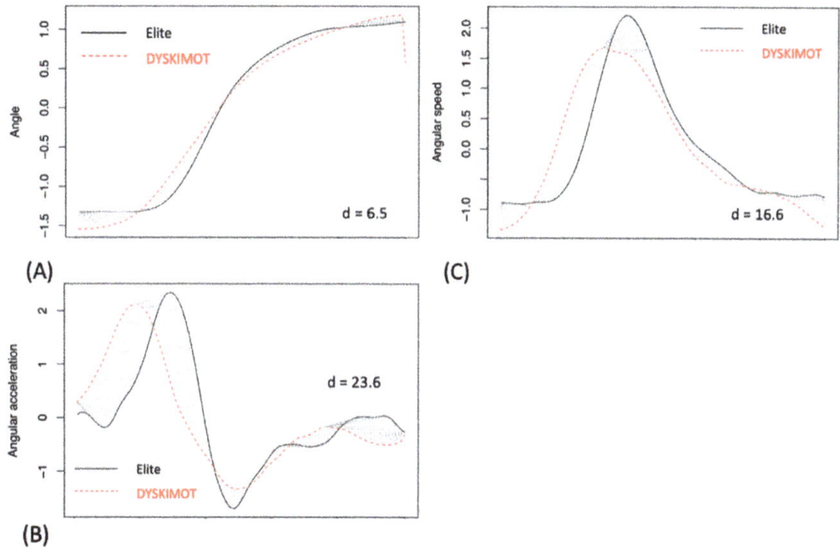

Figure 9. Typical plots of DTW matching between Elite (black solid lines) and DYSKIMOT (red dashed lines) analyzed for z-normalized data in one NDP (22 years, Male, NDI = 0, NRPS = 0) participant: (**A**) Angle, (**B**) Angular acceleration, (**C**) Angular speed. The DTW distance is added for completeness.

4. Discussion

Before comparing Elite and DYKIMOT measurements we recall their main features in Table 4.

Table 4. Summary of Elite and DYSKIMOT main features relative to the present study.

Elite		DYSKIMOT	
Infrared digital cameras	8	MARG sensor	IMU LSM9DS1
Resolution	1.5 Mpixel	Gyrometer range	$\pm 245\ °/s$
Sample frequency	200 Hz	Sample frequency	100 Hz
Accuracy/volume	<0.1 mm on $4 \times 3 \times 3$ m	Gyrometer sensitivity	$8.75\ 10^{-3}\ °/s$

The two-way ANOVA revealed that angle and average angular velocity were significantly different between Elite and DYSKIMOT systems. The difference in angle (1.76°) between the two systems is lower than the standard clinical angle evaluation of 5° reported via classical goniometry [9]. Such difference between the two systems is not clinically relevant, as an error of 2° is acceptable in most clinical situations [35]. Concerning the average angular velocity, the difference of $-5.50\ °\text{s}^{-1}$ may come from the errors induced by the derivation of the Elite position. This result is lower than the difference necessary to detect significant differences ($7.1\ °\text{s}^{-1}$) between adults and children [28]. Nevertheless, the clinical significance of a such difference is currently unknown. Apart from these differences, two other ANOVA results may be noted. First, as in [36], no signifiant difference between DP and NDP were observed for all variables studied. Second, the interactions effects (System x Status) did not induce significant differences. At this stage, DYSKIMOT and Elite give broadly similar results, but the computed parameters do not allow to distinguish between DP and NDP, at least in our population. Another point preventing the separation of DP and NDP is that the differences of the means were generally larger for the system factor than for the status factor: The accuracy of the DYSKIMOT device has to be improved, e.g., by appropriate filtering of the raw data and a better integration of gyrometer data, to reduce these discrepancies and improve the diagnostic ability of the sensor.

In a classic way [37–39], we have previously used Bland and Altman's method to evaluate the agreement between DYSKIMOT and Elite [40]. It appeared that the Bland and Altman's plots did not show a trend with the mean values of the measurements. The Bland and Altman plot for the angle parameter showed more points outside or close to the limits of agreement than the other plots, which is an indication that agreement between both systems is less obvious for the range of motion than for other parameters [40]. Since this method does not provide any quantified results on the comparison and leaves the user to decide whether this agreement is clinically acceptable or not, we analyzed the agreement between DYSKIMOT and Elite using Passing–Bablok regressions. The Passing–Bablok regression method is a non-parametric method for estimating the slope and the intercept of the linear relationship between two compared [34]. These two parameters are valued by medians and are less sensitive to extreme data and not making assumptions about errors distribution [41]. In our results, the Passing–Bablok indicated that the link between same parameters computed from both systems was well compatible with a linear shape ($r = 0.694$ to 0.922) for all parameters but angle, for which Pearson's coefficient was rather weak ($r = 0.431$) [42]. Nonzero offsets were observed but the 95% confidence intervals were large and always contained 0 value, while the slopes were close to 1 (up to 10% accuracy) with 95% confidence intervals always containing the value 1. Another advantage to this method is that by assuming that Elite results are gold-standard values, the Passing–Bablok regressions could be used to convert measured parameters with DYSKIMOT into "exact values" which are the Elite ones.

Although DTW has been known in the field of acoustic signal comparison [43], it has also been proposed for the purposes of similarity analysis during the functional pattern of gait [44], but never to compare motion neck signals obtained by two different devices. DTW is, by definition, sensitive for measuring two sequences with different lengths using dynamic programming [45]. In this work, the DTW distance between Elite and DYSKIMOT curves was adopted as an indicator of the similarity (up to an affine transformation) between the curves. In other words, the question was: Do both systems measures the same qualitative behaviors in position, angular velocity, and angular acceleration? Although angle measurements displayed a poor agreement between both systems, the DTW distance between DYSKIMOT and Elite angle was minimal: This result was expected since the structure of angle was simpler than angular velocity and angular acceleration. The DTW distance then increases between the angular velocity and the angular acceleration of the DYSKIMOT and Elite systems. This mostly results in the noise induced by the successive derivations, showing that qualitative features of these curves, especially the angular acceleration, should be interpreted carefully and might be artefacts of the sensor used.

The identification of particular kinematic events is relevant for the clinical assessment of patients, but the global shape of time series may contain more information of clinical interest. In our case for example, it is known that patients with neck pain have poorer sensory-motor control with open eyes, characterized by an increase in joint positioning error and a decrease in speed and acceleration during all movements [12]. The absence of difference in our kinematics data between patients and participants could seem unexpected as previous studies showed significant differences in terms of kinematics [5,46]. However this absence of difference could be explained by our sample size, resulted in low power, and by the difficulty for the DidRen laser to discriminate between such groups [47,48].

An obvious limitation of the present study is that we restricted our comparison of Elite and DYSKIMOT to cervical movements, while potential clinical applications may involve any other joints. Another limitation is that we used "naïve" drift correction following data acquisitions, which had to be implemented in real-time in the software. The Arduino prevented us to reach the desired frequency of 100 Hz with real-time complex filters like Kalman or Mahony. A future development would be the replacement of the Arduino by a slightly more expensive controller (ARM, 30 €), that will allow for real-time filtering and eventually for real-time angular data visualization without entailing too much the low-cost aspect of the DYSKIMOT project. It is therefore obvious that the presented experiments were carried out with a non-user-friendly interface, particularly because of the drift-related problems.

However, as the goal of this study was to evaluate the accuracy of a device that could be used by clinicians in clinical practice, we have chosen to leave this concern for future works. A user-friendly interface is currently under development.

In conclusion, the DYSKIMOT-based analysis system compares fairly well to a gold-standard optoelectronic system (Elite) up to linear errors. This ultra-low-cost sensor is recommended for clinical use as it provides more accurate information than the commonly used systems in clinical practice.

Author Contributions: Conceptualization, F.B., R.H. and F.D.; methodology, F.B., F.D. and R.H.; software, F.B., C.D., L.J., R.H. and W.E.; validation, W.E., L.J. and R.H.; formal analysis, F.B., R.H., F.D. and C.D.; investigation, R.H.; resources, R.H.; data curation, R.H. and F.B.; writing—original draft preparation, F.B., R.H. and F.D.; writing—review and editing, F.B., R.H., F.D. and C.D.; visualization, F.B.; supervision, L.P., C.D.; project administration, L.P., C.D.; funding acquisition, F.B. and F.D. All authors have read and agreed to the published version of the manuscript.

Funding: CeREF acknowledges financial support from the First Haute-Ecole programme, project no 1610401, DYSKIMOT, in partnership with OMT-Skills (http://omtskills.be/), https://www.cerisic.be/technique/projet-cerisic/developpement-dun-systeme-multitaches-immersif-et-low-cost-denregistrement-et-analyse-de-donnees-cinematiques-en-vue-de-levaluation-de-dyskinesies-motrices-et-de-leur-prise/.

Acknowledgments: We thank Stanley Lognoul and Alexandre Simeoni for their help in taking the measurements. We also thank Jean-Michel Brismée for his careful reading of the manuscript.

Conflicts of Interest: The authors declare no conflict of interest. The funders had no role in the design of the study; in the collection, analyses, or interpretation of data; in the writing of the manuscript, or in the decision to publish the results.

References

1. Blanpied, P.R.; Gross, A.R.; Elliott, J.M.; Devaney, L.L.; Clewley, D.; Walton, D.M.; Sparks, C.; Robertson, E.K. Neck Pain: Revision 2017. *J. Orthop. Sport. Phys. Ther.* **2017**, *47*, A1–A83. [CrossRef] [PubMed]
2. Vos, T.; Barber, R.M.; Bell, B.; Bertozzi-Villa, A.; Biryukov, S.; Bolliger, I.; Charlson, F.; Davis, A.; Degenhardt, L.; Dicker, D.; et al. Global, regional, and national incidence, prevalence, and years lived with disability for 301 acute and chronic diseases and injuries in 188 countries, 1990–2013: A systematic analysis for the Global Burden of Disease Study 2013. *Lancet* **2015**, *386*, 743–800. [CrossRef]
3. Coulter, I.D.; Crawford, C.; Vernon, H.; Hurwitz, E.L.; Khorsan, R.; Booth, M.S.; Herman, P.M. Manipulation and Mobilization for Treating Chronic Nonspecific Neck Pain: A Systematic Review and Meta-Analysis for an Appropriateness Panel. *Pain Physician* **2019**, *22*, E55–E70. [PubMed]
4. Bays, P.M.; Wolpert, D.M. Computational principles of sensorimotor control that minimize uncertainty and variability. *J. Physiol.* **2007**, *578 Pt 2*, 387–396. [CrossRef]
5. Roijezon, U.; Djupsjobacka, M.; Bjorklund, M.; Hager-Ross, C.; Grip, H.; Liebermann, D.G. Kinematics of fast cervical rotations in persons with chronic neck pain: A cross-sectional and reliability study. *BMC Musculoskelet. Disord.* **2010**, *11*, 222. [CrossRef]
6. Sjolander, P.; Michaelson, P.; Jaric, S.; Djupsjobacka, M. Sensorimotor disturbances in chronic neck pain–range of motion, peak velocity, smoothness of movement, and repositioning acuity. *Man. Ther.* **2008**, *13*, 122–131. [CrossRef]
7. Stenneberg, M.S.; Rood, M.; de Bie, R.; Schmitt, M.A.; Cattrysse, E.; Scholten-Peeters, G.G. To What Degree Does Active Cervical Range of Motion Differ Between Patients With Neck Pain, Patients With Whiplash, and Those Without Neck Pain? A Systematic Review and Meta-Analysis. *Arch. Phys. Med. Rehabil.* **2017**, *98*, 1407–1434. [CrossRef] [PubMed]
8. Lemeunier, N.; Jeoun, E.B.; Suri, M.; Tuff, T.; Shearer, H.; Mior, S.; Wong, J.J.; da Silva-Oolup, S.; Torres, P.; D'Silva, C.; et al. Reliability and validity of clinical tests to assess posture, pain location, and cervical spine mobility in adults with neck pain and its associated disorders: Part 4. A systematic review from the cervical assessment and diagnosis research evaluation (CADRE) collaboration. *Musculoskelet. Sci. Pract.* **2018**, *38*, 128–147. [CrossRef]
9. Youdas, J.W.; Carey, J.R.; Garrett, T.R. Reliability of measurements of cervical spine range of motion-comparison of three methods. *Phys. Ther.* **1991**, *71*, 98–104. [CrossRef]
10. Bonnechere, B.; Salvia, P.; Dugailly, P.M.; Maroye, L.; Van Geyt, B.; Feipel, V. Influence of movement speed on cervical range of motion. *Eur. Spine J.* **2014**, *23*, 1688–1693. [CrossRef]

11. Treleaven, J. Dizziness, Unsteadiness, Visual Disturbances, and Sensorimotor Control in Traumatic Neck Pain. *J. Orthop. Sport. Phys. Ther.* **2017**, *47*, 492–502. [CrossRef]

12. de Zoete, R.M.; Osmotherly, P.G.; Rivett, D.A.; Farrell, S.F.; Snodgrass, S.J. Sensorimotor control in individuals with idiopathic neck pain and healthy individuals: A systematic review and meta-analysis. *Arch. Phys. Med. Rehabil.* **2016**. [CrossRef]

13. Roijezon, U.; Clark, N.C.; Treleaven, J. Proprioception in musculoskeletal rehabilitation. Part 1: Basic science and principles of assessment and clinical interventions. *Man. Ther.* **2015**, *20*, 368–377. [CrossRef] [PubMed]

14. Sarig Bahat, H.; Weiss, P.L.; Sprecher, E.; Krasovsky, A.; Laufer, Y. Do neck kinematics correlate with pain intensity, neck disability or with fear of motion? *Man. Ther.* **2014**, *19*, 252–258. [CrossRef]

15. Dugailly, P.M.; Coucke, A.; Salem, W.; Feipel, V. Assessment of cervical stiffness in axial rotation among chronic neck pain patients: A trial in the framework of a non-manipulative osteopathic management. *Clin. Biomech.* **2018**, *53*, 65–71. [CrossRef] [PubMed]

16. Dugailly, P.M.; De Santis, R.; Tits, M.; Sobczak, S.; Vigne, A.; Feipel, V. Head repositioning accuracy in patients with neck pain and asymptomatic subjects: Concurrent validity, influence of motion speed, motion direction and target distance. *Eur. Spine J.* **2015**, *24*, 2885–2891. [CrossRef]

17. Michiels, S.; Hallemans, A.; Van de Heyning, P.; Truijen, S.; Stassijns, G.; Wuyts, F.; De Hertogh, W. Measurement of cervical sensorimotor control: The reliability of a continuous linear movement test. *Man. Ther.* **2014**, *19*, 399–404. [CrossRef]

18. Hage, R.; Dierick, F.; Roussel, N.; Pitance, L.; Detrembleur, C. Age-related kinematic performance should be considered during fast head-neck rotation target task in individuals aged from 8 to 85 years old. *PeerJ* **2019**, *7*, e7095. [CrossRef]

19. Nagai, T.; Clark, N.C.; Abt, J.P.; Sell, T.C.; Heebner, N.R.; Smalley, B.W.; Wirt, M.D.; Lephart, S.M. The Effect of Target Position on the Accuracy of Cervical-Spine-Rotation Active Joint-Position Sense. *J. Sport Rehabil.* **2016**, *25*, 58–63. [CrossRef]

20. Zhou, Y.; Loh, E.; Dickey, J.P.; Walton, D.M.; Trejos, A.L. Development of the circumduction metric for identification of cervical motion impairment. *J. Rehabil. Assist. Technol. Eng.* **2018**, *5*. [CrossRef]

21. Willemsen, A.T.; van Alste, J.A.; Boom, H.B. Real-time gait assessment utilizing a new way of accelerometry. *J. Biomech.* **1990**, *23*, 859–863. [CrossRef]

22. Picerno, P. 25 years of lower limb joint kinematics by using inertial and magnetic sensors: A review of methodological approaches. *Gait Posture* **2017**, *51*, 239–246. [CrossRef]

23. Robert-Lachaine, X.; Mecheri, H.; Larue, C.; Plamondon, A. Validation of inertial measurement units with an optoelectronic system for whole-body motion analysis. *Med. Biol. Eng. Comput.* **2017**, *55*, 609–619. [CrossRef]

24. Boissy, P.; Briere, S.; Hamel, M.; Jog, M.; Speechley, M.; Karelis, A.; Frank, J.; Vincent, C.; Edwards, R.; Duval, C. Wireless inertial measurement unit with GPS (WIMU-GPS)–wearable monitoring platform for ecological assessment of lifespace and mobility in aging and disease. In Proceedings of the 2011 Annual International Conference of the IEEE Engineering in Medicine and Biology Society IEEE Engineering in Medicine and Biology Society Annual Conference, Boston, MA, USA, 30 August–3 September 2011; pp. 5815–5819. [CrossRef]

25. Szczesna, A.; Skurowski, P.; Lach, E.; Pruszowski, P.; Peszor, D.; Paszkuta, M.; Stupik, J.; Lebek, K.; Janiak, M.; Polanski, A.; et al. Inertial Motion Capture Costume Design Study. *Sensors* **2017**, *17*, 612. [CrossRef] [PubMed]

26. Hage, R.; Ancenay, E. Identification of a relationship between cervical spine function and rotational movement control. *Ann. Phys. Rehabil. Med.* **2009**, *52*, 653–667. [CrossRef] [PubMed]

27. Sarig Bahat, H.; Weiss, P.L.; Laufer, Y. The effect of neck pain on cervical kinematics, as assessed in a virtual environment. *Arch. Phys. Med. Rehabil.* **2010**, *91*, 1884–1890. [CrossRef]

28. Hage, R.; Buisseret, F.; Pitance, L.; Brismee, J.M.; Detrembleur, C.; Dierick, F. Head-neck rotational movements using DidRen laser test indicate children and seniors' lower performance. *PLoS ONE* **2019**, *14*, e0219515. [CrossRef]

29. Vernon, H.; Mior, S. The Neck Disability Index: A study of reliability and validity. *J. Manip. Physiol. Ther.* **1991**, *14*, 409–415.

30. Meisingset, I.; Stensdotter, A.K.; Woodhouse, A.; Vasseljen, O. Neck motion, motor control, pain and disability: A longitudinal study of associations in neck pain patients in physiotherapy treatment. *Man. Ther.* **2016**, *22*, 94–100. [CrossRef]

31. Bulgheroni, M.V.; Antonaci, F.; Ghirmai, S.; Sandrini, G.; Nappi, G.; Pedotti, A. A 3D kinematic method for evaluating voluntary movements of the cervical spine in humans. *Funct. Neurol.* **1998**, *13*, 239–245.

32. Davis, R.B., III; Õunpuu, S.; Tyburski, D.; Gage, J.R. A gait analysis data collection and reduction technique. *Hum. Mov. Sci.* **1991**, *10*, 575–587. [CrossRef]

33. Theobald, P.S.; Jones, M.D.; Williams, J.M. Do inertial sensors represent a viable method to reliably measure cervical spine range of motion? *Man. Ther.* **2012**, *17*, 92–96. [CrossRef]

34. Bilic-Zulle, L. Comparison of methods: Passing and Bablok regression. *Biochem. Med.* **2011**, *21*, 49–52. [CrossRef] [PubMed]

35. Cuesta-Vargas, A.I.; Galan-Mercant, A.; Williams, J.M. The use of inertial sensors system for human motion analysis. *Phys. Ther. Rev.* **2010**, *15*, 462–473. [CrossRef] [PubMed]

36. Duc, C.; Salvia, P.; Lubansu, A.; Feipel, V.; Aminian, K. A wearable inertial system to assess the cervical spine mobility: Comparison with an optoelectronic-based motion capture evaluation. *Med. Eng. Phys.* **2014**, *36*, 49–56. [CrossRef] [PubMed]

37. Teufl, W.; Miezal, M.; Taetz, B.; Frohlich, M.; Bleser, G. Validity, Test-Retest Reliability and Long-Term Stability of Magnetometer Free Inertial Sensor Based 3D Joint Kinematics. *Sensors* **2018**, *18*, 1980. [CrossRef] [PubMed]

38. Saber-Sheikh, K.; Bryant, E.C.; Glazzard, C.; Hamel, A.; Lee, R.Y. Feasibility of using inertial sensors to assess human movement. *Man. Ther.* **2010**, *15*, 122–125. [CrossRef]

39. Bolink, S.A.; Naisas, H.; Senden, R.; Essers, H.; Heyligers, I.C.; Meijer, K.; Grimm, B. Validity of an inertial measurement unit to assess pelvic orientation angles during gait, sit-stand transfers and step-up transfers: Comparison with an optoelectronic motion capture system. *Med. Eng. Phys.* **2016**, *38*, 225–231. [CrossRef]

40. Hage, R.; Lognoul, S.; Siméoni, A.; Fourré, A.; Detrembleur, C.; Dierick, F. An Ultra low-Cost Inertial Sensor is Able To Assess Neck's Kinematics In Non-Disabled And Slighltly-Disabled Adults During The DidRen-Laser Test. In Proceedings of the Congrès ECMT Antwerpen, Antwerpen, Belgium, 20–21 September 2019.

41. Fuhrman, C.; Chouaid, C. Concordance between two variables: Numerical approaches (qualitative observations—The kappa coefficient-; quantitative measures. *Rev. des Mal. Respir.* **2004**, *21*, 123–125. [CrossRef]

42. Grenier, B.; Dubreuil, M.; Journois, D. Comparison of two measurement methods: the Bland and Altman assessment. *Ann. Fr. D'anesth. et de Reanim.* **2000**, *19*, 128–135. [CrossRef]

43. Rosa, M.; Fugmann, E.; Pinto, G.; Nunes, M. An anchored dynamic time-warping for alignment and comparison of swallowing acoustic signals. In Proceedings of the 2017 Annual International Conference of the IEEE Engineering in Medicine and Biology Society IEEE Engineering in Medicine and Biology Society Annual Conference, Seogwipo, Korea, 11–15 July 2017; pp. 2749–2752. [CrossRef]

44. Lee, H.S. Application of dynamic time warping algorithm for pattern similarity of gait. *J. Exerc. Rehabil.* **2019**, *15*, 526–530. [CrossRef] [PubMed]

45. Yang, C.Y.; Chen, P.Y.; Wen, T.J.; Jan, G.E. IMU Consensus Exception Detection with Dynamic Time Warping-A Comparative Approach. *Sensors* **2019**, *19*, 2237. [CrossRef] [PubMed]

46. Sarig Bahat, H.; Chen, X.; Reznik, D.; Kodesh, E.; Treleaven, J. Interactive cervical motion kinematics: Sensitivity, specificity and clinically significant values for identifying kinematic impairments in patients with chronic neck pain. *Man. Ther.* **2015**, *20*, 295–302. [CrossRef] [PubMed]

47. de Zoete, R.M.J.; Osmotherly, P.G.; Rivett, D.A.; Snodgrass, S.J. No Differences Between Individuals With Chronic Idiopathic Neck Pain and Asymptomatic Individuals on Seven Cervical Sensorimotor Control Tests: A Cross-Sectional Study. *J. Orthop. Sport. Phys. Ther.* **2019**, 1–37. [CrossRef] [PubMed]

48. Tsang, S.M.; Szeto, G.P.; Lee, R.Y. Movement coordination and differential kinematics of the cervical and thoracic spines in people with chronic neck pain. *Clin. Biomech.* **2013**, *28*, 610–617. [CrossRef] [PubMed]

Article

Low-Complexity Design and Validation of Wireless Motion Sensor Node to Support Physiotherapy

Jona Cappelle [1,*], **Laura Monteyne** [1], **Jarne Van Mulders** [1], **Sarah Goossens** [1], **Maarten Vergauwen** [2] **and Liesbet Van der Perre** [1]

[1] DRAMCO, Department of Electrical Engineering, Ghent Technology Campus, Katholieke Universiteit Leuven, B-9000 Ghent, Belgium; laura.monteyne@kuleuven.be (L.M.); jarne.vanmulders@kuleuven.be (J.V.M.); sarah.goossens@kuleuven.be (S.G.); liesbet.vanderperre@kuleuven.be (L.V.d.P.)

[2] Department of Civil Engineering, Geomatics Section, Ghent Technology Campus, Katholieke Universiteit Leuven, B-9000 Ghent, Belgium; maarten.vergauwen@kuleuven.be

* Correspondence: jona.cappelle@kuleuven.be

Received: 29 September 2020; Accepted: 4 November 2020; Published: 7 November 2020

Abstract: We present a motion sensor node to support physiotherapy, based on an Inertial Measurement Unit (IMU). The node has wireless interfaces for both data exchange and charging, and is built based on commodity components. It hence provides an affordable solution with a low threshold to technology adoption. We share the hardware design and explain the calibration and validation procedures. The sensor node has an autonomy of 28 h in operation and a standby time of 8 months. On-device sensor fusion yields static results of on average 3.28° with a drift of 2° per half hour. The final prototype weighs 38 g and measures ø6 cm × 1.5 cm. The resulting motion sensor node presents an easy to use device for both live monitoring of movements as well as interpreting the data afterward. It opens opportunities to support and follow up treatment in medical cabinets as well as remotely.

Keywords: physiotherapy; e-health; motion sensing; wireless charging; wireless connectivity; low power

1. Introduction

Context: evolution in physiotherapy. In the last few decades, physiotherapy has expanded from focusing on physical treatment solely with massage and stretching to a broader health context. Common treatments at a physiotherapist's practice nowadays are for example post-operative rehabilitation, neurological injury treatment, occupational injury prevention, etc. Not only the field of application has developed, but also physical treatment techniques and approaches have improved, thanks to general medical progress. In particular, the technological improvements in imaging have helped physiotherapists for example to locate injuries more precisely and adjust the patient's treatment [1]. The goal of the development reported in this paper is to introduce technological support at the patient's side to improve the treatment, both curative and preventive.

Focus: motion-sensing node. In this paper, we present an Inertial Measurement Unit (IMU) sensor node to support the tracking and visualization of a patient's execution of physical exercises or daily movements. The priorities for the sensor design were low-power, low-complexity, low-cost, and a small form factor. We achieved the goal to realize a sensor node with a diameter of maximum 6 cm, weighing less than 50 g, costing less than 30€, which can be lowered significantly for higher volumes. Considering the medical context, the sensor node must be hermetically sealed. Therefore, wireless charging is implemented. The measured data is transmitted wirelessly to a base station for further analysis. Calibration of the different sensors is done on-board to obtain measurements with higher precision than by using no calibration.

Progress with respect to the state of technology. Comparing the proposed sensor to currently available systems like [2,3], we focus on the raw output data rather than developing software that processes this automatically. Nonetheless, the raw data can be displayed graphically and present the data in a meaningful way. Secondly, the sensor design has features that contribute to the user-friendliness and accessibility for our target audience, the patient, and the physiotherapist. The presented motion sensor node thus exhibits a low complexity and user-friendly solution that can lower the cost with respect to available systems considerably, while preserving the same functionality. A smart watch, for example, is widely adopted to track overall activity of people. However, it is not fit to be attached anywhere on the body to monitor particular movements in physiotherapy, nor does it fulfill the low-cost and low-complexity requirements of the sensor nodes we aim for.

Contribution. We propose an innovative design, based on low-cost sensors, and the operation of the contactless sensor module, including automated calibration, which is in particular relevant to the targeted applications in e-treatment for physiotherapy. The novel contribution of this paper is threefold. First, we present the design and implementation of the wireless sensor node featuring wireless communication and charging and full filling the other requirements that were put forward. We share the open design it via GitHub [4]. Secondly, we elaborate on a simple, straightforward one-time sensor calibration procedure. This eases the operation of the system and ensures the reliable performance of the system. Lastly, we show how we performed the sensor validation with photogrammetry, which can be realized with inexpensive and widely available equipment in a real-life experiment. We further provide technological and application context.

Structure of this paper. This paper is further organized as follows: Section 2 presents the low-complexity design of the wireless sensor node. It zooms in on the calibration and wireless connectivity, as well as how the sensor node was optimized for low energy. The prototype is presented, meeting the initial requirements. In Section 3, the operation and accuracy of the sensor is validated using easily accessible equipment, avoiding expensive instruments. Next to this static validation, Section 4 elaborates on the dynamic behavior. This can be done with physical exercises. We explain the opportunities opened by the wireless sensor node for e-treatment in physiotherapy applications, and envisioned extensions to the system in Section 5. Section 6 summarizes the main conclusions of this paper and looks forward to potential future work.

2. Low Complexity Design of Wireless Motion Sensor Node

In the design of the sensor node, the following targets were set:

- **Accuracy.** The sensor node needs to be able to measure the human body movement with high precision. With proper calibration, it is possible to achieve a target accuracy of $\pm 2^{\circ}$ with a sampling frequency of 50 Hz [5].
- **User-friendly.** The device needs to be easy to use, capable of being operated by anyone, regardless of any medical or technical background. We opted to implement wireless charging to increase user-friendliness in operation and maintenance. The data is also wirelessly transferred to eliminate a mess of cables and thus providing freedom of movement.
- **Autonomy.** Users want to focus on the application rather than constantly thinking about charging the device. Therefore, an autonomy of at least 5 h and a charge time of less than 1.5 h is necessary.
- **Affordable.** To provide an appealing multi-purpose product for a wide range of applications, it needs to come at a low cost. That way, we want to reach a wide audience, both professionals as individuals.

The sensor node is built around an IMU. The data is wirelessly transmitted to a receiver and the internal battery can be wirelessly charged. Figure 1 shows an overview of the system. We discuss the main features of the sensor node here below.

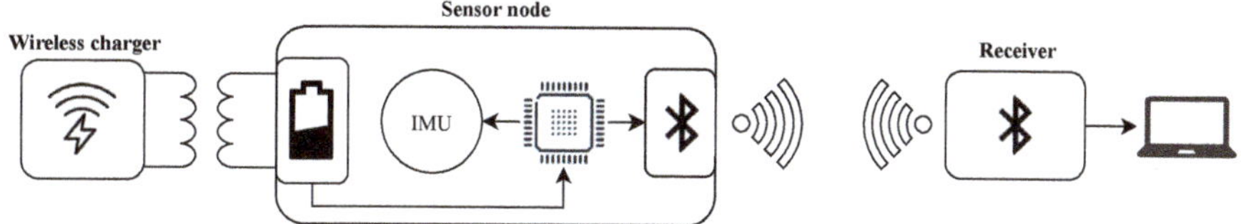

Figure 1. Overview of the hardware: The sensor node built around an Inertial Measurement Unit (IMU), wirelessly rechargeable and with wireless connectivity to a receiver base station.

2.1. Sensors

Motion can be monitored in several ways. A camera-based motion capturing system such as [6] can be used. These systems are highly accurate but expensive and cannot be used anywhere. Another method of monitoring movement that is more suited for our requirements, is by using an IMU. This type of sensor consists of several internal sensors. A 6 Degrees of Freedom (DoF) IMU is commonly used in recent works [7,8]. It has a constant drift in the resulting measurement data that cannot be corrected for. To eliminate this problem, our design uses a 9 DoF IMU in which the additional magnetometer provides a fixed reference. It consists of a Microelectromechanical Systems (MEMS) gyroscope, accelerometer, and compass. The IMU (ICM-20948 from Invensense) [9] was chosen for its ultra-low power operating current and high accuracy. The gyroscope is set to ±2000 dps full scale, the accelerometer is set to ±4 G full scale and the magnetometer is set to ±4900 µT. The sample rate of all sensors is set to 50 Hz.

To obtain accurate orientation data, sensor fusion is needed. As used in [10], a Digital Motion Processor (DMP) can be very efficient for running specialized sensor fusion algorithms. By offloading computationally heavy calculations from the main processor, the system can be more power-efficient. The lack of control of the sensor fusion and calibration is a significant drawback. By implementing our own sensor fusion and calibration, we can implement the most suited fusion algorithms and have full control over the calibration. Some systems use sensor fusion algorithms like a Kalman filter [7], which provide very accurate results but can be computationally intensive. A complementary filter, which is very easy to process but typically provides less accurate results than a Kalman filter, is sometimes used. It uses a high pass filter for the gyroscope values and a low pass filter for the accelerometer values. This method of sensor fusion is inaccurate during long measurements with a lot of movement. In [8] for example, [8] use a complementary filter for measuring static angles, thus primarily depending on the accelerometer values. We need high dynamic accuracy with low processing power thus implemented a Madgwick filter [11], which combines the best of both worlds.

The algorithm runs on the central microcontroller (ARM Cortex M0+ microcontroller (EFM32HG) from Silabs) [12]. It is designed by [11]. By combining the efficiency of this algorithm with a high accuracy, a bit of battery power is saved. Quaternions, a very good way of representing orientations, are used for the calculations. Figure 2 illustrates the functional block diagram of this filter with $\otimes$ a quaternion product, $\dot{q}$ a quaternion derivative and $\hat{q}$ a normalized vector. The algorithm has two adjustable parameters, β and f. β represents the error on the gyroscope measurements as the magnitude of a quaternion derivative. It determines the proportion of the correction value for the gyroscope. f represents the frequency of the measurements. The orientation is mainly calculated by integrating the changes in angular velocity from the gyroscope (1). At the same time, an orientation is calculated using the accelerometer and magnetometer values. A gradient descent algorithm, represented by ∇, is used to find the most likely solution in the set of infinite solutions. (3) represents the measured orientation from the earth's magnetic field. In (4), the measurements are normalized and mapped to the plane of the earth. In (5) = ∇f, the orientation from magnetometer and accelerometer values is calculated using a gradient descent algorithm. These values are normalized in (6) and used to correct gyroscope values with a factor β. These corrected gyroscope values are integrated in (2). Everything is further normalized in (7) to form unit quaternions and the results form is given as in Equation (1).

$$q = a + b \cdot i + c \cdot j + d \cdot k \qquad (1)$$

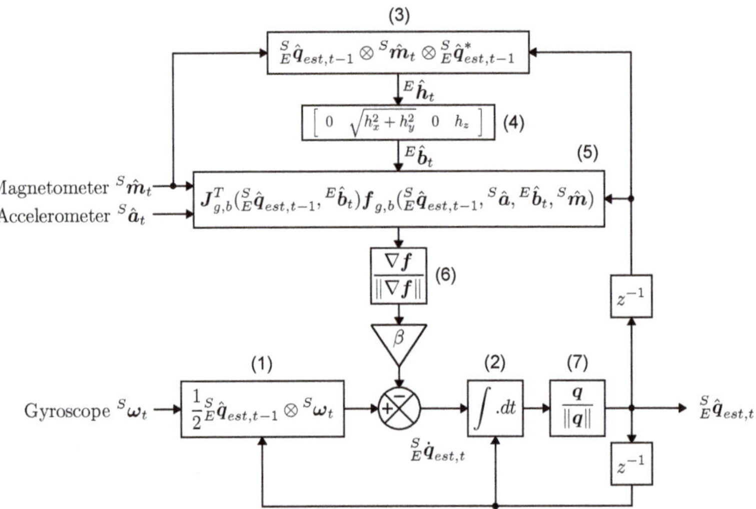

Figure 2. Block diagram Madgwick algorithm [11].

By using quaternions, we avoid the gimbal lock problem, the inability to uniquely represent an orientation, of Euler angles. When for example the pitch angle is 90°, yaw and roll cause the sensor to move in exactly the same fashion. Another problem is the inability to produce reliable estimates when an angle approaches 90° [13]. The benefit of the Madgwick algorithm is that we can run it at a very low speed and still get accurate results. At 50 Hz, the sample rate used by the sensor node, we get a static error of ±1° and a dynamic error of ±2° [11].

The sensor node goes to sleep as much as possible to conserve battery energy. When the sensor node is picked up, the always-on accelerometer generates an interrupt and wakes up the system. Figure 3 illustrates this procedure. When the Madgwick parameters are set correctly, just a small portion of the accelerometer and compass values are used to correct the gyroscope error. When the sensor node wakes from sleep, the gyroscope has no reference orientation thus it would take approximately 30 s to obtain a correct orientation, depending on how much the actual orientation, when the sensor is picked up, differs from the orientation in sleep. After wake-up, the parameters of the Madgwick filter are dynamically adjusted to obtain a correct orientation quicker. The accelerometer and compass are used in the first few seconds of activity to obtain a correct reference frame. After this, the parameters are automatically adjusted to a high accuracy mode. In this mode, the integration of changes in angular velocity from the gyroscope is mostly relied on for calculating the orientation of the node. The accelerometer and the compass are only used to make small corrections.

As a low power design consideration, inactivity is detected by checking the gyroscope values every second in an interrupt service routine, called from an Real Time Counter (RTC) interrupt. The gyroscope values are supposed to be zero when idle. This procedure automatically puts the sensor node in sleep when it is not used.

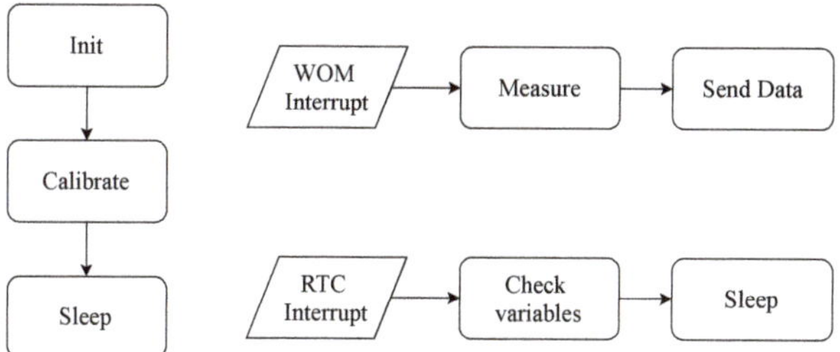

Figure 3. Flowchart code: The sensor node is calibrated once at initialization, a Wake On Motion (WOM) interrupt wakes up the system and measurements can start, an Real Time Counter (RTC) timer is used to periodically check the status of the sensor node to maximize autonomy.

2.2. Calibration

Calibration is an essential part of motion capturing systems. With calibration, the accuracy of the measurements can be drastically increased. In this design, a manual one-time calibration is used. The manual calibration allows the use of a very energy-efficient microcontroller and can yield calibration values with high accuracy. First, the gyroscope and the accelerometer are calibrated. This happens by simply putting the sensor node on a flat, leveled surface. During these measurements, no changes in angular velocity from the gyroscope or acceleration forces from the accelerometer are expected. The accelerometer and gyroscope are temporarily set to the most sensitive measurement range of ±250 dps full scale and ±2 G full scale to obtain the highest calibration accuracy possible. A few thousand measurements are taken by filling the First In First Out (FIFO) buffer of the IMU. From these measurements, a gyroscope and accelerometer bias offset is calculated and further subtracted from the actual measurements. After calibration of the gyroscope and accelerometer, the measurement ranges are changed back to ±2000 dps full scale and ±4 G full scale.

The compass is calibrated by rotating the device 360° around its three axes or performing a figure-8 movement. For these measurements, the maximum sampling frequency of 100 Hz is temporary used to have more data to work with and therefore obtain a better calibration. After calibration, the magnetometer sample rate is changed back to 50 Hz. The result of such a measurement is shown in Figure 4, a 2D visualization of the three planes of the 3D sphere after rotating the sensor node. Two types of distortions can occur on the IMU measurements: hard and soft iron distortions [14]. Hard iron distortions, caused by a permanent magnetic material, create a constant offset on the sphere. These offsets can be determined by calculating the center of the sphere and subtracting this value from the measurements. Soft iron distortions are caused by materials like iron. These materials do not create their own magnetic field but create a deformation on one or more axes. This will generally create an ellipse instead of a circle in a 2D plot. Soft iron distortions are more difficult to correct. Each axis is multiplied with a scale factor to calibrate the measurements. The minimal and maximal compass values captured in the calibration procedure of each axis are measured determined. The span of the compass values for all three axes is calculated, as well as the mean span for the three axes. The scale factor per axis is thus mean divided by the span of the axis that will be corrected. Equation (2) provides the equation for the x-axis scale factor, exemplary for the three axes. The result of these corrections, with the three circles perfectly round and centered, is given in Figure 5, showing that the calibration procedure operates correctly.

$$\text{Scalefactor} = \frac{max_x - min_x + max_y - min_y + max_z - min_z}{3 \cdot (max_x - min_x)} \qquad (2)$$

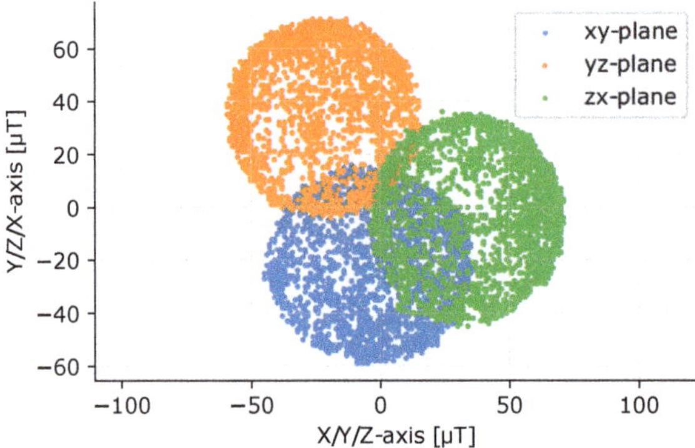

Figure 4. 2D plot of a sphere after rotating the sensor node around each axis before calibration. The circles are not perfectly round (elliptical sphere in 3D) caused by soft iron distortions. Also, offsets between the centers of the circles and the origin, caused by hard iron distortions, are present.

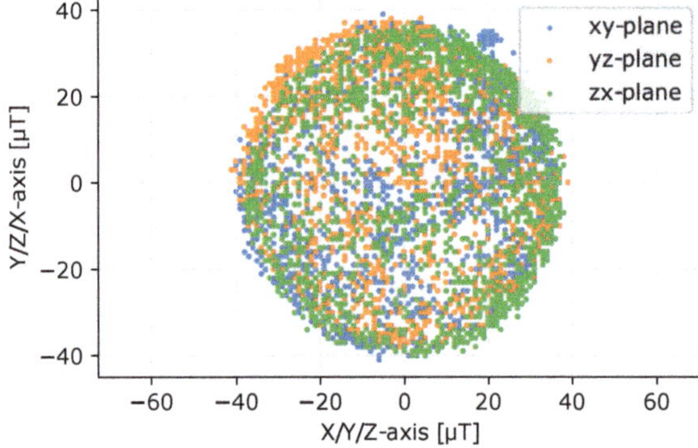

Figure 5. 2D plot of a sphere after rotating the sensor node around each axis after calibration. The circles are perfectly round (near perfect sphere in 3D) and no offsets between the center of the circles and the origin are visible.

2.3. Wireless Connectivity

Many wireless connectivity standards for Wireless Body Area Networks (WBAN) are available. We here briefly comment on the most considered technologies given the application focus of the presented design.

ZigBee operates with very low power usage. It works on top of the IEEE 802.14.4 standard, has a range of up to 100 m, and can be implemented as a mesh network. The low data rates of up to 250 kbps at 2.4 GHz make ZigBee less suited [15]. A second wireless standard is Z-Wave, a low data rate communication protocol with data rates of 40 kbps—100 kbps and a range of up to 30 m. Since it uses the 900 MHz band, it is not bothered by interference from 2.4 Ghz wireless communication like WiFi. It is commonly used in home automation for interconnecting energy efficient sensor nodes. The master-slave type network has a typical latency of 200 ms [16]. A third wireless standard is Bluetooth. It is based on the IEEE 802.15.1 standard, has a higher data rate of up to 2 Mbps and a range of up to 100 m. The more advanced Bluetooth protocol is widely used for data and audio transmission. It uses a master-slave model for communication [17]. For the design of the low power sensor node, Bluetooth Low Energy (BLE) is more appropriate. This special Bluetooth version is specifically designed for applications with very low power usage. A maximal data rate of 1 Mbps and a range of a few tens of meters can be achieved. BLE can use a master-slave model in a star topology or BLE devices can form a mesh network [16]. The advantage of BLE is its ability to directly connect to a smartphone or Bluetooth enabled device without the need for a separate receiving station. Following up with WiFi, based on the IEEE 802.11 standard with a very high data rate of 54 Mbps. The high power consumption makes WiFi less suited for a low power design [17]. We also studied the possibility of using a proprietary solution. The advantages are a possible further reduction in power consumption by packets with increased information density. Table 1 summarises the different wireless connectivity options in terms of power consumption, range, data rate and price. BLE is chosen for its low power consumption, sufficient range, relatively high data rate, low price, and high compatibility with existing devices.

Table 1. Comparison between available wireless technologies: ZigBee, Z-Wave, Bluetooth 5, BLE, and WiFi [15–17].

	ZigBee	Z-Wave	Bluetooth 5	BLE	WiFi
Power consumption (max)	100 mW	1 mW	100 mW	10 mW	>100 mW
Range (max)	100 m	30 m	100 m	<100 m	1000 m
Data rate (max)	250 kbps	100 kbps	2 Mbps	1 Mbps	54 Mbps
Price	Low	High	Very low	Very low	Average

A WBAN is necessary for transmitting the measured data. BLE is chosen for its high throughput, minimal power consumption, and interoperability with other devices [16]. The Proteus II module (AMB2623 module from WE based on an nRF52832) [18] is chosen for its small form factor and integrated PCB antenna. The data is transmitted at 0 dBm.

The data packet, sent out at 50 Hz, contains a preface, the module ID of the receiver, the RSSI, the data, and a checksum for error correction. This is clarified in Figure 6.

Preface				Module ID						RSSI	Data x				Data y				Data z				Batt	Checksum
02	84	0B	00	BC	04	20	DA	18	00	XX	x	x	x	x	y	y	y	y	z	z	z	z	%	CS

Figure 6. Bluetooth Low Energy (BLE) data packet structure: The data is composed of a preface, the module ID, three Euler angles, the remaining battery charge (percentage) and a checksum.

The quaternions from the Madgwick sensor fusion filter are converted to Euler angle floats. The three floats each take up four bytes in memory [19]. Exactly those bytes will be read from memory and transmitted wirelessly to ensure no loss in accuracy. One byte is added to transmit the battery status.

To guarantee a low power design, some software features are added. When the sensor node is picked up and cannot connect to a receiving device within five seconds, the sensor node enters sleep mode. The automatic reconnection of the sensor node with the receiving device is also built-in.

The receiving device is based on a development board (STM32L4+ microcontroller on an ST NUCLEO L45ZI development board) [20]. The same BLE module is chosen for this device. To be able to receive the transmitted data fast enough, an interrupt-based method is used together with a circular buffer [21]. The UART interrupt receives data and stores it in the buffer in the background. The received data is processed independently in the main program. A second UART transmits the data to a pc. A 3D representation of the orientation is written in VPython for visualization purposes.

2.4. Wireless Charging

Inductive wireless energy transfer is mainly used to recharge batteries of smartphones, wearables or, Internet of Things (IoT) devices. Implementation standards such as Qi, PMA, or AirFuel ensure a safe, efficient transfer of energy. Low power applications, below 5 W, often use e.g., proprietary solutions such as the "LinkCharge Low Power" technology from Semtech. Wearable devices, Electric toothbrushes, or LoRa based sensors are some of the many applications for the implementation of this technology [22]. ST Microelectronics also offers wireless power solutions for Smartwatches, or IoT battery-powered smart devices. The last option is to design your own Wireless Power Transfer (WPT) system without using existing standards. Building more efficient systems is time-consuming and not necessary since a lot of research has already been carried out in the 5 W WPT range.

In recent years, it has been generally accepted that the Qi is preferred over all other standards. The Wireless Power Consortium (WPC) manages and develops this standard. In the meantime, PMA, AirFuel and WPC have started a collaboration. All Qi-certified devices can communicate with each other. Charging a Qi-supported device can be performed by any Qi-certified charger. A series of functions in the standard ensures a safe charge cycle, such as thermal shutdown protection, foreign object detection, and overvoltage AC clamp protection [23].

The first wirelessly rechargeable smartwatches used proprietary WPT standards. New wearables switched to the Qi standard in contrast to wirelessly rechargeable smartphones, which were immediately equipped with the Qi standard. Recent smartphones are available with the option "Reverse Charging", which means that the internal smartphone coil can be used to charge devices that support Qi [24]. This new feature offers the possibility of recharging smartwatches with a smartphone. It makes sense that Qi was chosen above all other options for the sensor module. In most households, a Qi charger or a smartphone that supports reverse charging is available. Future measurements with this sensor can be used within families, as they can recharge their sensor modules at home.

We here further discuss the actual implementation of the battery charging circuit in the design of the sensor node presented in this paper. Since energy is transferred wirelessly via the Qi protocol, a Qi receiver IC was used. A TI Qi receiver IC (BQ51050) [23] was selected because of its high efficiency, wireless power receiver, integrated rectifier, and battery charger in a single package. The BQ51050A variant, combined with a Li-Ion battery is chosen because of its 4.20 V output voltage limitation. It is paired with an inductor coil (760308101214 coil from WE) [25], chosen for its very small size and a relatively decent Q-factor. The charging current is 200 mA with a termination current of 20 mA to ensure fast and safe charging. Temperature control with automatic cut-off functionality at 60° is implemented by using a Negative Temperature Coefficient (NTC) resistor. Because of the small coil, we implemented some extra shielding to ensure a more optimal WPT.

Figure 7 shows the two coils in the system with corresponding resonant circuits. A power transmitter coil is present in each charger pad and a receiver coil in each battery-powered device. Wireless charging achieves higher link efficiencies when implementing LC resonant circuits on both the receiver and transmitter. The coupling factor between the two coils is very low. Therefore implementing a resonant circuit can filter out the leak inductance and improve the link efficiency drastically [26]. A Qi charger pad has a built-in amplifier connected to an LC series resonant circuit. The energy receiver side consists of an LC resonant circuit with L, C_{s1}, and C_{s2}. These capacities can be calculated with the Equations (3) and (4). L_s' represents the inductance measured when the receiver coil is placed on top of a charger pad. L_s is the free-space inductance. f_s and f_D are fixed values respectively 100 kHz and 1 MHz [23].

$$C_1 = \frac{1}{(2\pi \cdot f_s)^2 \cdot L_s'} \tag{3}$$

$$C_2 = \left((f_D \cdot 2\pi)^2 \cdot L_s - \frac{1}{C_1} \right)^{-1} \tag{4}$$

Filling in the formula and converting to values for which actual hardware components are commercially available gives 100 nF for C_{s1} and 1 nF for C_{s2}. Three other types of capacitors have an important function in the circuit. The BOOT, COMM, and CLAMP capacitors. The BOOT or bootstrap capacitors are used for driving the high-side FETs of the synchronous rectifier. The COMM capacitors allow communication with the charger pad. Here, capacitive load modulation is used. An extra capacitance is connected to the resonance circuit, which changes the resonance frequency. This change is visible on the charger pad side. Load modulation allows communication between the power receiver charging circuit and the power delivery pad circuit. Guidance values for resistive load modulation can be found in the datasheet. The CLAMP capacitors ensure overvoltage protection. Above the rectified voltage of 15 V, the CLAMP capacitors are switched to change the resonance frequency and protect the circuit against high voltages. The datasheet provides suggestions for these values. Values of 10 nF, 470 nF and 47 nF were used for the BOOT, CLAMP and COMM capacities, respectively [23].

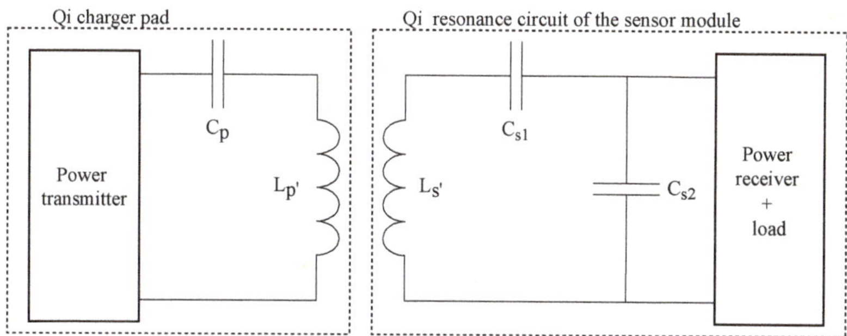

Figure 7. Wireless Power Transfer (WPT) setup. A Qi power transmitter with a Qi power receiver and load, based on LC resonant circuits.

2.5. Optimization for Low Energy

One of the main focuses of this work is the realization of a node with a convenient autonomy. A Li-Ion battery is chosen for its high energy density and low weight. The round battery with a capacity of 200 mAh is ideal for this prototype. This battery is rechargeable. With compatibility and ease of use in mind, Qi-compatible wireless charging is implemented. The whole system is powered at 2 V with an ultra-low Iq buck converter. In this configuration, a buck converter is much more efficient than a Low-dropout (LDO) regulator, even in sleep mode. The IMU works at 1.8 V. Here, the use of an LDO for the voltage drop of 0.2 V is more efficient. By running the whole system at 2 V instead of the traditional 3.3 V, a theoretical power difference of 9.610 mW is calculated when quiescent currents are neglected. This translates to a gain in the autonomy of 29.3 %. The sensor node consumes 0.102 mW in sleep mode and 25.839 mW in active measurement mode. This is reflected in an autonomy of 28 h in operation and of 261 days in sleep mode, which is well above the five hours put forward. An active power consumption of 25.839 mW is very low for this kind of system and can't be significantly improved with the hardware we are currently using. This power consumption in combination with a 200 mAh battery allows for a long enough time between charges. The sleep current of 0.102 mW can possibly be improved by disabling the Qi-wireless charger completely when it's not being used, thus eliminating quiescent currents. This can be done by using a MOSFET.

2.6. Prototype

A small physical design that is easy to place on the body is crucial. The sensor node features a round design with no sharp edges. The final prototype weighs 38 g and has dimensions ⌀6 cm × 1.5 cm. The structure of the case is shown in Figure 8. The wireless charging coil is positioned at the bottom (1). It is held in place by some offsets in the case (2). On top of that is the battery (3). Above the battery is the PCB (4) which is supported by four pins in the case (5). Everything is fastened nicely by the cover (6), which can be attached with a twist top. We did not yet hermetically seal the case for the initial experiments. By applying some sealant on the twist top, one can make the case more waterproof. Figure 9 shows the assembled prototype of the sensor node. The total cost of components is 28€ with case and 22€ without the case.

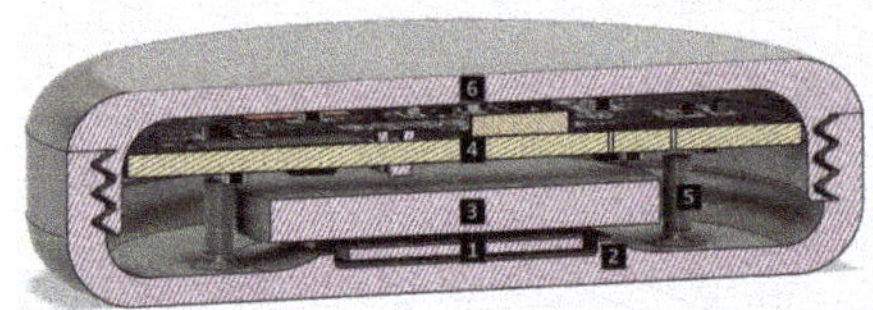

Figure 8. Cross-section of the sensor node. 1: wireless charging coil, 2: offsets, 3: battery, 4: PCB, 5: support pins, 6: twist top.

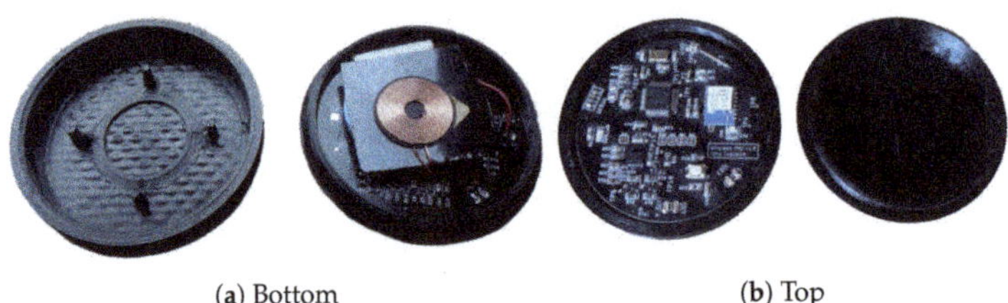

(a) Bottom   (b) Top

Figure 9. Result: sensor node in the 3D printed case. (**a**) Bottom side: wireless charging coil and battery. (**b**) Topside: IMU, microcontroller, and BLE chip.

3. Validation with Easily Accessible Equipment

For the validation of the accuracy of the motion measurements realized by the sensor node, it is common to use professional equipment. The static verification process of the IMU has already been performed by using a computer monitored pan-tilt unit to place the sensor node in specific angles or by using a Vicon motion capturing system [6,8,27]. In the validation of the Madgwick filter for example, a Vicon motion capture system is also used [28]. Sensor validation on this equipment in general yields very accurate results but it is less accessible, expensive and time-consuming.

We propose an alternative, very accessible way of validation using convenient equipment in the context of designing a low-cost system that is user-friendly. With photogrammetry, one can get a fairly accurate representation of the performance of the sensor node. In this method, we take and interpret photographic images of positions of the sensor. By comparing the data from the IMU with the data extracted from images, the static error on the measurements can be derived. The advantages are that this method can be performed almost anywhere and can be used with consumer off-the-shelf equipment. Since, in contrast to professional cameras, lower-cost equipment, such as a smartphone camera, suffers from lens distortions and lower quality recordings, some measures must be taken. To minimize the effect of the lower quality equipment, the camera is placed horizontally and perpendicular to the wall. This way, foreshortening effects are eliminated. Furthermore, the sensor is positioned such that its projection lies near the center of the image where radial distortion is minimal. This eliminates the need for a camera calibration procedure. Finally, we add several markers to the scene as shown in Figure 10. The relative position of these markers is measured up to ±2 mm.

Since all we need is angles, we can perform measurements in the image and transfer them to the reference system of the sensor node. By attaching a lever to the sensor node, the accuracy of the readings in the image increases. The angle of the sensor can easily be measured by indicating front and endpoints of the lever (red and green points in Figure 10) and mapping these points in the image to points on the wall, using the coordinate system defined by the surrounding markers. By comparing the data from the IMU with the data extracted from the images, the static error on the measurements can be derived for the pitch and roll axis. In our experiments, only static measurements are performed. Dynamic measurements are possible as well, in which case video instead of images should be recorded and the video frames must be synchronized with the output data of the sensor node. Doing so, one can obtain angles at frame level. Instead of manually indicating points in each video frame, this process can be automated using image tracking [29,30].

Figure 10. Method for sensor validation based on photogrammetry using convenient, commercial off-the-shelf equipment. By comparing the data from the IMU with the data extracted from images, the static error on the measurements can be derived for the pitch and roll axis.

Table 2 gives an overview of the measurements. For roll and pitch angles, the setup as shown in Figure 10 is used with the sensor node rotated 90° between roll and pitch measurements. Since the yaw values have no real fixed orientation, relative measurements are taken by using the setup as shown in Figure 11 where the sensor and markers are positioned on the floor instead of against the wall. Several static measurements were performed. The static sensor drift is 2° per half hour. The average error on the pitch axis is 3.06°, the average error on the roll axis is 2.75° and the average error on the yaw axis is 4.04°.

Table 2. Result of pitch, roll, and yaw static measurements with their respective error at different angles.

	Target Angle [°]	Reference [°]	Sensor [°]	Error [°]
Pitch	0	0.08	−3.2	3.28
	45	44.76	42.5	2.26
	90	90.19	95.04	−4.85
	180	178.45	176.6	1.85
Roll	0	0.47	1.8	−1.33
	45	48.41	44.8	3.61
	90	90.15	87	3.15
	180	180.01	177.1	2.91
Yaw	45	48.03	45.1	2.93
	90	95.49	88.9	6.59
	180	182.18	185.2	−3.02
	270	274.12	270.5	3.62

Alternatively, it is possible to measure all three (roll, pitch, yaw) angles at once by measuring the position of the lever endpoints in 3D using a stereo or multi-camera setup. However, drawbacks of such a method are the much higher complexity, the need for calibration and synchronization, and the lower accuracy in the depth dimension.

There are some irregularities in the measurements. The yaw value at 90° seems to be off. A root cause could be the influence of a nearby magnetic object. The sensor can get disturbed in the near proximity of magnetic objects such as speakers and smartphones. These magnets create a distortion in the magnetic field which isn't fixed to the reference frame of the sensor node, thus can't be corrected for in calibration. The user can perform reliable measurements when staying half a meter away from these objects to obtain accurate measurements. The pitch error at 90° is also too large. The reason is that Euler angles are not good at representing orientations in the neighborhood of 90° [13].

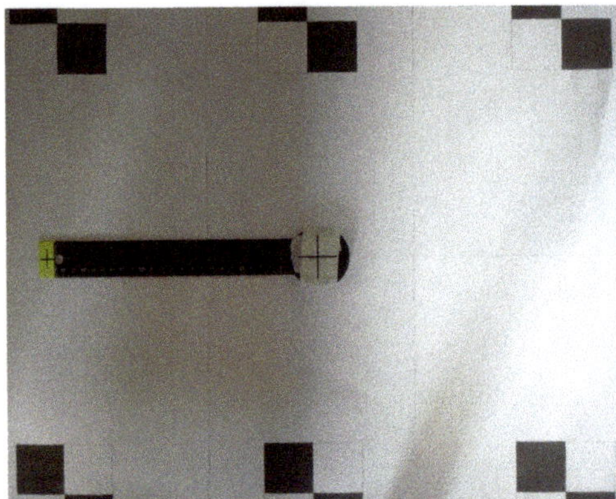

Figure 11. Photogrammetry-based method for yaw axis sensor validation.

4. Validation with Real-Life Exercises

To evaluate and validate the dynamic behavior of the sensor node and real-life operation, two back exercises are performed. The first exercise starts with a person kneeling with hands on the ground. The back is periodically rounded and made hollow, thus demonstrating the periodic concavity of the spine. This is illustrated in Figure 12.

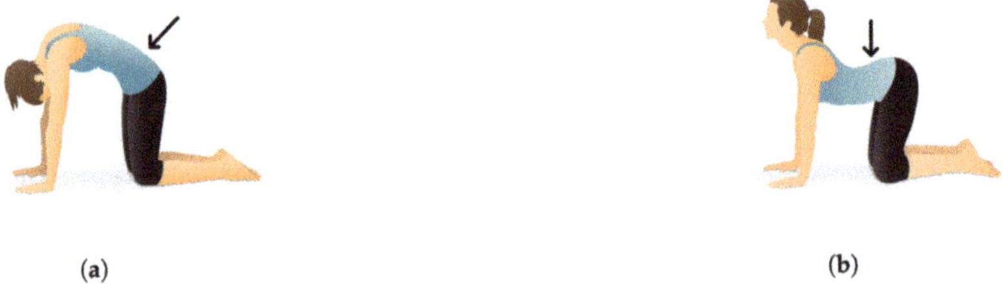

(a)
(b)

Figure 12. Illustration of the first exercise: periodic concavity of the spine (Images provided by Pocket Yoga (www.pocketyoga.com)). The arrows indicate the position of the sensor node. (**a**) Start position. (**b**) End position.Illustration of the first exercise: periodic concavity of the spine

Figure 13 presents the result of the measurements. The exercise has been performed in a set of 3 repetitions. A periodic movement with a variation of ±45° on the roll axis can be observed. The pitch axis shows a little bit of sideways rotation in the lower back. The yaw axis is stable, which is to be expected. A second captured exercise is the lateral rotation of the back, illustrated in Figure 14. The patient should rotate the hull sideways, while maintaining stable lower limbs. The measured result is represented in Figure 15. An angular deviation of ±50° is present in the yaw axis data. Small changes in roll and pitch values are also observed. These two exercises provide a first evaluation of the dynamic characteristics of the sensor node. We clearly see that the amount of samples taken is appropriate to acquire accurate results. However, more testing, either by dynamic photogrammetry or with specialized equipment, is needed before a firm conclusion on accuracy can be made.

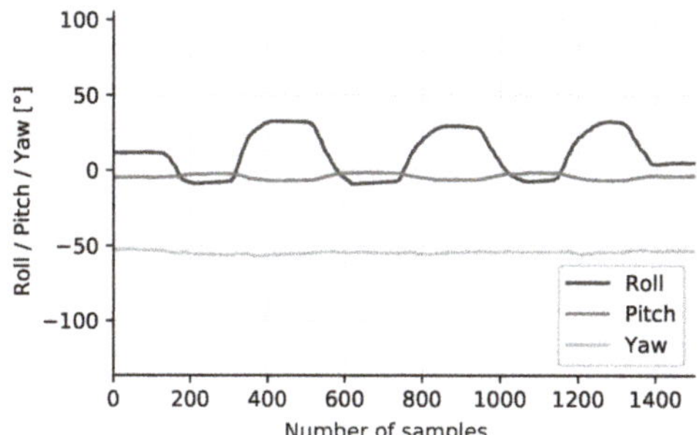

Figure 13. Exercise: Rounded back—hollow back. A periodic movement with a variation of ±45° on the roll axis can be observed. The pitch and yaw axis are stable.

Figure 14. Illustration of the second exercise: lateral rotation of the back (Images provided by Pocket Yoga (www.pocketyoga.com)). The arrow indicates the position of the sensor node.

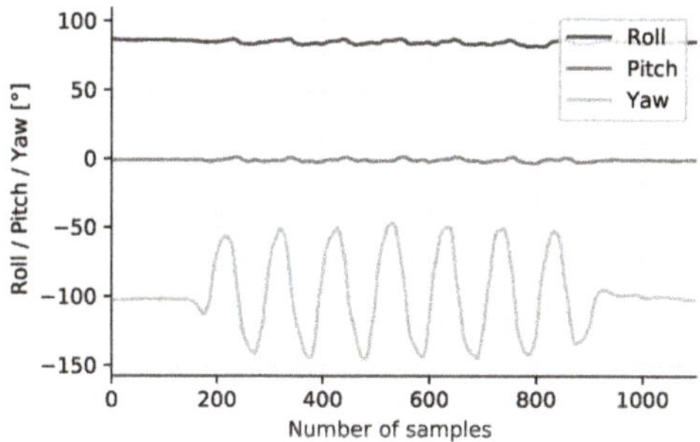

Figure 15. Exercise: Rotation of the back. A periodic movement with a variation of ±50° on the yaw axis can be observed. The roll and pitch axis are stable.

5. Opportunities in e-Treatment Applications and Extended Functionalities

We here first explain the opportunities opened up by stand-alone low-cost and low-complexity sensor nodes in physio-therapeutic e-treatment. We benchmark the current solution and introduce further extensions of the system that can bring interesting features for both private and professional users.

5.1. Opportunities in Supporting e-Treatment in Physiotherapy

The presented wireless sensor node has been designed to meet the particular needs to support physiotherapy treatment. We wish to introduce technical support at the patient's side to improve both curative and preventive treatment. The sensor thus enables *e-treatment*, which we define as (remote) physical therapy that is supported by measurements made by wireless sensors. In a curative treatment, the patient can wear the sensor to assist the physiotherapist in the evaluation of (eventual take-home) rehabilitation exercises. A preventive treatment could consist of monitoring a person's daily movements or measuring a patient's flexibility. We specifically expect measurements at work to be interesting, knowing that the large majority of neuromusculoskeletal disorders result from repetitive movements and bad posture at work [31].

Also important in our definition of e-treatment, is the word *remote*. In the case of remote treatment, the patient is not physically present in the physiotherapist's practice, but for example at home and possibly assisted with one or more sensors. Especially because of the increasing cost of healthcare in our ageing society, it is important to look at efficient and low-cost alternatives. The connection is then real-time through a conference call, or non real-time by exchanging exercises over a manual for example. There are several reasons why a remote session can be preferred over a conventional consultation:

- The patient can perform the session more or less independently.
- The patient is abroad and wants to continue the treatment with the same physiotherapist. For example, elite athletes who have to travel a lot.
- A patient is not allowed to leave the house. The COVID-19 pandemic proved this to be a realistic scenario.

The effectiveness of e-treatment in a remote sense is exhaustively discussed in [32]. A last important field of application is the education of physiotherapists. With the help of our technology in a bigger ecosystem, we want to reach physiotherapist with e-learning and help them train and improve. In summary, the sensor can enable remote treatment, as well as support conventional consultations or even acquire measurement data for preventive purposes.

5.2. Extension to Multiple Sensor Nodes

Richer information and support in rehabilitation and e-treatment could be offered by the combination of multiple sensor nodes, either of the same type or using heterogeneous sensors. An especially relevant type is an Surface Electromyography (sEMG) sensor module for measuring muscle activity. While we have designed the first prototype for this sensor type, in a future version we will combine the IMU and the sEMG sensor into one module. By combining these sensors, we can capture a more complete picture of what the human body is doing. However, this generates extra technological challenges, especially with respect to synchronization, both intra- and inter-module, required to ensure concurrent measurements. Synchronization between the sEMG and the IMU can be implemented using a shared clock. Both sensors will experience the same clock drift. BLE beacon packets from a central node, in this case the receiver, or a custom protocol can be used to synchronize the clocks between sensor nodes [33]. The data can be transmitted using unidirectional beacon packets without re-transmission. This type of data transfer is very simple but does not guarantee the packet arrives at the receiver. A better way would be to use the BLE re-transmission functionality to ensure the packets are received properly. Time synchronization beacon packets could be sent in between. It is evident that both the electrical and the mechanical design will be more complicated, not in the least because of the need to integrate the functions in a small space.

6. Conclusions and Future Work

Conclusion. In this paper, a wireless on-body sensor node for measuring movement is presented. The careful choice of components, software optimizations, and overall low power design considerations lead to a sensor node with an autonomy of 28 h. An 'always-on' buttonless design, with a standby time of 8 months is developed that is ready to measure whenever it is picked up. We explained the calibration of the sensor node and zoomed in, in particular on a photogrammetric procedure to validate the sensor with easily accessible, low-cost equipment. On-device sensor fusion by using a Madgwick filter yields static results of on average 3.28° with a drift of 2° per half hour. The final prototype weighs 38 g and measures ø6 cm × 1.5 cm. The result of this work can be used in a broad range of applications. It allows doctors and physiotherapists to have an easy to use device to pass along with patients and afterward interpreting the results, it can be used for live monitoring of rehabilitation exercises or anything motion tracking related.

Future work. We see multiple opportunities in future work to both the current sensor node, and to extend it with new functionality and features. Firstly, we plan to further examine the accuracy of the sensor node by checking it against specialized equipment. We will add other sensors to get a more in-depth view of the human body. We also designed a sEMG sensor for measuring muscle activity. These two sensors could be integrated into one module to perform simultaneous measurements. Synchronization, both inter- and intra-sensor node, will be implemented to ensure precise, simultaneous measurements. A future upgrade could also implement a real-time calibration by using artificial intelligence [34]. This could well be implemented on a low power microcontroller with an ARM Cortex M4 chip (nRF52832 from Nordic Semiconductor), which is already used in the BLE module. By running the Bluetooth stack and the peripheral code on the same chip, we could eliminate the central Cortex M0+ microcontroller and further reduce the power consumption. We could also design our own PCB antenna. In the current design, the data is, other than being visualized, not further processed. To detect and analyze complex movements, further data analysis as well as learning algorithms can be implemented. Another extension to the system is a direct communication between the sensor nodes and a smartphone through an app. This eliminates the need for a separate receiver.

Author Contributions: Conceptualization, Formal analysis, Investigation, J.C.; Methodology, J.C. and M.V.; Supervision, L.V.d.P.; Writing—review & editing, J.C., L.M., J.V.M., S.G., M.V. and L.V.d.P. All authors have read and agreed to the published version of the manuscript.

Funding: NOMADe project of the EU Interreg Program France-Wallonie-Vlaanderen under grant agreement 4.7.360.

Acknowledgments: This research has been supported by the NOMADe project of the EU Interreg Program France-Wallonie-Vlaanderen under grant agreement 4.7.360. We thank our colleagues in the NOMADe project for the valuable inputs leading to the design of our wireless sensor node that is adequate for usage in physiotherapy treatment.

Conflicts of Interest: The authors declare no conflict of interest.

Abbreviations

The following abbreviations are used in this paper:

BLE	Bluetooth Low Energy
DMP	Digital Motion Processor
DoF	Degrees of Freedom
FIFO	First In First Out
IMU	Inertial Measurement Unit
IoT	Internet of Things
LDO	Low-dropout
MEMS	Microelectromechanical Systems
NTC	Negative Temperature Coefficient
RTC	Real Time Counter
sEMG	Surface Electromyography
WBAN	Wireless Body Area Networks
WOM	Wake On Motion
WPC	Wireless Power Consortium
WPT	Wireless Power Transfer

References

1. Porciuncula, F.; Roto, A.V.; Kumar, D.; Davis, I.; Roy, S.; Walsh, C.J.; Awad, L.N. Wearable Movement Sensors for Rehabilitation: A Focused Review of Technological and Clinical Advances. *PM&R* **2018**, *10*, S220–S232. [CrossRef] [PubMed]
2. Inertial Motion Capture System|Shop|Eliko. Available online: https://www.eliko.ee/shop/inertial-motion-capture-system/ (accessed on 27 October 2020).
3. IMU Sensor Development Kit|Wireless IMU Sensor|9DOF Motion Sensor. Available online: https://www.shimmersensing.com/products/shimmer3-development-kit (accessed on 27 October 2020).
4. Cappelle, J. Wireless Motion Sensor Node. Available online: https://github.com/DRAMCO/NOMADe-Wireless-Motion-Sensor-Node (accessed on 4 November 2020). [CrossRef]
5. Brodie, M.; Walmsley, A.; Page, W. The static accuracy and calibration of inertial measurement units for 3D orientation. *Comput. Methods Biomech. Biomed. Eng.* **2008**, *11*, 641–648. [CrossRef] [PubMed]
6. Vicon|Award Winning Motion Capture Systems. Available online: https://www.vicon.com/ (accessed on 9 September 2020).
7. Kadir, K.; Yusof, Z.M.; Rasin, M.Z.M.; Billah, M.M.; Salikin, Q. Wireless IMU: A Wearable Smart Sensor for Disability Rehabilitation Training. In Proceedings of the 2018 2nd International Conference on Smart Sensors and Application (ICSSA), Kuching, Malaysia, 24–26 July 2018; pp. 53–57.
8. Petropoulos, A.; Sikeridis, D.; Antonakopoulos, T. Wearable Smart Health Advisors: An IMU-Enabled Posture Monitor. *IEEE Consum. Electron. Mag.* **2020**, *9*, 20–27. [CrossRef]
9. ICM-20948—TDK. Available online: https://www.invensense.com/products/motion-tracking/9-axis/icm-20948/ (accessed on 26 October 2019).
10. Fei, Y.; Song, Y.; Xu, L.; Sun, G. Micro-IMU based Wireless Body Sensor Network. In Proceedings of the 33rd Chinese Control Conference, Nanjing, China, 28–30 July 2014; pp. 428–432.
11. Madgwick, S.O.H.; Harrison, A.J.L.; Vaidyanathan, R. Estimation of IMU and MARG orientation using a gradient descent algorithm. In Proceedings of the 2011 IEEE International Conference on Rehabilitation Robotics, Zurich, Switzerland, 29 June–1 July 2011; pp. 1–7.
12. EFM32 Happy Gecko Family EFM32HG Data Sheet. Available online: https://www.silabs.com/documents/public/data-sheets/efm32hg-datasheet.pdf (accessed on 6 November 2020).
13. Understanding Euler Angles—CH Robotics. Available online: http://www.chrobotics.com/library/understanding-euler-angles (accessed on 6 May 2020).
14. Tuupola, M. How to Calibrate a Magnetometer? Available online: https://appelsiini.net/2018/calibrate-magnetometer/ (accessed on 10 April 2020).
15. Ergen, S. ZigBee/IEEE 802.15.4 Summary. *UC Berkeley Sept.* **2004**, *10*, 11.
16. Bin Ab Rahman, A. Comparison of Internet of Things (IoT) Data Link Protocols. Available online: https://www.semanticscholar.org/paper/Comparison-of-Internet-of-Things-(-IoT-)-Data-Link-Rahman/1cf94e2ebb27aaecdae3742e444ca9e87314216b (accessed on 7 November 2020).
17. Danbatta, S.J.; Varol, A. Comparison of Zigbee, Z-Wave, Wi-Fi, and Bluetooth Wireless Technologies Used in Home Automation. In Proceedings of the 2019 7th International Symposium on Digital Forensics and Security (ISDFS), Barcelos, Portugal, 10–12 June 2019; pp. 1–5.
18. Proteus-II—Bluetooth Smart 5.0 Module (AMB2623). Available online: https://katalog.we-online.de/en/wco/WIRL_BTLE_5 (accessed on 26 October 2019).
19. Cx51 User's Guide: Floating-Point Numbers. Available online: http://www.keil.com/support/man/docs/c51/c51_ap_floatingpt.htm (accessed on 18 March 2020).
20. NUCLEO-L4R5ZI—STM32 Nucleo-144 Development Board with STM32L4R5ZI MCU— STMicroelectronics. Available online: https://www.st.com/en/evaluation-tools/nucleo-l4r5zi.html (accessed on 17 March 2020).
21. UART Receive Buffering—Simply Embedded. Available online: http://www.simplyembedded.org/tutorials/interrupt-free-ring-buffer/ (accessed on 17 March 2020).
22. Semtech Releases Next-Generation LinkCharge®LP (Low Power) Wireless Charging Platform. Available online: https://www.semtech.com/company/press/semtech-releases-next-generation-linkcharge-lp-low-power-wireless-charging-platform (accessed on 6 November 2020).
23. BQ5105xB High-Efficiency Qi v1.2-Compliant Wireless Power Receiver and Battery Charger. Available online: ti.com/lit/ds/symlink/bq51051b.pdf?ts=1604672421028 (accessed on 6 November 2020).

24.   Galaxy S10 Reverse Wireless Charging Feature: How to Use It. Available online: https://www.valuewalk.com/2019/03/galaxy-s10-reverse-wireless-charging/ (accessed on 6 November 2020).

25.   WE-WPCC Wireless Power Charging Receiver Coil 760308101214. Available online: https://www.we-online.com/catalog/datasheet/760308101214.pdf (accessed on 6 November 2020).

26.   Puers, R. *Inductive Powering: Basic Theory and Application to Biomedical Systems*, 1st ed.; Springer: Dordrecht, The Netherlands, 2009.

27.   Computer Controlled Pan-Tilt Unit Model PTU-D46. Available online: www.imagelabs.com/wp-content/uploads/2011/01/Specs-PTU-D46.pdf (accessed on 6 November 2020).

28.   Madgwick, S.O.H. An Efficient Orientation Filter for Inertial and Inertial/Magnetic Sensor Arrays. *Rep. x-io Technol. Univ. Bristol (UK)* **2010**, *25*, 113–118.

29.   Shi, J.; Tomasi, C. Good Features to Track. In Proceedings of the IEEE Conference on Computer Vision and Pattern Recognition, Seattle, WA, USA, 21–23 June 1994; pp. 593–600.

30.   Isard, M.; Blake, A. CONDENSATION—Conditional Density Propagation for Visual Tracking. *Int. J. Comput. Vis.* **1998**, *29*, 5–28. [CrossRef]

31.   de Kok, J.; Vroonhof, P.; Snijders, J.; Roullis, G.; Clarke, M.; Peereboom, K.; van Dorst, P.; Isusi, I. *Work-Related Musculoskeletal Disorders: Prevalence, Costs and Demographics in the EU*; European Agency for Safety and Health at Work: Bilbao, Spain, 2019.

32.   Dierick, F.; Buisseret, F.; Brismée, J.M.; Fourré, A.; Hage, R.; Leteneur, S.; Monteyne, L.; Thevenon, A.; Thiry, P.; Van der Perre, L.; et al. Opinion on the Effectiveness of Physiotherapy Management of Neuro-Musculo-Skeletal Disorders by Telerehabilitation. Available online: https://www.ifompt.org/site/ifompt/Telerehab_EN.pdf (accessed on 27 October 2020).

33.   Coviello, G.; Avitabile, G.; Florio, A. A Synchronized Multi-Unit Wireless Platform for Long-Term Activity Monitoring. *Electronics* **2020**, *9*, 1118. [CrossRef]

34.   Claesson, E.; Marklund, S. Calibration of IMUs using Neural Networks and Adaptive Techniques Targeting a Self-Calibrated IMU. Master's Thesis, Chalmers University of Technology, Gothenburg, Sweden, 2019.

Publisher's Note: MDPI stays neutral with regard to jurisdictional claims in published maps and institutional affiliations.

 *sensors*

Article

Lower Limb Kinematics Using Inertial Sensors during Locomotion: Accuracy and Reproducibility of Joint Angle Calculations with Different Sensor-to-Segment Calibrations

Julien Lebleu [1,*], Thierry Gosseye [2,3], Christine Detrembleur [1], Philippe Mahaudens [1,4], Olivier Cartiaux [1,5] and Massimo Penta [2,3]

[1] Neuro Musculo Skeletal Lab (NMSK), Institut de Recherche Expérimentale et Clinique, Université Catholique de Louvain, Secteur des Sciences de la Santé, Avenue Mounier 53, B-1200 Brussels, Belgium; christine.detrembleur@uclouvain.be (C.D.); philippe.mahaudens@uclouvain.be (P.M.); crt@ecam.be (O.C.)

[2] Institue of Neurosciences (IONS), Université Catholique de Louvain, Secteur des Sciences de la Santé, Place Pierre de Coubertin 1, B-1348 Louvain-la-Neuve, Belgium; thierry@arsalis.com (T.G.); massimo@arsalis.com (M.P.)

[3] Arsalis SPRL, Chemin du Moulin Delay 6, B-1473 Glabais, Belgium

[4] Cliniques Universitaires St-Luc, Service D'orthopédie, avenue Hippocrate 10, B-1200 Brussels, Belgium

[5] ECAM, Brussels Engineering School, Promenade de l'Alma 50, B-1200 Brussels, Belgium

* Correspondence: julien.lebleu@uclouvain.be; Tel.: +32-2-764-53-75

Received: 11 December 2019; Accepted: 22 January 2020; Published: 28 January 2020

Abstract: Inertial measurement unit (IMU) records of human movement can be converted into joint angles using a sensor-to-segment calibration, also called functional calibration. This study aims to compare the accuracy and reproducibility of four functional calibration procedures for the 3D tracking of the lower limb joint angles of young healthy individuals in gait. Three methods based on segment rotations and one on segment accelerations were used to compare IMU records with an optical system for their accuracy and reproducibility. The squat functional calibration movement, offering a low range of motion of the shank, provided the least accurate measurements. A comparable accuracy was obtained in other methods with a root mean square error below 3.6° and an absolute difference in amplitude below 3.4°. The reproducibility was excellent in the sagittal plane (intra-class correlation coefficient (ICC) > 0.91, standard error of measurement (SEM) < 1.1°), good to excellent in the transverse plane (ICC > 0.87, SEM < 1.1°), and good in the frontal plane (ICC > 0.63, SEM < 1.2°). The better accuracy for proximal joints in calibration movements using segment rotations was traded to distal joints in calibration movements using segment accelerations. These results encourage further applications of IMU systems in unconstrained rehabilitative contexts.

Keywords: inertial sensor; gait; validity; functional calibration; accuracy; wearable electronic devices

---

1. Introduction

Quantitative assessment of lower limb kinematics is required in various applications, such as motion analysis, sports science, and rehabilitation. Although opto-electronic motion capture systems are considered the gold standard for this assessment, their widespread use is limited by their restricted area of measurement, their optical limitations due to marker occlusion or reflection, and their cost. Moreover, opto-electronic trackers are generally used in a restricted lab environment, which further limits the exploration of real-life movements and exercises. Wearable sensors, such as inertial measurement units (IMUs), have been developed to overcome these limitations [1], allowing for human motion analysis in unconstrained real-life conditions [2,3].

Although they generally contain a 3D accelerometer, a 3D gyroscope, and an optional 3D magnetometer, an IMU does not measure joint angles perfectly. Joint angles obtained via signal integration typically drift over time [4,5] and their accuracy varies with the joint assessed and the movement complexity [6,7]. While an error under 5° is generally accepted for most clinical gait applications [8], the measurement error typically ranges from 5° to 18.8° depending on the joint and the plane of motion [6]. Another major challenge in IMU-based human motion analysis that the IMUs' local coordinate systems are not aligned with physiologically meaningful axes [9]. Such alignment, required to compute joint angles, can be performed via a "sensor-to-segment" calibration procedure [10].

The first approach to ensure this alignment consists of a rigorous positioning of the sensor in relation to the anatomy [11]. This method assumes that the segment axes are parallel to the IMU axes, is approximate, and requires user expertise to locate the sensor axes relative to the joint axes for both segments around each joint. The second approach consists of placing an IMU on each segment and aligning the IMU and joint axes via a set of calibration postures [12] and/or movements around physiological motion axes. The latter functional method [13,14] consists of making the subject stand upright with straight legs for a few seconds to define the vertical axis for each IMU or segment, while the other axes are defined via active or passive movements [13,14]. Since the movements are generally human-controlled, the accuracy of the axes definition essentially relies on the subjects' ability to precisely hold a given posture and on the execution of a given movement [9]. The third approach consists of exploiting the kinematic constraints of the joints and use almost arbitrary movements to perform the sensor-to-segment calibration [9,15]. This method is particularly adapted to single axis joints that can be satisfactorily modelled as a hinge joint like the knee; however, the modeling of spherical joints requires the execution of movements mostly around one axis to identify the joint axes [9], which resemble the functional method.

The IMU-based tracking of lower limb spherical joints, using one of the two aforementioned methods, therefore requires the execution of a functional calibration movement. While the accuracy of the sensor-to-segment calibration is determined by the quality of the functional calibration movement, to our knowledge, it has only been investigated in one study for upper limb motion tracking [16] and no study has compared functional calibration movements for the tracking of lower limb joint angles.

This study aimed to (1) assess the accuracy of different functional calibration methods in order to compute the lower limb joint angles during walking, (2) assess the reproducibility of different functional calibration movements, and (3) compare the accuracy provided by functional calibration movements in different gait movements.

2. Materials and Methods

2.1. Participants

Seven healthy young adults participated in this study (6 females, 1 male, mean (SD) age = 22.6 (1.5) years, height = 1.67 (0.08) m, body mass = 65.4 (11.6) kg). Participants were included in the study if they were between 20 and 25 years old and free of any injury at the time of participation. The study protocol was approved by the ethics committee of our university (agreement number: B403201523492) and each patient provided written informed consent to the use of their anonymized data.

2.2. Experimental Setup and Recordings

To assess the lower limb joint kinematics, seven wearable IMUs; (x-IMU, x-io Techologies, Bristol, UK) were fixed in matched 3D-printed ABS (acrylonitrile butadiene styrene) enclosures and attached by means of a semi-elastic belt to seven lower body segments, as shown in Figure 1: the waistline at the level of the fifth lumbar vertebra (L5), the middle of the thighs, the middle of the shanks, and at the dorsal side of the feet. The IMUs were firmly strapped on the skin or clothes. Although this could lead to undesirable artifacts, it is more representative of records in an unconstrained context, such as outdoor conditions. Each IMU included a tri-axial accelerometer (full scale ±6 g), a gyroscope

(±2000°/s), and a magnetometer (±8.1 G) that were sampled at a frequency of 128 Hz. The IMUs were connected to a computer by means of a Bluetooth connection using a custom application based on open source software [17] (C# program, github.com/xioTechnologies). Each movement was recorded independently. The synchronization between the IMUs was ensured by a custom-built magnetic coil that sent a magnetic impulse at the beginning of each recording.

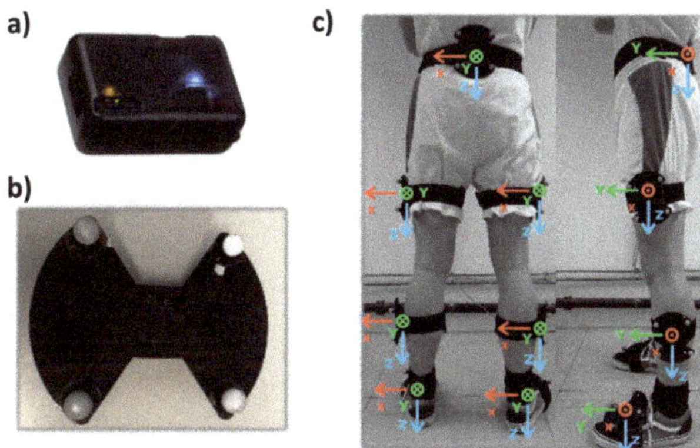

Figure 1. Sensor locations: (**a**) x-IMU from xi-o Technologies, (**b**) 3D-printed acrylonitrile butadiene styrene (ABS) enclosures (4 markers of 14 mm diameter) with the inertial measurement unit (IMU) reference frame, and (**c**) segment reference frame of the 7 IMUs on the subject.

Four reflective markers were fixed at each corner of the ABS enclosures to define clusters for each segment (Figure 1). Motion capture data were collected at a rate of 200 Hz using an eight-camera motion analysis system (Vicon V5 Motion Systems, Oxford Metrics Ltd., Oxford, UK)) and processed using Nexus 2.5 software. The position of each marker on the cluster allowed for the orientation of each segment to be computed in the lab reference frame.

2.3. Functional Calibration and Test Movements

The experimental protocol is illustrated in Figure 2. Four functional calibration movements were performed to assess their reproducibility and accuracy regarding lower limb joint angle measurement with the IMUs. These movements were designed to include a rotation in the sagittal plane of each lower body segment, including the pelvis, while being easy to explain and reproduce. Each functional calibration movement included (1) an upright static posture with the arms alongside the body and the feet parallel beside each other that was used to define the segment vertical axis and (2) a functional movement spanning a range of orientations for each segment in the sagittal plane that was used to define a second segment axis. In the static posture, the segment was supposed to have a zero angle in all three planes such that the segment reference frames were aligned with the lab reference frame. The X axis was defined as the medio-lateral axis, pointing to the left of the subject; the Y axis as the anterio-posterior axis, pointing in front of the subject; and the Z axis as the vertical axis, pointing downward (Figure 1c).

The instructions were as follows:

- Calibration movement 1: "Tilted to stand": Start in a leaned-back position with extended legs on the chair, bend the knees, bend the trunk forward, get up from the chair, and stop moving.
- Calibration movement 2: "Extension stand up": Sit on the chair, extend the knees in front of you, bend the knees, bend the trunk forward, get up from the chair, and stop moving.
- Calibration movement 3: "Squat": Stand in front of the chair, rise on your heels, squat deeply, get up, and stop moving.

- Calibration movement 4: "Walking" 5 m:
- 4a. Walk at your pace to the red line on the floor, then stop moving.
- 4b. Fast: walk five meters as fast as you can without running to the red line on the floor as if you were late to a meeting.
- 4c. Slow: walk five meters slowly to the red line on the floor but keep moving.

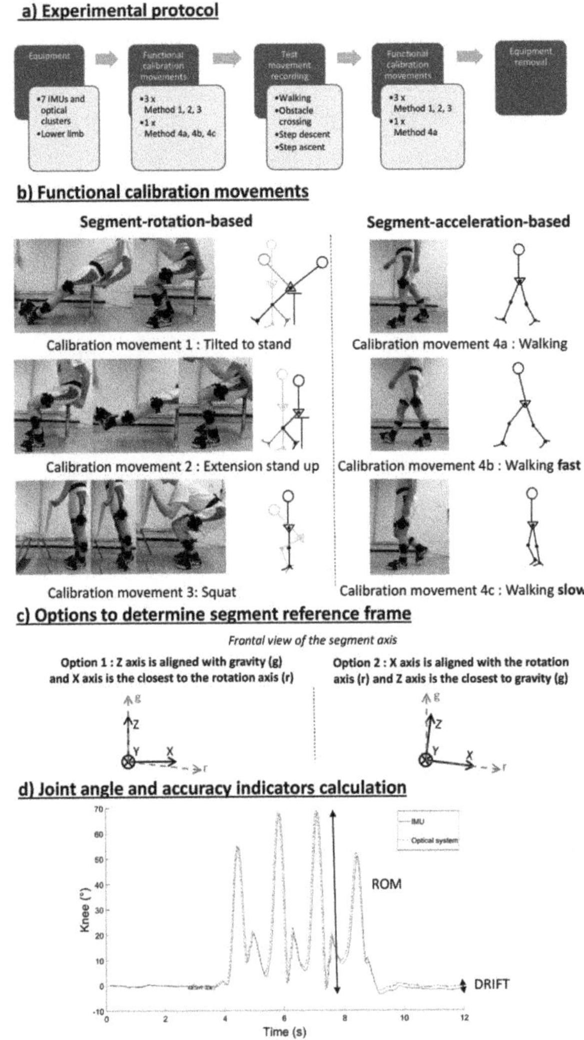

Figure 2. (**a**) Experimental protocol. (**b**) Functional calibration movements: for movements 1, 2, and 3, where the second axis was defined as the principal rotational axis as determined by a principal component analysis (PCA) on gyroscope signals; for movements 4a, 4b, and 4c, the second axis was defined as the principal acceleration axis through a PCA on accelerometer signals. (**c**) Two options to determine the segment reference frames, as shown in segment frontal views. (**d**) The accuracy was assessed using the root mean square error (RMSE), the absolute difference in the range of motion (ΔROM) between both systems, and the absolute drift accumulated during the movement (DRIFT).

The functional calibration movements were demonstrated by the operator and each participant received practice trials to get used to each movement. Each functional calibration movement was recorded three times before and three times after the execution of the test movements. The walking movement at self-selected speed was only performed two times, before and after the test movements.

Four test movements were performed in the same order:

- walking five meters at a self-selected speed;
- stepping over an obstacle 28 cm in height while walking 5 m at a self-selected speed;
- ascend a step 20 cm in height;
- descend a step 20 cm in height.

The mean recorded times for test movements were 11 s for walking, 11 s for stepping over an obstacle, 14 s for the step ascent, 16 s for the step descent.

2.4. Signal Processing

An open source attitude and heading reference system (AHRS) algorithm was used for sensor fusion between the accelerometer and gyroscope sensor data of the IMU (Mahony's AHRS algorithm) [18]; the magnetometer signals were omitted. The four calibration movements were used to compute the orientation of each segment relative to the lab reference frame in different ways. The gravity vector during the static upright posture was used to define the vertical axis for each segment. A second segment axis was defined in one of two ways depending on the method of functional calibration. For functional calibration movements 1, 2, and 3, it was defined as the principal rotational axis as determined by a principal component analysis (PCA) on gyroscope signals. Two options were used to determine the segment reference frame (see frontal views of the segment reference frame in Figure 2c): either the gravity vector (g) was defined as the vertical axis and the lateral axis was forced to be the orthogonal axis closest to the rotation axis of the functional calibration movement (r), or the lateral axis was defined as the functional calibration movement rotation axis and the vertical axis was the orthogonal axis closest to the gravity vector (and thus the transversal plane was not perfectly horizontal in the static upright posture). For functional calibration movements 4a, 4b, and 4c, the second axis was defined as the principal acceleration axis through a PCA on accelerometer signals transformed in a lab-fixed reference frame. The 3D orientation of the pelvis and joint angles for the hips, knees, and ankles were calculated from the segment orientations based on the recommendations of the International Society of Biomechanics [6] for the different functional calibration movements. Flexion-extension were rotations around the X axis, abduction-adduction was around the Y axis, and internal-external rotations were around the Z axis.

The lower body 3D kinematics derived from the optical system were computed in two different ways. They were either computed in the lab frame or computed through the same functional calibration procedures described above, using the principal axis of rotation or acceleration determined from optical records.

For each participant, a static period (about 5 s) in a standing position was captured at the beginning of each test to define the segment's initial orientation for the IMU AHRS algorithm.

2.5. Data Analysis

Joint angles of the walking test movement were calculated for all functional calibration procedures. The accuracy of the IMU kinematics was computed for each calibration procedure as the difference in joint angle between the IMU and optical measurements. The accuracy was assessed using the root mean square error (RMSE) during the movement period, the absolute difference in the range of motion (ROM) between both systems (ΔROM), and the absolute drift accumulated during the movement due to the error in the angular rate integration (DRIFT). The RMSE, ΔROM, and DRIFT parameters were computed using Matlab 2018 (Mathworks Inc, Natick, MA, USA) and are expressed in degrees.

A generalized linear model was used to assess the effect of (1) the functional calibration movement, (2) the option used to determine the segment's reference frame to compute the IMU orientation, and (3) the functional calibration method for the optical system on the amplitude of the RMSE, ΔROM, and DRIFT parameters for each joint angle and plane of motion. This analysis was performed with SPSS (version 25, IMB Corporation, Amonk, NY, USA) and the significance level was set to $\alpha = 0.05$.

The reproducibility of each functional calibration movement was assessed as the difference in joint angle computed from each repetition of the functional calibration movement. The reproducibility of the ROM parameter in all movement planes and joints was determined based on the intra-class correlation coefficient (ICC) [19] and standard error of measurement (SEM) [19]. Values of ICC ≥ 0.90 were considered as excellent, 0.70–0.89 as good, 0.40–0.69 as acceptable, and <0.40 as low [20]. The SEM estimates the non-systematic variance and reflects the within-subject variability among repeated calibrations. A proportional SEM (SEM%) was calculated by expressing the SEM relative to the mean ROM (SEM% = (SEM/mean) × 100%)) [21]. An SEM% above 10% was considered as high.

Once the most accurate and reproducible functional calibration method was selected for a walking test, the accuracy was determined for the other test movements, namely the step ascent, descent, and stepping over an obstacle, using the RMSE and ΔROM parameters. The parameters were calculated for the front leg (i.e., the first leg to touch the step in the step ascent, the first leg to touch the floor leg in the step descent, and the first leg to touch the ground in obstacle stepping) and for the back leg in the different test movements. Differences in the RMSE and ΔROM parameters between test movements were assessed with a one-way ANOVA. Tukey's post hoc test was used to reveal which groups differed in the case of significant p-values. The significance level was set to $\alpha = 0.05$.

3. Results

The pelvis orientation and the joint angles during a time-normalized typical gait stride at a self-selected speed of 1 m·s^{-1} are illustrated in Figure 3 for measurements with the IMUs and with the optical system using the walking functional calibration (Calibration movement 4a). The traces displayed a classical movement pattern as indicated by the similarity between both set of measurements, with a mean RMSE value lower than 5° for all joints in all planes.

3.1. Assessment of the IMU Accuracy

The mean differences between measurements from the IMUs and from the optical system, as well as the factors affecting this difference, are presented in Table 1. The highest mean error was observed at the knees and at the ankles, bilaterally, as shown by the RMSE between 3.0° and 4.1° and by the ΔROM between 1.9° and 5.1° in any plane of motion. The generalized linear model also indicated that these differences were significantly linked to the functional calibration movement; see the details in Figure 4. These errors were also significantly associated with the methods of determination of the IMUs reference frame, although to a lesser extent: the mean (SD) RMSE was 2.3° (1.7°) when the vertical axis was aligned to the gravity vector and 3.0° (3.1°) when the lateral axis was aligned to the segment rotation axis. The errors were not significantly linked to the two different methods used to compute the 3D kinematics from the optical system (lab frame or computed through the functional calibration). Table 1 also shows that the DRIFT was largely independent of the factors considered in the generalized linear model.

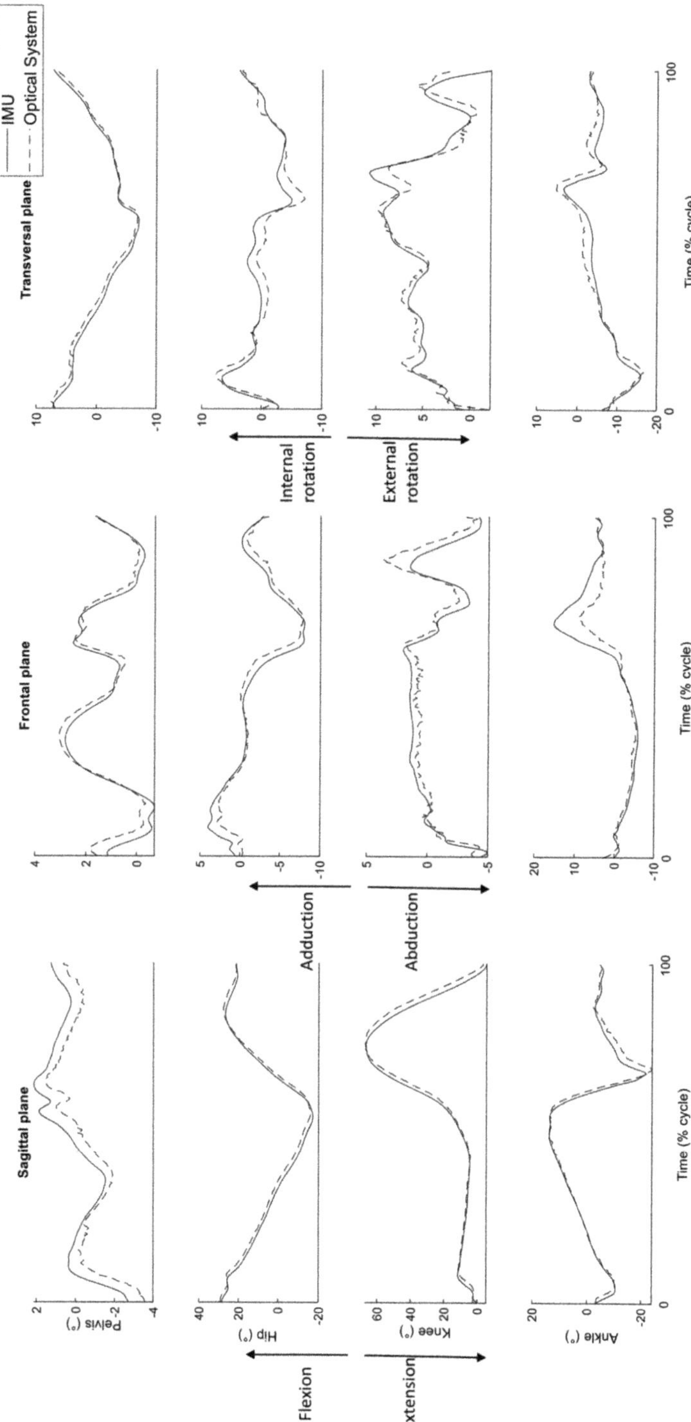

Figure 3. Typical trace of the 3D pelvis orientation and joint angles during a gait stride for measurements with the IMUs (plain lines) and with the optical system (short dash lines) using the walking functional calibration.

Table 1. Generalized linear model: effect of functional calibration movement, segment reference frame, and optical functional calibration on the accuracy.

Accuracy Indicator			RMSE (°)				ΔROM (°)				DRIFT (°)		
Calibration Factor			Functional Calibration Movement	Segment Reference Frame	Optical Functional Calibration		Functional Calibration Movement	Segment Reference Frame	Optical Functional Calibration		Functional Calibration Movement	Segment Reference Frame	Optical Functional Calibration
		Mean (SD)	Mean Difference between Methods			Mean (SD)	Mean Difference between Methods			Mean (SD)	Mean Difference between Methods		
Pelvis	Sagittal	0.9 (0.5)	-	-	-	0.4 (0.6)	1.0	-	-	0.4 (0.6)	-	-	-
	Frontal	1.1 (0.9)	1.3	-	-	0.5 (0.7)	1.1	-	-	0.5 (0.7)	-	-	-
	Transverse	1.5 (1.8)	-	-	-	0.2 (0.2)	-	-	-	0.2 (0.2)	-	-	-
Hip right	Sagittal	2.0 (1.2)	-	-	-	1.6 (1.6)	-	1.0	-	1.6 (1.6)	-	-	-
	Frontal	2.7 (2.1)	-	-	-	2.3 (1.5)	-	-	-	2.3 (1.5)	-	-	-
	Transverse	2.4 (1.5)	-	-	-	2.2 (1.3)	-	-	-	2.2 (1.3)	-	-	-
Knee right	Sagittal	**4.1 (3.1)**	-	1.5	-	2.4 (3.8)	3.9	-	-	2.4 (3.8)	1.2	-	-
	Frontal	**3.6 (2.3)**	**4.0**	-	-	**4.7 (6.4)**	9.2	-	-	**4.7 (6.4)**	-	-	-
	Transverse	**3.3 (2.1)**	-	2.4	-	**3.2 (5.2)**	4.3	3.1	-	**3.2 (5.2)**	-	-	-
Ankle right	Sagittal	2.5 (1.7)	-	0.9	-	2.7 (5.1)	-	-	-	2.7 (5.1)	-	-	-
	Frontal	**3.3 (2.5)**	**3.7**	-	-	**4.2 (5.6)**	7.2	-	-	**4.2 (5.6)**	-	-	0.5
	Transverse	2.4 (4.3)	**4.3**	2.6	-	**5.1 (6.6)**	5.6	2.8	-	**5.1 (6.6)**	2.0	1.0	-
Hip left	Sagittal	2.5 (0.8)	-	-	-	0.8 (0.7)	-	-	-	0.8 (0.7)	-	-	-
	Frontal	2.1 (1.6)	-	-	-	1.9 (1.7)	-	-	-	1.9 (1.7)	-	-	-
	Transverse	2.1 (1.6)	-	-	-	2.0 (1.8)	1.9	0.6	-	2.0 (1.8)	-	-	-
Knee left	Sagittal	**3.7 (3.5)**	-	1.9	-	1.9 (3.6)	-	-	-	1.9 (3.6)	-	-	-
	Frontal	**3.5 (1.7)**	1.9	-	-	**4.3 (4.0)**	4.9	-	-	**4.3 (4.0)**	-	-	-
	Transverse	3.0 (1.6)	-	1.4	-	2.7 (3.9)	3.0	2.6	-	2.7 (4.0)	-	-	-
Ankle left	Sagittal	3.2 (1.5)	-	-	-	2.3 (4.0)	4.6	2.2	-	2.3 (4.0)	-	-	-
	Frontal	2.9 (1.6)	2.8	-	-	**3.6 (3.9)**	3.8	-	-	**3.6 (3.9)**	-	-	-
	Transverse	3.2 (4.3)	**4.8**	2.4	-	**3.6 (5.1)**	6.7	2.0	-	**3.6 (5.1)**	-	-	-

Numerical values indicate significant effects, dashes indicate non-significant effects. Mean differences between calibration factors are presented in degrees for significant effects ($p < 0.05$) and in bold for angles >3°. For the functional calibration movement factor, the highest mean difference between the six movements is presented. RMSE: root mean square error, ΔROM: absolute difference in range of motion, DRIFT: absolute difference between IMU and optical system at end of recording, SD: standard deviation.

The variation of the RMSE across different functional calibration movements is illustrated in Figure 4. While the RMSE varied between approximately 1° and 6° for most joints, planes of motion, and functional calibration movements, it was larger for the "squat" calibration movement, especially at the knee and ankle where the upper confidence limit of the RMSE reached up to 8° and the upper confidence limit ΔROM reached up to 10° or more. Tilted, extension, and walking functional calibration movements tended to provide less accurate measurements at the knee. Walking functional calibration movements tended to report more accurate angles for distal lower limb joints, although without a clear visible impact on the ROM accuracy. After excluding the "squat" movement, the mean accuracy for all other calibration movements is summarized in Table 2, indicating a mean RMSE of less than 2° at the pelvis, less than 3° at the hip and ankle, and less than 4° at the knee with a trend for larger errors in the frontal plane. The mean errors were smaller in ΔROM. DRIFT values at the hips and at other lower limb joints in the sagittal plane were on average under 2.7°, while higher mean values up to 4.9° were observed at the knee and at the ankle in the frontal and transverse planes.

Table 2. Summary of the accuracy without the squat calibration movement and with reference frames aligned with gravity.

		RMSE (°)	ΔROM (°)	DRIFT (°)
		Mean (SD)		
Pelvis	Sagittal	1.0 (0.7)	0.7 (0.8)	0.4 (0.6)
	Frontal	1.2 (1.1)	0.7 (0.9)	0.5 (0.4)
	Transverse	1.5 (1.8)	0.2 (0.2)	0.2 (0.2)
Hip right	Sagittal	2.1 (1.3)	1.0 (1.0)	1.6 (1.6)
	Frontal	2.9 (2.2)	2.0 (1.7)	2.3 (1.6)
	Transverse	2.2 (1.5)	1.7 (1.1)	2.1 (1.3)
Knee right	Sagittal	**3.6 (2.5)**	1.1 (0.6)	1.9 (1.8)
	Frontal	2.8 (1.6)	2.5 (2.3)	**4.0 (5.3)**
	Transverse	2.2 (1.2)	1.8 (1.9)	2.9 (4.8)
Ankle right	Sagittal	2.0 (1.3)	1.6 (1.1)	2.7 (5.4)
	Frontal	2.4 (1.6)	2.6 (2.7)	**3.5 (4.6)**
	Transverse	2.2 (1.0)	**3.4 (2.9)**	**4.9 (6.6)**
Hip left	Sagittal	2.5 (1.7)	1.1 (0.7)	0.8 (0.7)
	Frontal	2.4 (1.9)	1.8 (1.8)	2.0 (1.7)
	Transverse	2.0 (1.5)	1.5 (1.3)	2.0 (1.8)
Knee left	Sagittal	**3.2 (2.6)**	1.0 (0.7)	1.4 (1.1)
	Frontal	**3.2 (1.7)**	**3.1 (3.0)**	**4.0 (3.9)**
	Transverse	2.3 (0.9)	1.5 (1.1)	2.5 (3.9)
Ankle left	Sagittal	2.9 (1.4)	0.9 (0.8)	2.1 (4.1)
	Frontal	2.3 (1.2)	2.3 (1.9)	**3.1 (3.3)**
	Transverse	1.9 (0.8)	1.7 (1.7)	**3.1 (5.1)**

Bold values indicate angles >3°.

3.2. Assessment of the IMU Reproducibility

The mean ROM recorded for each joint in each plane of motion during one walking test movement are presented in Table 3, together with the reproducibility indices computed across repetitions of each calibration movement. The mean ROM displayed symmetrical values for both limbs and classical movement amplitudes for walking at 1 m·s⁻¹. Overall, the mean reproducibility was excellent for all calibration movements in the sagittal plane (ICC: 0.96–0.99) and it was good to excellent in the transverse plane (ICC: 0.87–0.93). In the frontal plane, the mean reproducibility was good for all calibration movements (ICC: 0.79–0.86), except for walking, which had an acceptable reproducibility on average (ICC: 0.63) due to the low ICC observed for the hip and knee. The reproducibility was uniformly good to excellent across the calibration movements, except for the walking movement, which reported slightly lower reproducible movement amplitudes for proximal joints (as low as ICC = 0.76 for the hip compared to ICC = 0.94 for any other movement) and slightly higher reproducibility in distal joints (as high as ICC = 0.91 for the ankle compared to ICC = 0.82 for other movements). For all functional calibration movements, whatever the joint, the mean SEM was within 1.2° in all planes, although the mean SEM remained generally higher at the knee (0.9°) and ankle (1.8°) compared to the hip (0.6°) and pelvis (0.1°). This resulted in acceptable variations between movements, as shown by a mean SEM% of 1.8% in the sagittal plane, 4.8% in the transverse plane, and 7.9% in the frontal plane, where articular amplitudes were smaller for all lower limb joints.

3.3. Assessment of Accuracy in Different Test Movements

The functional calibration movements that provided the highest accuracy were the tilted and extension movements, as well as walking at a self-selected speed (mean RMSE for all joints, respectively: 2.5°, 2.3°, 2.2°). Since these calibration movements provided a comparable performance for the functional calibration of the IMU, the mean accuracy with the walking functional calibration was computed for the measurement of the four gait test movements, namely walking, ascending or descending one step, and stepping over an obstacle, when considering the lateral axis of the reference frame perpendicular to the gravity vector (as similar accuracies were obtained when considering it parallel to the axis of the segment rotation during the functional calibration). The mean accuracy obtained with the functional calibration movements retained is presented in Table 4. The RMSE and ΔROM were both smaller than 6° for ascending a step and smaller than 13° for descending it. For stepping over an obstacle, the RMSE reached 13° but the ΔROM had maximum values of only 4°. Notably, the accuracy of the IMU measurements were higher in walking than in other gait movements, where larger inaccuracies were observed, especially for a step ascent (both ankles in the sagittal plane and back leg hip in the frontal plane), for a step descent (all joints of both legs), and for stepping over an obstacle (all joints of both legs excluding the pelvis), although the error was generally within 5° and only rarely exceeded 10°.

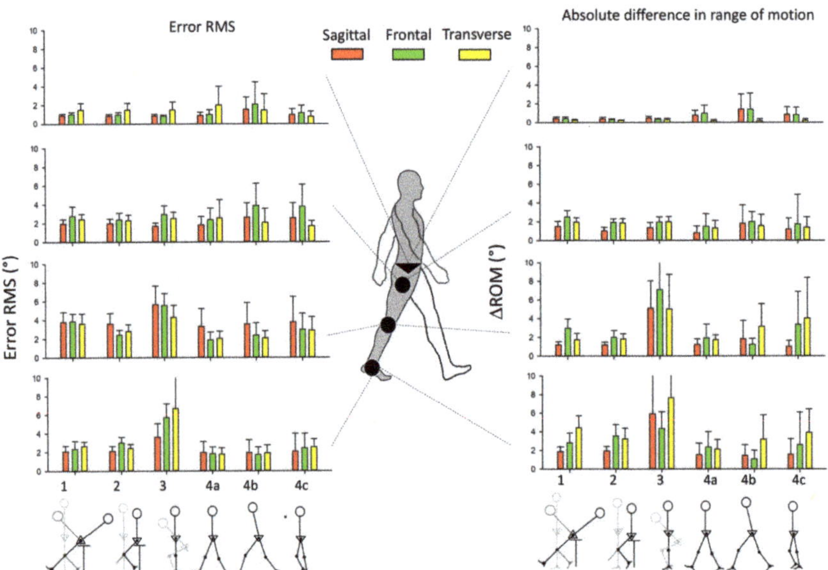

Figure 4. Accuracy across functional calibration movements (1. Tilted to stand, 2. Extension, 3. Squat, 4a. Walking (1 ± 0.1 m·s−1), 4b. Walking fast (1.5 ± 0.1 m·s−1), 4c. Walking slow (0.6 ± 0.1 m·s−1)). Error bars are the confidence interval means at 95%.

Table 3. Joint angle measurement reproducibility for one walk task across four functional calibration movements.

		ROM (°)		Tilted (Movement 1)			Extension (Movement 2)			Squat (Movement 3)			Walking (Movement 4a)	
		Mean (SD)	ICC	SEM (°)	SEM%	ICC	SEM (°)	SEM%	ICC	SEM (°)	SEM%	ICC	SEM (°)	SEM%
Pelvis	Sagittal	12.8 (2.1)	1.00	0.1	0.7%	1.00	0.1	0.5%	1.00	0.1	0.5%	0.98	0.3	2.5%
	Frontal	6.8 (0.8)	0.98	0.1	1.7%	0.98	0.1	1.6%	1.00	0.1	0.8%	0.76	0.4	6.1%
	Transverse	17.9 (4.1)	1.00	0.0	0.0%	1.00	0.0	0.0%	1.00	0.0	0.0%	1.00	0.0	0.0%
Hip right	Sagittal	50.4 (2.4)	0.98	0.3	0.7%	1.00	0.2	0.4%	1.00	0.3	0.6%	0.90	0.9	1.8%
	Frontal	14.0 (1.9)	0.81	0.8	5.8%	0.90	0.7	5.1%	0.90	0.6	4.2%	0.50	1.3	9.3%
	Transverse	18.4 (2.7)	0.98	0.4	2.0%	1.00	0.4	1.9%	1.00	0.2	1.1%	0.90	0.7	4.0%
Knee right	Sagittal	73.3 (3.2)	1.00	0.2	0.2%	1.00	0.2	0.3%	0.92	0.9	1.3%	1.00	0.2	0.3%
	Frontal	11.8 (1.9)	0.68	1.1	9.4%	0.72	1.0	8.7%	0.83	0.8	6.8%	0.15	1.8	**15.3%**
	Transverse	18.9 (2.8)	0.93	0.7	4.0%	0.91	0.8	4.4%	0.77	1.4	7.2%	0.78	1.3	6.9%
Ankle right	Sagittal	46.3 (16.0)	0.95	3.4	7.4%	0.97	2.9	6.2%	0.93	4.3	9.3%	1.00	1.0	2.2%
	Frontal	18.5 (3.2)	0.69	1.8	9.8%	0.56	2.1	**11.5%**	0.80	1.4	7.7%	0.83	1.3	7.2%
	Transverse	21.3 (4.8)	0.83	2.0	9.4%	0.81	2.1	9.7%	0.63	2.9	**13.6%**	0.81	2.1	9.9%
Hip left	Sagittal	46.9(4.7)	1.00	0.2	0.5%	1.00	0.2	0.5%	1.00	0.1	0.3%	0.90	1.2	2.6%
	Frontal	15.0 (1.6)	0.91	0.5	3.3%	0.90	0.5	3.2%	0.90	0.5	3.3%	0.40	1.3	8.5%
	Transverse	17.2 (3.5)	0.98	0.5	2.8%	1.00	0.8	4.5%	1.00	0.3	1.6%	0.90	0.9	5.2%
Knee left	Sagittal	75.2 (9.3)	1.00	0.5	0.7%	1.00	0.6	0.8%	0.98	1.2	1.6%	0.99	0.9	1.2%
	Frontal	10.4 (2.8)	0.60	1.8	**17.1%**	0.70	1.5	**14.9%**	0.86	1.0	**10.0%**	0.84	1.1	**10.9%**
	Transverse	19.7 (4.5)	0.99	0.4	2.1%	0.99	0.4	2.2%	0.97	0.7	3.8%	0.95	1.0	5.3%
Ankle left	Sagittal	41.0 (3.7)	0.92	1.0	2.5%	0.98	0.6	1.4%	0.96	0.7	1.8%	0.98	0.5	1.2%
	Frontal	17.4 (4.5)	0.87	1.6	9.1%	0.86	1.7	9.6%	0.70	2.4	**14.0%**	0.95	1.0	5.9%
	Transverse	19.8 (3.6)	0.68	2.0	**10.3%**	0.83	1.5	7.5%	0.75	1.8	9.2%	0.88	1.3	6.3%
	Mean Sagittal	49.4 (5.9)	0.98	0.8	1.8%	0.99	0.7	1.4%	0.97	1.1	2.2%	0.96	0.7	1.7%
	Mean Frontal	13.4 (2.4)	0.79	1.1	8.0%	0.80	1.1	7.8%	0.86	1.0	6.7%	0.63	1.2	9.0%
	Mean Transverse	19.0 (3.7)	0.91	0.9	4.4%	0.93	0.8	4.3%	0.87	1.0	5.2%	0.90	1.1	5.4%
	Mean Pelvis	12.5 (2.4)	0.99	0.1	0.8%	0.99	0.1	0.7%	1.00	0.0	0.4%	0.91	0.2	2.8%
	Mean Hip	26.9 (2.8)	0.94	0.5	2.5%	0.95	0.5	2.6%	0.96	0.3	1.8%	0.76	1.1	5.2%
	Mean Knee	34.8 (4.1)	0.86	0.8	5.6%	0.89	0.8	5.2%	0.89	1.0	5.1%	0.78	1.1	6.6%
	Mean Ankle	27.4 (5.9)	0.82	2.0	8.1%	0.84	1.8	7.7%	0.80	2.3	9.3%	0.91	1.2	5.5%

Bold values indicate SEM% > 10%, ROM: range of motion, SD: standard deviation, ICC: intraclass correlation coefficient, SEM: standard error of measurement, SEM%: standard error of measurement in proportion to the mean.

Table 4. Accuracy obtained for each joint, in the 3D, in the four test movements: walking, stair ascent, stair descent, and obstacle crossing.

			Mean ROM (°)			Mean RMSE (°)				Mean ΔROM (°)			
		Walking	Stair Ascent	Stair Descent	Obstacle Crossing	Walking	Stair Ascent	Stair Descent	Obstacle Crossing	Walking	Stair Ascent	Stair Descent	Obstacle Crossing
Front	Pelvis Sagittal	12.8	18.4	24.4	18.9	0.9	1.2	1.7 *	1.4	0.5	0.8	1.3	1.3
	Pelvis Frontal	6.8	13.7	20.0	13.3	0.9	2.2*	2.6 *	1.3	0.5	0.7	1.9 *	1.0
	Pelvis Transverse	17.9	20.0	47.6	27.1	1.4	1.5	3.5 *	2.0	0.2	0.6	0.8 *	0.6
	Hip Sagittal	50.4	83.5	69.7	104.5	2.1	2.8	7.8 *	7.3 *	1.2	1.0	2.6 *	1.5
	Hip Frontal	14.0	18.1	30.8	17.4	2.4	4.3 *	4.9 *	3.5 *	2.1	3.8	3.1	2.9
	Hip Transverse	18.4	20.6	30.0	21.4	2.2	3.1	4.2 *	4.5 *	2.0	2.2	3.0	2.8
	Knee Sagittal	73.3	98.7	101.6	119.2	3.6	5.3	12.8 *	10.2 *	1.3	4.3	2.3	5.3 *
	Knee Frontal	11.8	13.6	15.3	14.4	2.7	4.2	4.2	4.4	2.5	4.7	3.0	6.5 *
	Knee Transverse	18.9	19.1	25.3	27.8	2.4	3.8	5.4 *	5.1 *	1.9	4.2	4.0	5.6 *
	Ankle Sagittal	46.3	45.4	76.5	48.4	2.1	5.7 *	7.9 *	4.3 *	1.6	5.4	4.6	6.8
	Ankle Frontal	18.5	21.6	25.1	21.9	2.5	2.4	4.0 *	3.3	2.7	2.3	3.8	3.0
	Ankle Transverse	21.3	18.4	24.5	21.2	2.4	2.3	3.1	3.6 *	4.1	4.4	1.9	3.0
Back	Hip Sagittal	46.9	83.2	65.6	65.9	2.3	3.5	5.9 *	5.1 *	0.9	1.1	2.1	1.1
	Hip Frontal	15.0	20.2	29.9	19.6	1.8	5.2 *	4.0 *	3.3 *	1.5	4.8 *	1.9	3.6
	Hip Transverse	17.2	23.7	29.8	22.0	2.0	3.2 *	4.7 *	3.1 *	1.7	2.5	2.7	1.4
	Knee Sagittal	75.2	94.1	103.5	129.6	3.2	5.3	11.3 *	12.3 *	1.0	3.6	2.6	5.1 *
	Knee Frontal	10.4	16.2	16.3	15.3	2.7	3.7	4.4 *	4.0	2.7	2.8	4.0	4.9
	Knee Transverse	19.7	19.8	24.0	27.5	2.3	4.1 *	5.3 *	4.9 *	1.3	2.9	4.1 *	3.3
	Ankle Sagittal	41.0	50.8	70.8	43.9	2.9	4.5	9.0 *	7.2 *	1.0	6.4 *	3.5	4.6
	Ankle Frontal	17.4	18.4	22.1	22.7	2.2	2.3	3.5 *	3.9 *	2.4	2.2	3.2	2.8
	Ankle Transverse	19.8	17.8	22.1	31.8	1.9	2.3	3.9 *	5.0 *	1.8	3.6	2.2	10.5 *

ROM: range of motion, RMSE: root mean square error, ΔROM: absolute difference in range of motion, *: significant difference in the post hoc test against the walking task.

4. Discussion

The main objective of this study was to compare the accuracy and reproducibility of lower limb joint angles computed from IMUs following different functional calibration methods. The study showed that applying a functional calibration movement before IMU-based lower limb kinematic assessment allowed for a fairly accurate measurement of gait movements. Except for the squat calibration movement, only small discrepancies were observed between functional calibration movements during a walking task, with a peak mean error of 3.6° for any joint in any plane of movement. Overall, the absolute reproducibility was similar for the three planes, but relative reproducibility was higher in the sagittal plane, with a mean standard error of measurement of less than 1.1° observed between multiple repetitions of the same functional calibration movement. A comparable overall performance was observed for different calibration movements, although each movement reported variable merits for different joints and planes of movement. Although the highest accuracy was observed in straight walking with a mean error of 2.2°, more complex gait movements tended to provide larger but limited errors, with a mean error of 3.5° for a step ascent, 5.4° for a step descent, and 4.7° for crossing an obstacle.

4.1. Accuracy of Different Calibration Methods during Straight Walking

The accuracy reported in this study during straight walking ranged from 1.1° to 3.6° for RMSE and from 0.2° to 3.4° for ΔROM, which is comparable to the mean error below 3° reported when using marker clusters on segments [22] rather than markers on anatomical landmarks [23–25]. Indeed, both methods reported different joint kinematics and accounted differently for errors of markers placement, soft tissue artefacts, and biomechanical model calculations [22]. The functional calibration of the optical system did not influence the accuracy, indicating that the reference frame obtained for each segment with the functional calibration movements were close to the optical reference frame. As the magnetometer was not used in the AHRS algorithm, the DRIFT was controlled to be acceptable (mean of 2.3°) for such short experiments. The drift was slightly higher in the more distal joints, probably due to the higher speed of the movements [26,27]. The drift was slightly lower in the sagittal plane, probably because the drift in this plane was better compensated by the sensor fusion algorithm. Although a mean error under 2° has been obtained on a single-joint movement [13], the accuracy obtained with our multi-joint model is acceptable for most clinical gait applications [8].

While the tilted and extension calibration movements provided a higher accuracy in the hip and ankle kinematics compared to the knee, walking calibration movements reported a higher accuracy for distal joints, whatever the walking speed. This observation can be supported by (1) a greater variability in the knee kinematics during the tilted and extension movements compared to straight walking and (2) higher accelerations of the distal relative to the proximal segments during walking. This observation also showed that the reference frame for each segment can be equally determined via a rotational movement recorded by the gyroscopes or via a translational movement recorded by the accelerometers contained in each IMU. The accuracy obtained in slow walking also validates the use of this functional calibration movement in similar conditions, which is often encountered in pathological gaits or in older adults [28].

The lower accuracy reported for the knee and ankle via the squat calibration movement could be explained by the lower movement amplitude of the shank and foot segments during this calibration movement. Indeed, the smaller amplitude-to-noise ratio probably resulted in an erroneous definition of the reference frame, leading to kinematic crosstalk [29]. This result also showed that functional movements exploring a wide range of segment orientation tended to provide more accurate segment reference frames.

4.2. Reproducibility of Calibration Movements

Reproducibility was excellent in 65% of the tested joints and motion planes, good in 24%, acceptable in 10%, and poor in 1 observation out of 84. Concerning differences between calibration movements, the walking calibration movement produced the highest reproducibility and SEM% for the ankle, while the other functional calibration movements produced higher reproducibility indices for the pelvis and hip joints. The lower reproducibility at the ankle for the segment-rotation-based movements could be explained by the difficulty in reproducing movements purely in the sagittal plane. This observation also supports previous results showing that the variable position of the foot affects the functional calibration when using different static postures [12]. The use of more guidance or more repetitions of the calibration movements could improve the reproducibility by (1) avoiding parasitic movements of the feet out of the sagittal plane and (2) decreasing the impact of any parasitic movement on the definition of the rotation axis. However, in order to limit the complexity and burden of the functional calibration movements, the walking calibration movement remains a remarkably convenient alternative since it offers a good to excellent reproducibility (though lower than other movements for the proximal joints), with a very simple and ecological movement. Caution may be needed for subjects having an impaired walking pattern, e.g., a subject walking with the feet pointing outwards.

A higher reproducibility was observed in the sagittal plane compared to the frontal plane. This could be explained by the higher range of motion in the sagittal plane during walking, leading to more kinematic crosstalk in the other planes measured and/or by the fact that the functional calibrations mainly generated segment movements in the sagittal plane. The combination of the higher variability and smaller ROM in the frontal plane during walking led to a higher SEM% in this plane, as also shown in upper limb anteroposterior reaching tasks [19]. Higher SEM% values inevitably require higher changes to detect meaningful functional changes, e.g., after therapy. The reproducibility of calibration movements in the frontal plane should be explored for the assessment of functional outcomes involving larger movements in the frontal plane.

4.3. Accuracy across Different Gait Movements

More complex gait movements tended to provide larger errors than a peak mean RMSE of 3.6° and a peak mean ΔROM of 3.4° for straight walking. Indeed, the peak mean errors obtained in the sagittal plane for a step ascent of 3°, 5°, and 5° for hip, knee, and ankle, respectively, correspond to errors in elevation angles of 5°, 4°, and 4° previously reported for the same joints [30]. Similarly, the peak mean ΔROM of 6.4° obtained for the stair ascent and of 4.6° for the stair descent are comparable to the errors previously reported for healthy subjects (peak error of 4.1° for a stair ascent and 4.8° for a stair descent) [31]. Therefore, before implementing inertial sensors in a complex, real-life context, the accuracy should be established in such a context rather than extrapolated from simpler gait movements recorded in controlled lab conditions.

4.4. Limitations and Perspectives

This study focused on healthy adults and this could be a limitation in case the functional calibration movements proposed here would be used with patients with a limited range of motion or who have parasitic movements that may hinder an accurate and reproducible calibration movement. The transferability to the elderly or to patients with motion disabilities should be assessed in further studies.

The IMU magnetometer was voluntarily omitted in this study in order to avoid ferromagnetic disturbances. The recordings in this study were limited in time due to the short time required to execute the investigated movements. The drift resulting from longer records [32] could be limited by using the IMU magnetometer or algorithms that constantly fuse the segment's angular velocity and linear acceleration via known kinematic relations between segments [33].

Although a high accuracy for the lower limb joint angles has been obtained by using only the gyroscope signals, our methods could be improved by also accounting for the segment accelerations [9,34], which can be used to locate the joint centers and improve the robustness of the segment orientations [35]. Another approach consists in using a hinge joint model and kinematic constraints to develop automatic or so-called "plug and play" calibrations [9,36]. This less restrictive method may facilitate clinical applications where patients with motion disabilities cannot be expected to perform precise prescribed calibration movements.

5. Conclusions

This study documents the high accuracy of IMU-derived lower limb joint angles during walking using several functional movements for the sensor-to-segment calibration. Functional movements requiring larger segmental angular amplitudes provided more accurate segmental reference frames and led to a higher accuracy regarding the kinematics of the adjacent joints. Alternatively, the higher linear accelerations generated at distal segments during a walking functional calibration also led to a higher reproducibility for distal joints, in comparison to the functional calibration based on the principal rotational axes. The walking, tilted, and extension functional calibration movements were shown to be three equivalent options for the gait movement examined in this study. In addition, for examining walking, the walking functional calibration may be superior because it involved very limited material and instruction complexity, which strengthens its interest in uncontrolled environments.

Author Contributions: Conceptualization, J.L., T.G., C.D., O.C., and M.P.; data curation, J.L.; formal analysis, J.L.; funding acquisition, C.D.; investigation, J.L. and T.G.; methodology, J.L., T.G., C.D., O.C., and M.P.; project administration, J.L., C.D., and M.P.; resources, J.L., C.D., and M.P.; software, J.L. and T.G.; supervision, C.D., O.C., and M.P.; validation, J.L., T.G., C.D., P.M., O.C., and M.P.; visualization, J.L., T.G., and M.P.; writing—original draft, J.L. and M.P.; Writing—review and editing, J.L., T.G., C.D., P.M., and M.P. All authors have read and agreed to the published version of the manuscript.

Funding: This research received no external funding.

Acknowledgments: We wish to express our gratitude to Julie Lamsens and Guillaume Gaudet for proofreading the paper.

Conflicts of Interest: The authors declare no conflict of interest.

References

1. Figueiredo, J.; Santos, C.P.; Moreno, J.C. Automatic recognition of gait patterns in human motor disorders using machine learning: A review. *Med. Eng. Phys.* **2018**, *53*, 1–12. [CrossRef] [PubMed]
2. Boissy, P.; Blamoutier, M.; Briere, S.; Duval, C. Quantification of Free-Living Community Mobility in Healthy Older Adults Using Wearable Sensors. *Front. Public Health* **2018**, *6*, 216. [CrossRef] [PubMed]
3. Del Din, S.; Godfrey, A.; Galna, B.; Lord, S.; Rochester, L. Free-living gait characteristics in ageing and Parkinson's disease: Impact of environment and ambulatory bout length. *J. Neuroeng. Rehabil.* **2016**, *13*, 46. [CrossRef] [PubMed]
4. Bergamini, E.; Ligorio, G.; Summa, A.; Vannozzi, G.; Cappozzo, A.; Sabatini, A.M. Estimating orientation using magnetic and inertial sensors and different sensor fusion approaches: Accuracy assessment in manual and locomotion tasks. *Sensors* **2014**, *14*, 18625–18649. [CrossRef] [PubMed]
5. Lebel, K.; Boissy, P.; Hamel, M.; Duval, C. Inertial measures of motion for clinical biomechanics: Comparative assessment of accuracy under controlled conditions—Changes in accuracy over time. *PLoS ONE* **2015**, *10*, e0118361. [CrossRef] [PubMed]
6. Poitras, I.; Dupuis, F.; Bielmann, M.; Campeau-Lecours, A.; Mercier, C.; Bouyer, L.J.; Roy, J.S. Validity and Reliability of Wearable Sensors for Joint Angle Estimation: A Systematic Review. *Sensors* **2019**, *19*, 1555. [CrossRef]
7. Van der Straaten, R.; De Baets, L.; Jonkers, I.; Timmermans, A. Mobile assessment of the lower limb kinematics in healthy persons and in persons with degenerative knee disorders: A systematic review. *Gait Posture* **2018**, *59*, 229–241. [CrossRef]
8. Chèze, L. *Analyse Cinématique du Mouvement Humain*, 1st ed.; ISTE Group: London, UK, 2014.

9. Seel, T.; Raisch, J.; Schauer, T. IMU-Based Joint Angle Measurement for Gait Analysis. *Sensors* **2014**, *14*, 6891–6909. [CrossRef]

10. Choe, N.; Zhao, H.; Qiu, S.; So, Y. A sensor-to-segment calibration method for motion capture system based on low cost MIMU. *Measurement* **2019**, *131*, 490–500. [CrossRef]

11. Cutti, A.G.; Giovanardi, A.; Rocchi, L.; Davalli, A.; Sacchetti, R. Ambulatory measurement of shoulder and elbow kinematics through inertial and magnetic sensors. *Med. Biol. Eng. Comput.* **2008**, *46*, 169–178. [CrossRef]

12. Palermo, E.; Rossi, S.; Marini, F.; Patanè, F.; Cappa, P. Experimental evaluation of accuracy and repeatability of a novel body-to-sensor calibration procedure for inertial sensor-based gait analysis. *Measurement* **2014**, *52*, 145–155. [CrossRef]

13. Favre, J.; Aissaoui, R.; Jolles, B.M.; de Guise, J.A.; Aminian, K. Functional calibration procedure for 3D knee joint angle description using inertial sensors. *J. Biomech.* **2009**, *42*, 2330–2335. [CrossRef] [PubMed]

14. Cutti, A.G.; Ferrari, A.; Garofalo, P.; Raggi, M.; Cappello, A.; Ferrari, A. 'Outwalk': A protocol for clinical gait analysis based on inertial and magnetic sensors. *Med. Biol. Eng. Comput.* **2010**, *48*, 17–25. [CrossRef] [PubMed]

15. Laidig, D.; Müller, P.; Seel, T. Automatic anatomical calibration for IMU-based elbow angle measurement in disturbed magnetic fields. *Curr. Dir. Biomed. Eng.* **2017**, 3. [CrossRef]

16. Bouvier, B.; Duprey, S.; Claudon, L.; Dumas, R.; Savescu, A. Upper Limb Kinematics Using Inertial and Magnetic Sensors: Comparison of Sensor-to-Segment Calibrations. *Sensors* **2015**, *15*, 18813–18833. [CrossRef]

17. Technologies, x.-o. x-Imu GUI. Available online: https://github.com/xioTechnologies/x-IMU-GUI (accessed on 4 December 2017).

18. Technologies, x.-o. Open-Source-AHRS-With-x-IMU. Available online: https://github.com/xioTechnologies/Open-Source-AHRS-With-x-IMU/tree/master/x-IMU%20IMU%20and%20AHRS%20Algorithms/x-IMU%20IMU%20and%20AHRS%20Algorithms/AHRS (accessed on 4 December 2017).

19. Wagner, J.M.; Rhodes, J.A.; Patten, C. Reproducibility and minimal detectable change of three-dimensional kinematic analysis of reaching tasks in people with hemiparesis after stroke. *Phys. Ther.* **2008**, *88*, 652–663. [CrossRef]

20. Shrout, P.E.; Fleiss, J.L. Intraclass correlations: Uses in assessing rater reliability. *Psychol. Bull.* **1979**, *86*, 420–428. [CrossRef]

21. Lebleu, J.; Mahaudens, P.; Pitance, L.; Roclat, A.; Briffaut, J.B.; Detrembleur, C.; Hidalgo, B. Effects of ankle dorsiflexion limitation on lower limb kinematic patterns during a forward step-down test. *J. Back Musculoskelet. Rehabil.* **2018**. [CrossRef]

22. Teufl, W.; Miezal, M.; Taetz, B.; Fröhlich, M.; Bleser, G. Validity of inertial sensor based 3D joint kinematics of static and dynamic sport and physiotherapy specific movements. *PLoS ONE* **2019**, *14*, e0213064. [CrossRef]

23. Al-Amri, M.; Nicholas, K.; Button, K.; Sparkes, V.; Sheeran, L.; Davies, J.L. Inertial Measurement Units for Clinical Movement Analysis: Reliability and Concurrent Validity. *Sensors* **2018**, *18*, 719. [CrossRef]

24. Tadano, S.; Takeda, R.; Miyagawa, H. Three dimensional gait analysis using wearable acceleration and gyro sensors based on quaternion calculations. *Sensors* **2013**, *13*, 9321–9343. [CrossRef] [PubMed]

25. Nuesch, C.; Roos, E.; Pagenstert, G.; Mundermann, A. Measuring joint kinematics of treadmill walking and running: Comparison between an inertial sensor based system and a camera-based system. *J. Biomech.* **2017**. [CrossRef] [PubMed]

26. Lebel, K.; Boissy, P.; Nguyen, H.; Duval, C. Inertial measurement systems for segments and joints kinematics assessment: Towards an understanding of the variations in sensors accuracy. *Biomed. Eng. Online* **2017**, *16*, 56. [CrossRef] [PubMed]

27. Lebel, K.; Boissy, P.; Hamel, M.; Duval, C. Inertial measures of motion for clinical biomechanics: Comparative assessment of accuracy under controlled conditions—Effect of velocity. *PLoS ONE* **2013**, *8*, e79945. [CrossRef]

28. Fried, L.P.; Tangen, C.M.; Walston, J.; Newman, A.B.; Hirsch, C.; Gottdiener, J.; Seeman, T.; Tracy, R.; Kop, W.J.; Burke, G.; et al. Frailty in older adults: Evidence for a phenotype. *J. Gerontol. Ser. A Biol. Sci. Med. Sci.* **2001**, *56*, M146–M156. [CrossRef]

29. Brennan, A.; Zhang, J.; Deluzio, K.; Li, Q. Quantification of inertial sensor-based 3D joint angle measurement accuracy using an instrumented gimbal. *Gait Posture* **2011**, *34*, 320–323. [CrossRef]

30. Bergmann, J.H.; Mayagoitia, R.E.; Smith, I.C. A portable system for collecting anatomical joint angles during stair ascent: A comparison with an optical tracking device. *Dyn. Med.* **2009**, *8*, 3. [CrossRef]

31. Laudanski, A.; Brouwer, B.; Li, Q. Measurement of lower limb joint kinematics using inertial sensors during stair ascent and descent in healthy older adults and stroke survivors. *J. Healthc. Eng.* **2013**, *4*, 555–576. [CrossRef]
32. Picerno, P. 25 years of lower limb joint kinematics by using inertial and magnetic sensors: A review of methodological approaches. *Gait Posture* **2016**, *51*, 239–246. [CrossRef]
33. Slajpah, S.; Kamnik, R.; Munih, M. Kinematics based sensory fusion for wearable motion assessment in human walking. *Comput. Methods Programs Biomed.* **2014**, *116*, 131–144. [CrossRef]
34. Olsson, F.O.; Seel, T.; Lehmann, D.; Halvorsen, K. Joint axis estimation for fast and slow movements using weighted gyroscope and acceleration constraints. *arXiv* **2019**, arXiv:1903.07353.
35. Seel, T.; Schauer, T.; Raisch, J. Joint Axis and Position Estimation from Inertial Measurement Data by Exploiting Kinematic Constraints. In Proceedings of the IEEE International Conference on Control Applications, Dubrovnik, Croatia, 3–5 October 2012; pp. 45–49.
36. Nowka, D.; Kok, M.; Seel, T. On Motions That Allow for Identification of Hinge Joint Axes from Kinematic Constraints and 6D IMU Data. In Proceedings of the 18th European Control Conference (ECC), Naples, Italy, 25–28 June 2019; pp. 4325–4331.

Article

SteadEye-Head—Improving MARG-Sensor Based Head Orientation Measurements Through Eye Tracking Data

Lukas Wöhle * and Marion Gebhard

Group of Sensors and Actuators, Department of Electrical Engineering and Applied Sciences,
Westphalian University of Applied Sciences, 45877 Gelsenkirchen, Germany; marion.gebhard@w-hs.de
* Correspondence: lukas.woehle@w-hs.de

Received: 14 April 2020; Accepted: 9 May 2020; Published: 12 May 2020

Abstract: This paper presents the use of eye tracking data in Magnetic AngularRate Gravity (MARG)-sensor based head orientation estimation. The approach presented here can be deployed in any motion measurement that includes MARG and eye tracking sensors (e.g., rehabilitation robotics or medical diagnostics). The challenge in these mostly indoor applications is the presence of magnetic field disturbances at the location of the MARG-sensor. In this work, eye tracking data (visual fixations) are used to enable zero orientation change updates in the MARG-sensor data fusion chain. The approach is based on a MARG-sensor data fusion filter, an online visual fixation detection algorithm as well as a dynamic angular rate threshold estimation for low latency and adaptive head motion noise parameterization. In this work we use an adaptation of Madgwicks gradient descent filter for MARG-sensor data fusion, but the approach could be used with any other data fusion process. The presented approach does not rely on additional stationary or local environmental references and is therefore self-contained. The proposed system is benchmarked against a Qualisys motion capture system, a gold standard in human motion analysis, showing improved heading accuracy for the MARG-sensor data fusion up to a factor of 0.5 while magnetic disturbance is present.

Keywords: data fusion; MARG; IMU; eye tracker; self-contained; head motion measurement

---

1. Introduction

Measuring head motion for medical diagnostics, virtual or augmented reality applications or human-machine collaboration usually involves multimodal sensorized interfaces to estimate motion and generate an appropriate control input for the desired application. In the context of human-machine collaboration and direct interaction with assistive technologies these interfaces are often designed to be used hands-free [1–3]. Such an interface needs to be precise, robust and fail safe. The research and development of new interaction technologies is in demand. A promising hands-free sensing modality involves Magnetic AngularRate Gravity (MARG)-sensors to estimate orientation of the operators head to interact with or teleoperate a robotic system [1]. In general MARG-sensors consist of a tri-axis accelerometer, tri-axis gyroscope as well as a tri-axis magnetometer. Such a sensor can also be termed inertial measurement unit (IMU) if it does not feature the tri-axis magnetometer. The orientation estimation from these sensors is based on the integration of the gyroscope raw angular rate data. This raw signal suffers from various noise terms that need to be taken care of, especially the dc-bias [4]. Even if the dc-bias of the gyroscope is set to zero at the MEMS-fab, there is a dc-bias that depends on sensor type, packaging and temperature, which in turn leads to a drift of the integrated gyroscope data [5]. This drift is usually corrected by the underlying data fusion process. Most common algorithms use the absolute references, direction of gravity and geomagnetic field to reduce orientation drift introduced through the gyroscope measurements [4,6]. The influence of permanent magnets, that is, hard iron effect, and magnetizable materials, that is, soft iron effect, in the direct neighborhood of the MARG-sensor will cause superpositions of the surrounding magnetic field. The distorted magnetic field can no longer be used to correct for dc-bias errors in the heading estimate (yaw-axis). During magnetic disturbances the MARG-sensor relies on gyroscope data only and will over time result in a drift in the heading estimate because of the accumulated errors if not corrected somehow [7,8].

This paper presents a novel approach to reduce heading estimation errors of head movement measurements by functional combination of mobile eye tracking and a head worn MARG-sensor. The approach utilizes the physiological connection between eye and head movements to identify static and dynamic motion phases. The eye tracking glasses are used to track visual fixations which indicate static phases. This indication is used for zero orientation change updates in a MARG-sensor data fusion algorithm. The approach relies on an infrared based eye camera only and does not need a scene or world camera and therefore no eye to world camera calibration. The presented approach is decoupled from most surrounding environmental conditions.

Human-robot collaboration in industrial production as well as rehabilitation robotics are applications that benefit from the proposed approach. These applications are mostly indoor and introduce potential magnetic interference at the location of the MARG-sensor. The work presented here not only enhances robustness of heading estimate of the head orientation measurements, it also presents a self-contained device mostly decoupled from varying visual markers and lightning conditions in the surrounding scenery.

1.1. State of the Art in MARG Based Orientation Measurements

Many algorithms exist to fulfill orientation estimation for head motion tracking. The Kalman filter has become the de facto standard for industrial and research applications [9,10]. However, various fast and lightweight data fusion methods for orientation estimation have been developed to reduce computational load while keeping orientation errors at a reasonable level. A famous method was introduced by Madgwick et al. in 2011 [11]. This method is based on the gradient descent algorithm and has gained popularity due to its fast and effective implementation on a microcontroller. Unfortunately MARG-sensors are exposed to magnetic field disturbance, termed soft- or hard-iron effects, resulting in incorrect orientation estimation [7]. This error scales with respect to the distance between the sensor and source of magnetic disturbance and magnetic properties of the source, for example, 35–50° error near large ferromagnetic metal objects or the floor (indoor) [12]. Recent approaches try to overcome these magnetic disturbances by software, for example, online gyroscope bias estimation [13] or fast online magnetometer calibration [14]. These software based corrections do not require additional hardware nor other sources of reference but might require certain motion conditions. For example online magnetometer calibration approaches are usually based on a sampling of sparse 3D magnetometer data points to adapt the calibration matrix, which might not be possible in every situation due to fast changing magnetic field values or the rather small motion space of the head. The use of other hardware, for example, visual odometry, usually provides a decent source of reference for the heading estimate but depends on the surrounding environmental conditions, for example, structured environment, reasonable lighting conditions and small relative motion in the scenery [15]. These conditions can not always be guaranteed especially in the context of human-machine collaboration which will feature a lot of relative motion from the robotic system and heavy dynamic magnetic disturbances due to the robots metal-links and motor-joints.

1.2. State of the Art in Eye Tracking

The analysis of the direction or the location of interest of a user through eye tracking is key for many applications in various fields, that is, human-computer interfaces (gaze mouse), human-machine interfaces, medical diagnostic and many more [16]. Therefore, fast and reliable eye tracking devices and software have been heavily researched [17]. Eye trackers can be separated into stationary or mobile devices. Stationary eye trackers are fixed in position referenced to the world frame. The devices' camera observes the users eyes and maps the tracked gaze to a defined surface (e.g., screen monitor) [18]. Mobile devices on the other hand usually consist of a frame, that is worn like a pair of glasses, a mono- or binocular eye camera fixed to the frame monitoring the pupil and a world camera to merge calibrated pupil positions to a gaze point in the world frame [17]. Furthermore, modern mobile eye tracking devices either feature a MARG-sensor or an IMU-sensor or the eye tracker can be extended by a custom or third-party sensor board. In this work, we use a monocular mobile eye tracker that gained popularity in the research community over the past years due to the open source software and affordable pricing [19].

2. Working Principle

The proposed work is based on the physiological relationship between eye and head rotations during visual fixations of stationary objects. The eyes of a human are centered in a fixed axis of rotation inside the head and are therefore naturally affected by head rotations. Visual fixations of objects will result in small or nonsignificant rotation of the eyeball during stationary motion phases, see Figure 1a. A rotation of the head during visual fixation of a stationary object however, will result in an opposite rotation of the eyeball due to the vestibo-ocular reflex, stabilizing the visual scenery [20], see Figure 1b. The physiological relationship between head and eye rotations therefore represents a natural indicator for head rotation and can be used to support head orientation measurements.

A mobile eye tracker is used to measure the above mentioned relationship and utilize this indicator for MARG-sensor based head orientation measurements. The method assumes that the MARG-sensor is fixed in position with respect to the eye tracker frame, for example, attached to it. The mobile eye tracker is worn by the user and should be adjusted in a way that prevents heavy slippage during head motion, for example, through an eye wear strap which is common practice in mobile eye tracking. From the setup given in Figure 1, the following constraints can be derived.

The coordinate system of the mobile eye tracker and MARG-sensor share a common reference frame with the users head and rotate conjointly, see Figure 1. Every rotation of the head is directly coupled with the rotation of the eye-tracker frame and MARG- or IMU-sensor. This rotation will result in a change of the pupil position in the eye camera image. This is either due to a voluntary change in the visual fixation or a head movement. In contrast to head rotation based changes of the pupil position, a change of visual fixation is usually coupled with high angular velocity of the eyeball due to a significant pupil position change of consecutive eye camera images. If the visual target is moving while the head is stationary or the fixation changes to another visual target, there will be a significant change in pupil position, due to the given coordinate system setup. The eye tracker camera is in a fixed position with respect to the head coordinate system. If the eyes follow a moving target or switch the visual fixation the pupil position changes within consecutive frames because the target changes its position with respect to the head and eye tracker coordinate system. Changes in the pupil position will therefore always indicate motion, whether it is introduced through head motion or voluntary eye motion. Near zero changes in the pupil position between eye tracker camera images however indicate near zero head rotation with one possible exception from this assumption. If a visual fixation stays on a moving target while the head is rotating at the same rotational speed at which the target is moving, all local coordinate systems do keep their relative positions between each other. This would result in a no motion classification from the pupil position change criteria. In this kind of situation the pupil position does not change with respect to the eye tracker coordinate system since the head and target coordinate system do not change their relative positions and orientations between each other. A MARG- or IMU-sensor however does measure motion related to the world coordinate system and will therefore measure a change in the orientation between the world and eye tracker coordinate system. The rotational velocity of the motion, or in other words the change of orientation between coordinate systems, needs to exceed a minimum threshold to distinguish the motion from gyroscope noise during these special phases.

The approach presented here uses this pupil motion description under visual fixation of an object to reduce the drift effect at stationary phases and therefore improve the MARG-sensor based heading estimate. Since all local coordinate systems are in a fixed position towards each other, every simultaneous measured movement or motion is caused by head rotations. Every significat pupil position change indicates motion, either from head rotation or eye rotatation. Zero or no rotation however, is indicated by every visual fixation, independent of the total fixation time, that results in near zero change in the pupil position and angular velocity.

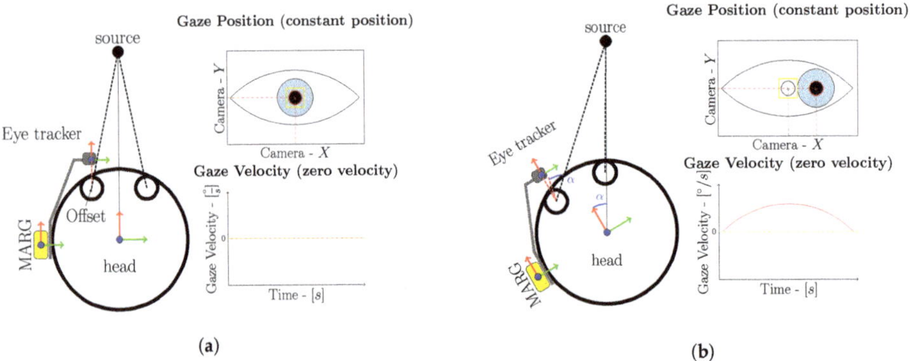

Figure 1. Coordinate systems between eye tracker, Magnetic AngularRate Gravity (MARG)-sensor and head when fixating an object (**a**) without and (**b**) with head motion. The condition of fixation of the object results in a stable gaze position. Possible motion of the pupil during fixations itself, so called microsaccades, is very small (0.2° at a duration of 20–30 ms) and can be neglected (yellow boundary box).

3. Data Fusion Process

In this work it is proposed to support MARG-sensor based heading estimates by zero rotation updates measured and indicated by visual fixation detection. The detection of visual fixation (given in subsection A) is used to feed the previous estimate of heading from the MARG-sensor fusion process (given in subsection B) recursively to the filter itself to reduce accumulation of gyroscope bias related heading errors. We calculate an IMU heading vector $^{N}\vec{N}_{IMU,k}$ based on the previous estimate of the heading that represents the direction of a horizontalized heading vector in the North East Down (NED) frame. This heading vector can be used as a complete substitute to the magnetometer based horizontalized north direction vector in the MARG-equation of an adapted form of Madgwick's Gradient Descent filter stage.

Synchronization of both systems is achieved through timestamp based message filtering. The data of the mobile eye tracker as well as the used MARG- or IMU-sensor should be accessible in real time by the manufacturers application programming interface (API) and provide a timestamp that can be used for synchronization processes, for example, using the message filter package from the robot operating system (ROS) framework. An angular rate threshold based switching can be either implemented on the MARG-sensor or host computer to account for possible latency issues between both systems. This threshold is based on the median head motion noise in static motion phases indicated by visual fixations. If the gyroscope raw signal exceeds the median noise level, the zero orientation update is turned off. This median noise threshold is also used to address the special case that the pupil position does not change while the head and eyes are following a moving target at the same rotational speed. During these special motions the magnitude of the measured angular rate will exceed the median gyroscope threshold which in turn disables the zero orientation update.

3.1. Visual Zero Rotation Detection

The trigger signal for the zero rotation update is based on an online visual fixation detection algorithm that utilizes dispersion (spatial movement, th_d) and duration (th_t) thresholds to identify fixations. These thresholds define the total amount of allowed gaze or pupil position differences (Δp) between time successive eye camera images (Δt). The algorithm utilizes a sliding window which length is determined by the duration threshold th_t and sampling frequency. Dispersion p is calculated as the sum of the differences between consecutive pupil positions

$$\Delta p = [(max(x) - min(x) + (max(y) - min(y)], \tag{1}$$

where x and y are the eye tracker cameras pixel positions. The dispersion is compared to the maximum dispersion threshold th_d. Fixations are identified as such if the dispersion stays below th_d. This results in an expansion of the window to the right until the dispersion exceeds this threshold. If no fixation is detected the window does not expand but moves forward in time [21].

This kind of algorithm has proven to be very accurate and robust regarding online fixation identification but needs careful parameter setting [21]. While the visual fixation stays on a target and inside the dispersion threshold boundaries, the head is assumed to be stationary. The threshold parameter ratings for the magnitude of dispersion in time is given due to involuntary movement, for example, microsaccades and tremor. However, these involuntary movements usually consist of rather small duration in the range of 20–30 ms and amplitudes peaking in a visual angle of $0.2°$ [22]. The fixation detection parameters should be chosen in a way that fixations are still detected even in the presence of microsaccades and tremor. A fixation is identified and labeled as such, as soon as the fixation duration threshold is reached. Upon this a trigger signal (S_t) is emitted indicating a zero orientation update cycle for the MARG-sensor data fusion process.

$$S_t = \begin{cases} 1, \ \Delta p \leq th_d \wedge \Delta t \leq th_t \\ 0 \end{cases} \tag{2}$$

The trigger starts an acquisition cycle that stores gyroscope raw data while the fixation holds true. When a sufficient amount of gyroscope samples has been recorded, updated motion noise parameters are sent to the MARG-sensor to update the threshold to account for desynchronization and latency issues between both systems and their different sampling rates. This procedure ensures adaptive and individual noise parameterization for the current user and use case and enables a real-time support.

3.2. MARG-Sensor Datafusion

In general the approach can be used independently of the underlying MARG-sensor data fusion process, since it indicates whether the users head is in dynamic or static motion phases. In this work, we exploit the approach on an adaptation of Madgwick's gradient descent algorithm (GDA) based filter. Figure 2 depicts the complete filter fusion approach that will be explained in detail in the following subsection.

As proposed by Madgwick et al. a quaternion ${}^N_B\mathbf{q}$ is computed by solving a minimization problem

$$\min_{{}^N_B\mathbf{q} \in \mathbb{R}^4} \ f({}^N_B\mathbf{q}, {}^N\vec{d}, {}^B\vec{s}), \tag{3}$$

that rotates a vector ${}^N\vec{d}$ into the orientation of a reference vector ${}^B\vec{s}$

$$f({}^N_B\mathbf{q}, {}^N\vec{d}, {}^B\vec{s}) = {}^N_B\mathbf{q} \bullet \begin{pmatrix} 0 \\ {}^N\vec{d} \end{pmatrix} \bullet {}^N_B\dot{\mathbf{q}} - \begin{pmatrix} 0 \\ {}^B\vec{s} \end{pmatrix}, \tag{4}$$

where ^N_B denotes the orientation of the global navigation frame relative to the body frame and $^N_B\mathbf{q}$ is the four component quaternion

$$^N_B\mathbf{q} = \begin{pmatrix} q_1 & q_2 & q_3 & q_4 \end{pmatrix}^T. \tag{5}$$

A possible solution to the optimization problem in Equation (3) can be given by gradient descent based solving for the obtained magnetometer and accelerometer vector measurements respectively

$$^N_B\mathbf{q}_{k+1} = {^N_B}\mathbf{q}_k - \mu_t \frac{\nabla f(^N_B\mathbf{q}_k, {^N}\vec{d}, {^B}\vec{s})}{||\nabla f(^N_B\mathbf{q}_k, {^N}\vec{d}, {^B}\vec{s})||}, \ k = 0,1,2,\ldots n, \tag{6}$$

where μ_t denotes the stepsize of the gradient function. For a complete mathematical explanation of the filter see Reference [11] or Reference [8]. The GDA filter stage computes a complete quaternion $^N_B\mathbf{q}_k$ either based on gyroscope, magnetometer and accelerometer (MARG-case) or gyroscope and accelerometer only data (IMU-case). This is to reduce errors in the heading estimate from magnetic disturbances but requires two different sets of equations [8]. This is due to the missing magnetometer measurement vector in the IMU case set of equations and therefore needs an adapted objective function and Jacobian respectively.

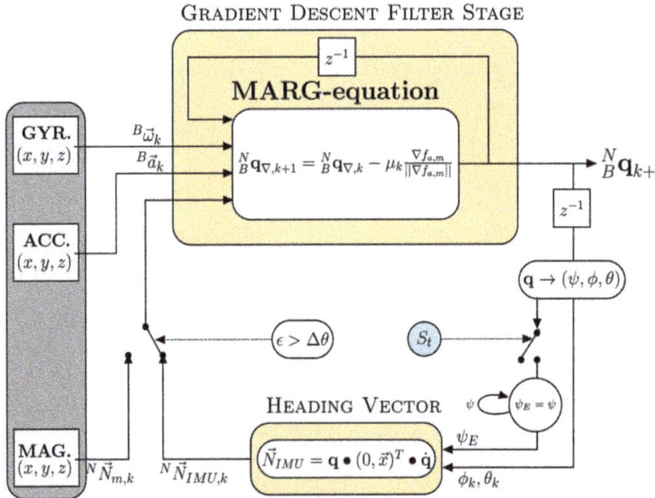

Figure 2. Block diagram of the proposed fusion approach. Upon detection of a zero rotation update through the eye tracker ($S_t = 1$), the current orientation is used to define an inertial measurement unit (IMU) heading vector ($^N\vec{N}_{IMU}$). This IMU heading vector will be updated with accelerometer measurements and fixed heading information as long as $S_t = 1$ is triggered and magnetic disturbance is present. If magnetic disturbance is present and the trigger is zero ($S_t = 0$) the IMU heading vector is iterativly updated. Switching between magnetometer based north direction vector ($^N\vec{N}_m$)and IMU heading vector ($^N\vec{N}_{IMU}$) is based on the deviation angle ϵ between both vectors.

In this work, we propose calculating an IMU heading vector that substitutes the magnetometer vector while magnetic disturbance is present to reduce the needed sets of equations as well as to guarantee convergence and a continuous quaternion solution to the minimization problem. We use the north direction vector $^N\vec{N}_m$ from the NED formulation through accelerometer and magnetometer measurements while no disturbance is present. The north direction vector is defined as the cross product between the down and east vector,

$$^{N}\vec{N}_m =^{N} \vec{D} \times^{N} \vec{E},\tag{7}$$

where the down vector is defined as the inverse of the acceleration measurement vector ($^{B}\vec{a}$)

$$^{N}\vec{D} = -^{B}\vec{a},\tag{8}$$

and the east vector is defined as the cross product between the down vector and the magnetometer measurement vector ($^{B}\vec{m}$)

$$^{N}\vec{E} =^{N} \vec{D} \times^{B} \vec{m}.\tag{9}$$

During rotation the acceleration vector will be subject to motion acceleration and therefore does not accurately reflect the direction of gravity. This effect however is typically reduced by low pass filtering the acceleration vector. Most modern sensors provide onboard low pass filter banks that can be configured by the user to the appropriate needs. Furthermore, the rotational accelerations will be rather small compared to the dominant acceleration originating from gravity. This is due to the small distance (in this case 0.1 m) between the rotational center of the head and the position of the MARG-sensor resulting in minor inaccuracies during dynamic motion. The influence of these incaccuracies during dynamic motion is further reduced by the subsequent data fusion filter. The data fusion filter usually does emphasize gyroscope data integration during fast dynamic motion to reduce inaccuracies from the motion acceleration on the orientation estimation.

We calculate a substitute to the north direction vector, termed IMU heading vector $^{N}\vec{N}_{IMU}$, based on the orientation estimation from the gyroscope and accelerometer measurements. This is achieved through the following process.

We extract the heading information of the output quaternion $^{N}_{B}\mathbf{q}_k$ by calculating a three component vector ($^{N}\vec{N}_{IMU,k}$) describing heading information in the NED frame. First the heading information (yaw angle, ψ_E) from the quaternion $^{N}_{B}\mathbf{q}_k$ is converted to Euler angle representation

$$
\begin{aligned}
a &= (q_{k,1}{}^2 + q_{k,2}{}^2 - q_{k,3}{}^2 - q_{k,4}{}^2)\\
b &= 2 \cdot (q_{k,2} \cdot q_{k,3} + q_{k,1} \cdot q_{k,4})\\
\psi_E &= atan2(b,a).
\end{aligned}\tag{10}
$$

When a zero rotation update is triggered, the fusion process samples the current output angle ψ_E from the last output quaternion $^{N}_{B}\mathbf{q}_k$ of the GDA filter stage and holds it while the trigger S_t is true. The subscript E indicates that the angle ψ is not updated if the sample and hold mechanism is activated. If the trigger signal is false, indicating head motion, the fusion process updates the angle ψ_E with every new output quaternion $^{N}_{B}\mathbf{q}_k$.

Second we convert the iterative updated roll (ϕ_k) and pitch (θ_k) angles derived from the current quaternion $^{N}_{B}\mathbf{q}_k$ by the following process

$$
\begin{aligned}
a &= 2 \cdot (q_{k,3} \cdot q_{k,4} + q_{k,1} \cdot q_{k,2})\\
b &= (q_{k,1}{}^2 - q_{k,2}{}^2 - q_{k,3}{}^2 + q_{k,4}{}^2)\\
c &= 2 \cdot (q_{k,2} \cdot q_{k,4} + q_{k,1} \cdot q_{k,3})\\
\phi_k &= atan2(b,a)\\
\theta_k &= asin(-c).
\end{aligned}\tag{11}
$$

From the yaw (ψ_E) angle and the current roll (ϕ_k) and pitch (θ_k) angles we build a new temporary quaternion shown in Equation (12),

$$
{}_B^N\mathbf{q}_{E,k} =
\begin{bmatrix}
c(\frac{\phi_k}{2})c(\frac{\theta_k}{2})c(\frac{\psi_E}{2}) + s(\frac{\phi_k}{2})s(\frac{\theta_k}{2})s(\frac{\psi_E}{2}) \\[2mm]
s(\frac{\phi_k}{2})c(\frac{\theta_k}{2})c(\frac{\psi_E}{2}) - c(\frac{\phi_k}{2})s(\frac{\theta_k}{2})s(\frac{\psi_E}{2}) \\[2mm]
c(\frac{\phi_k}{2})s(\frac{\theta_k}{2})c(\frac{\psi_E}{2}) + s(\frac{\phi_k}{2})c(\frac{\theta_k}{2})s(\frac{\psi_E}{2}) \\[2mm]
c(\frac{\phi_k}{2})c(\frac{\theta_k}{2})s(\frac{\psi_E}{2}) - s(\frac{\phi_k}{2})s(\frac{\theta_k}{2})c(\frac{\psi_E}{2})
\end{bmatrix},
\tag{12}
$$

where c and s are sine and cosine functions respectively.

This quaternion is now applied to a x-axis unit vector because the north direction vector defines the sensors body x-axis, resulting in

$$
\vec{x} = \begin{pmatrix} 1 & 0 & 0 \end{pmatrix}^T
$$
$$
{}^N\vec{N}_{IMU,k} = {}_B^N\mathbf{q}_{E,k} \bullet \begin{pmatrix} 0 \\ \vec{x} \end{pmatrix} \bullet {}_B^N\dot{\mathbf{q}}_{E,k},
\tag{13}
$$

where $\bullet$ indicates quaternion multiplication and $\dot{\mathbf{q}}$ represents the conjugate quaternion to $\mathbf{q}$. The vector ${}^N\vec{N}_{IMU,k}$ now represents the direction as a substitute to the magnetometer based north direction vector in the NED frame, as can be seen in Figure 3.

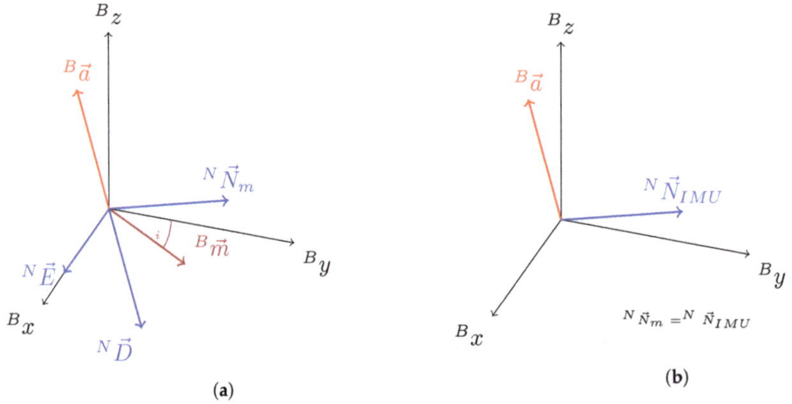

Figure 3. Depiction of the north direction vector substitutes, (**a**) north direction vector, termed ${}^N\vec{N}_m$ from accelerometer (${}^B\vec{a}$) and magnetometer (${}^B\vec{m}$) measurements in the case of undisturbed magnetic field measurement, and (**b**) IMU heading vector, termed ${}^N\vec{N}_{IMU}$ calculated based on quaternion vector multiplication from the gradient descent algorithm (GDA) filter without magnetometer data in the case of disturbed magnetic field measurements.

Since the vectors ${}^N\vec{N}_m$ and ${}^N\vec{N}_{IMU}$ lie in the same plane it is possible to calculate a deviation angle (ϵ) that can be used to determine magnetic disturbance due to sudden changes in the direction of the north direction vector in contrast to the IMU heading vector. The deviation angle is calculated as follows

$$
\epsilon = \cos^{-1}\left({}^N\vec{N}_m \bullet {}^N\vec{N}_{IMU}\right),
\tag{14}
$$

where $\bullet$ represents the dot product respectively.

If magnetic disturbance is present, the deviation angle ϵ will increase. If it exceeds a threshold $\Delta\theta$, the filter switches towards the virtual sensor vector based quaternion calculation and vice versa if it vanishes. This procedure enables the calculation of a complete and continuous quaternion solution that involves current sensor measurements from the accelerometer and the extracted heading information from the previous quaternion. It is possible to use the same set of equations without any adaptation and switch from magnetometer based north direction vector to the IMU heading vector without divergence of the quaternion. While the zero rotation trigger is enabled, the fusion process holds the recent calculated yaw angle ψ_E. This ensures that the GDA based calculation of the new quaternion ${}^{N}_{B}\mathbf{q}_k$ is less affected by possible drift in the heading direction due to uncorrected gyroscope bias but will however be corrected in the remaining axes through accelerometer updates and preserves a continuous solution and convergence. While no trigger is emitted, the fusion approach simply updates the measurement quaternion with every iteration based on either magnetic north direction vector when no disturbance is present or the IMU heading vector from the current orientation. It is known that Euler angle representation is subject to gimbal lock if two rotation axis align. This effect can be dealt with in two different ways. Either by designing the experiment in a way that does not include head rotations around the pitch exceeding $\pm 90°$, which causes gimbal lockin the chosen rotation order (Z-Y-X), or by formulating a quaternion based heading orientation estimation method. The quaternion based solution can be found in the following paragraph.

The heading information (yaw angle, ψ_E) from the quaternion ${}^{N}_{B}\mathbf{q}_k$ is converted to a quaternion representing only the yaw rotation (${}^{N}_{B}\mathbf{q}_{\psi,E}$) by deriving it from the corresponding Euler angle representation [23]. A unit quaternion representing heading information (${}^{N}_{B}\mathbf{q}_{\psi}$) is expressed as a rotation ψ around the z-axis

$$
\mathbf{q} = \left(cos(\psi/2) \quad 0 \quad 0 \quad sin(\psi/2))^{T} \right),
$$
$$
{}^{N}_{B}\mathbf{q}_{\psi} = \frac{\mathbf{q}}{\|\mathbf{q}\|}.
\tag{15}
$$

We can express the heading quaternion ${}^{N}_{B}\mathbf{q}_{\psi,E}$ without trigonometric functions by substituting the corresponding Euler angle Equation (10) with (15) and normalize it, resulting in

$$
\mathbf{q} = \left(((q_{k,1}{}^2 + q_{k,2}{}^2 - q_{k,3}{}^2 - q_{k,4}{}^2)) \quad 0 \quad 0 \quad (2 \cdot (q_{k,2} \cdot q_{k,3} + q_{k,1} \cdot q_{k,4})) \right)^{T},
$$
$$
{}^{N}_{B}\mathbf{q}_{\psi} = \frac{\mathbf{q}}{\|\mathbf{q}\|}.
\tag{16}
$$

To get the half rotation angle from Equation (10) we add a unit quaternion and normalize the result

$$
\mathbf{q} = {}^{N}_{B}\mathbf{q}_{\psi} + \left(1 \quad 0 \quad 0 \quad 0 \right)^{T},
$$
$$
{}^{N}_{B}\mathbf{q}_{\psi,E} = \frac{\mathbf{q}}{\|\mathbf{q}\|}.
\tag{17}
$$

Likewise, to the Euler angle solution the zero rotation update trigger samples the current output quaternion ${}^{N}_{B}\mathbf{q}_{\psi,E}$ from the last output quaternion ${}^{N}_{B}\mathbf{q}_k$ of the GDA filter stage and holds it while it is activated. If the trigger signal is deactivated, the fusion process updates the heading quaternion ${}^{N}_{B}\mathbf{q}_{\psi,E}$ with every new output quaternion ${}^{N}_{B}\mathbf{q}_k$.

We calculate a quaternion (${}^{N}_{B}\mathbf{q}_{\phi,\theta,k}$) representing the iterative updated roll (ϕ_k) and pitch (θ_k) angles based on the current quaternion ${}^{N}_{B}\mathbf{q}_k$. This is achieved by removing the yaw component of the current quaternion ${}^{N}_{B}\mathbf{q}_k$ through conjugate quaternion multiplication. We calculate a yaw quaternion ${}^{N}_{B}\mathbf{q}_{\psi,k}$ based on the Equation (16) and apply the conjugate to the current quaternion ${}^{N}_{B}\mathbf{q}_k$

$$
{}^{N}_{B}\mathbf{q}_{\phi,\theta,k} = {}^{N}_{B}\dot{\mathbf{q}}_{\psi,k} \bullet {}^{N}_{B}\mathbf{q}_k,
\tag{18}
$$

where $\bullet$ indicates quaternion multiplication and $\dot{\mathbf{q}}$ represents the conjugate quaternion to $\mathbf{q}$.

The final quaternion ${}^{N}_{B}\mathbf{q}_{E,k}$ can be computed by combining the heading quaternion ${}^{N}_{B}\mathbf{q}_{\psi,E}$ and the iterative updated quaternion representing only roll and pitch ${}^{N}_{B}\mathbf{q}_{\phi,\theta,k}$ through quaternion multiplication,

$$
{}^{N}_{B}\mathbf{q}_{E,k} = {}^{N}_{B}\mathbf{q}_{\psi,E} \bullet {}^{N}_{B}\mathbf{q}_{\phi,\theta,k}.
\tag{19}
$$

The quaternion ${}^{N}_{B}\mathbf{q}_{E,k}$ now represents a complete orientation expressed as quaternion and combines heading information from the sample and hold mechanism with current updates regarding roll and pitch information from the filter's output quaternion. This solution does not suffer from gimbal lock and can be used as the input quaternion to Equation (13). Both methods, Euler angle conversion or the complete quaternion based heading calculation, are valid and can be chosen based on the desired application and design of experiment.

4. Interface Setup

We use a custom designed MARG-sensor board running the GDA based sensor data fusion on an Espressif 32 bit dual-core microcontroller unit (MCU). The sensor board features a low power 9-axis ICM 20948 InvenSense MARG-sensor, see Figure 4. The MCU is running the FreeRTOS realtime operating system on both cores at 1 kHz scheduler tick rate [24]. The standalone implementation is designed to simultaneous calculate orientation data from two copies of the data fusion process at 250 Hz, while their only difference is the active eye tracking trigger update. The two data fusion filters run in real time on the MCU and publish the two sets of fused orientation data and the calibrated 9-axis sensor data are at 100 Hz via micro-USB over UART. The sensor is attached via a custom casing to a low cost monocular eye tracker from Pupil Labs [19]. The tracker frame is secured via an eyewear strap on the users head. The eye tracker features 120 Hz frame rate of the eye tracking process at a resolution of 400×400 pixels. It is connected to a host computer running the Pupil Labs open source capture tool to acquire and preprocess the data as well as taking care of online fixation detection. The data are accessible in real-time through ZeroMQ. Two custom c++ ROS (Robot Operating System) nodes handle the synchronization and inter device communication. Synchronization between the MARG and eye tracking data is achieved through comparison of their corresponding timestamp upon arriving at the host system with a maximum lag of 3 ms between the timestamps. While fixation is true, the trigger signal is broadcasted to the MARG-sensor system indicating zero rotation. Furthermore, the trigger starts the gyroscope raw data capture process on the host computer. When the visual fixation is released the trigger is set to false which stops the gyroscope capture process as well as the zero orientation update cycle. The median gyroscope noise for stationary motion phases is sent to the MARG-sensor, when sufficient amount of data has been captured. To reduce latency impact on the orientation calculation, a movement threshold based on this median gyroscope noise is implemented on the MARG-sensor to ensure that the trigger will be set to false without latency drops or desynchronization.

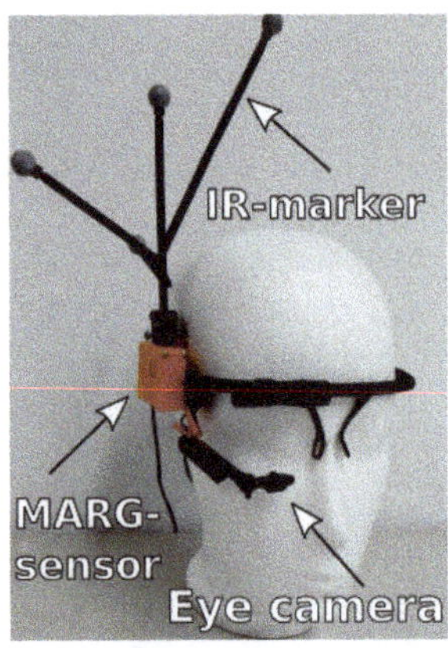

(a)

(b)

Figure 4. Proposed head interface and custom designed MARG-sensor board. The head interface (**a**) consists of a pupil core headset, a passive IR-marker tree that is in line to the custom MARG-sensor board (**b**) which is attached to the eye tracker frame.

5. Experimental Setup

The accuracy of the proposed interface is benchmarked against a Qualisys motion capture system (Qualisys Miqus Camera M3, Qualisys AB, Kvarnbergsgatan 2, Göteborg, Sweden). The interface is worn by a user alongside a leightweight medium density fiberboard based rigid body passive IR-marker tree connected to the MARG-sensor casing, see Figure 4. The capture process of the Qualisys motion capture system broadcasts data at 120 Hz over a real-time client, allowing for timestamp based synchronization via the before mentioned ROS-nodes. The threshold $\Delta\theta$ is chosen based on the 3σ standard deviation of the north direction vector under static conditions for a short period of time (10 s). The first calculated north direction vector of this series is the initial vector. This initial vector is used to calculate the standard deviation of this series of deviation angles ϵ based on Equation (14). For the ICM20948 on the custom sensor board the threshold $\Delta\theta$ results in $3°$. After a warm up phase, the magnetometer data is turned off to simulate a magnetically disturbed environment and examine the eye tracking supported zero orientation change trigger update mechanism as a proof of concept. The proof of concept of the proposed orientation estimation update mechanism can be provided either by using real magnetometer data or by turning off the magnetometer data measurement completely. A difference is not evident. This is due to the switching from north heading vector to IMU heading vector when a magnetic disturbance is present. In such a case magnetometer data is not used in the orientation estimation algorithm and is therefore not dependent on real magnetometer data input. Thus, we simulate disturbance by switching magnetometer data off to investigate the performance of the proposed data fusion during periods of non usable magnetometer input. To compare rotations, the coordinate system of the Qualisys data is transformed into the body coordinate frame of the MARG-sensor by calculating an alignment matrix from six stationary positions through least square method described in Reference [25]. The user is instructed to freely move his head and eyes with some static or no-rotation phases spanning between 2–5 s in duration. The total duration for one trial was limited to 15 min. A total of six trials was gathered for one individual user as a proof of concept for the proposed method. Visual fixation detection parameters were chosen based on experimental pretests that minimize latency drops when changing from stationary to dynamic head motion and were set to the following: $th_d = 0.21°$, and $th_t = 220$ ms.

Two pretests were conducted to demonstrate the interchangeability of the north direction vector substitute calculations described in Section 3.2 and the filter's capability to detect interference based on the deviation angle ϵ. The three axis magnetometer was calibrated based on the process described in Reference [26]. The MARG-sensor is rotated arbitrarily in all dimensions. The tri-axis magnetometer data is sampled during this period. After recording the magnetometer data is calibrated through least square fitting of the ellipsoid data into a unit sphere and scaled to the surrounding field strength afterwards. During a 10 min warm up phase, the filter uses magnetometer data to converge towards the direction of magnetic north and gravity respectively. To demonstrate the interchangibility of the vectors we switch off the filter to use the IMU heading vector instead of the magnetometer based north direction vector and move the sensor arbitrarily for a short period of time (50 s). Both vector measurements are recorded throughout the trial. The second pretest covers the validation of magnetic interference detection and the switching from north to IMU heading vector. The sensor is set up according to the previous mentioned calibration and warm up routines. We record two sets of orientation: (a) The proposed filter with magnetic interference detection and switching and (b) the same filter without the switching mechanism. After 44 s an iron bar is brought close to the sensor (15 cm) to introduce magnetic interference.

6. Results and Discussion

In this work it is proposed to support MARG-sensor based heading estimates by zero rotation updates measured and indicated by visual fixation detection. An interchangeable north direction vector substitute is used for a gradient descent based orientation estimation. Section 6.1 gives an overview of the pretest to show the interchangeability of the heading vector substitutes calculation described in Section 3.2 whereas Section 6.2 the filters capability to detect magnetic disturbance and switch towards IMU heading vector. Section 6.3 presents the experimental results from the full fusion approach using visual fixations for zero rotation update.

6.1. Interchangeable North Direction Vector Substitutes

Figure 5 depicts normalized individual x-, y- and z-axis results for north direction vector $^N\vec{N}_m$ from calibrated magnetometer data through Equations (7)–(9) and the IMU heading vector $^N\vec{N}_{IMU}$ based on the process given by Equations (10)–(13). Both vectors show similar results during the whole trial with maximum deviations of ±0.1 normalized units. The north direction vector from magnetometer data has a larger spread of measurement values compared to the IMU heading vector. This originates from the different noise characteristics and computations of the vector components. The north direction vector is directly calculated from raw accelerometer and magnetometer data and will directly reflect raw sensor noise, whereas the IMU heading vector is based on smoothed quaternion fusion from gyroscope and accelerometer readings from the GDA filter. The noise spreading level, however, does stay at a reasonable level during the trial. This pretest shows the interchangeability of the different north direction vector substitutes which guarantees a continuous quaternion solution and convergence of the filter.

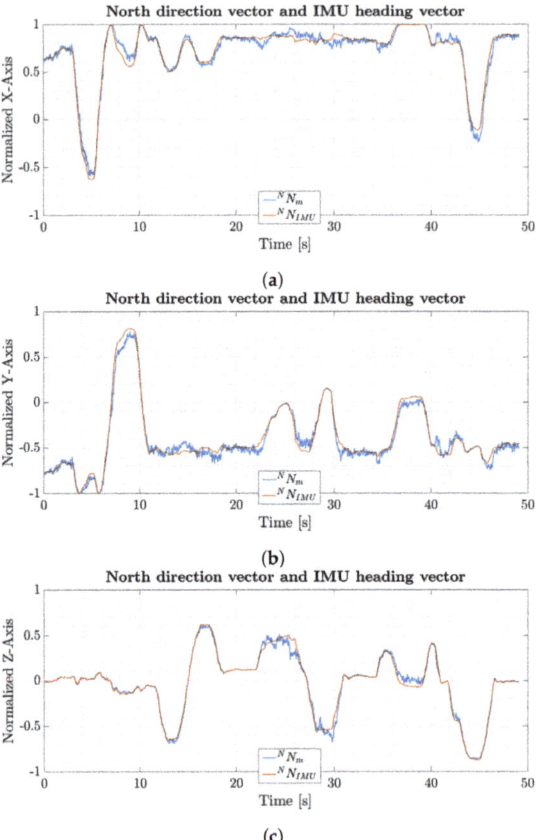

Figure 5. Comparison between (**a**) magnetometer based normalized x-axis north direction vector and normalized x-axis IMU heading vector, (**b**) magnetometer based normalized y-axis north direction vector and normalized y-axis IMU heading vector and (**c**) magnetometer based normalized z-axis north direction vector and normalized z-axis IMU heading vector.

The IMU heading vector will drift apart from the north direction vector with respect to time. This is due to uncorrected gyroscope biasresulting in drift in the heading estimate of the quaternion used for determining the IMU heading vector. For short periods of time and under the same initial conditions however both vectors are almost identical. The length of the time period in which both vectors are mostly identical depends on the individual noise characteristics of the used gyroscope and computational errors from the discrete implementation. High grade navigation gyroscopes will experience less drift compared to consumer based gyroscopes used in this work. The maximum time before gyroscope errors accumulate more than 1° drift in the heading estimate is 50 s for the custom MARG-sensor board used in this work.

6.2. Magnetic Disturbance Detection

Figure 6 depicts yaw angle results as well as the corresponding yaw angle errors for the magnetic disturbance detection and switching from north direction to IMU heading vector based on Equation (14). The Figure 6a presents three different yaw angle estimations over time: ground truth yaw data (Qualisys, yellow), yaw estimations from the proposed filter with deviation detection and heading vector substitutes (M_E, blue) as well as yaw estimations of a version of the filter without heading vector switching (M_O, orange). The figure also presents values for the magnetic deviation angle ϵ (black) over time. Figure 6b presents the corresponding yaw angle errors referenced to the ground truth yaw data. Magnetic disturbance is introduced for a short period of time starting at 44 s and ending at 66 s by bringing an iron bar close to the sensor (15 cm).

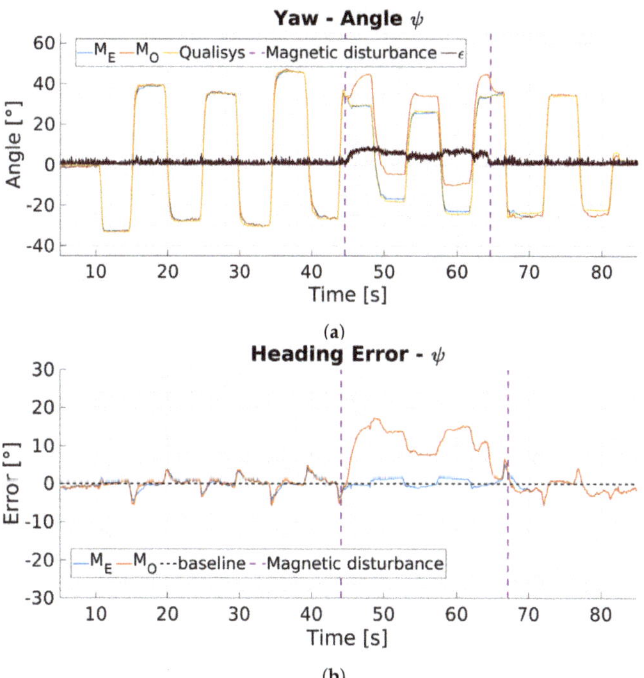

Figure 6. Influence of magnetic disturbance on yaw angle estimation: (**a**) yaw angle comparisons between ground truth (Qualisys, yellow), the proposed filter with magnetic disturbance detection and heading vector substitute switching (M_E, blue), a filter version without heading vector switching (M_O, orange) as well as the magnetic deviation angle ϵ (black). (**b**) depicts the corresponding heading error referenced to the Qualisys system. Magnetic interference is introduced for a short period of time (43 s to 66 s) by bringing and iron bar close to the MARG-sensor. The proposed filter detects the interference and switches towards IMU heading vector usage resulting in less error.

The proposed filter (M_E) detects the disturbance when it is introduced because the deviation angle ϵ exceeds the threshold $\Delta\theta$, see Figure 6a. The filter switches towards IMU heading vector substitute and is not affected by the disturbance, resulting in a maximum error of 2° during this phase. In contrast, the filter without switching mechanism experiences large yaw angle erros and results in up to 17° total error (see Figure 6b). This pretest demonstrates the filters capability of magnetic disturbance detection based on the deviation angle calculation between north direction and IMU heading vector. After the filter detects a disturbance it uses the IMU heading vector for orientation estimation. In this mode the filter furthermore enables visual zero rotation updates mechanism to reduce heading error accumulation over time.

6.3. MARG-Sensor Data Fusion Approach Using Visual Fixations

Figure 7 shows typical data for yaw angle measurements from a 30 s sequence of one 900 s trial. The yaw angles are presented in degrees over time in seconds. The figure depicts yaw angle estimation data from the ground truth motion capture system (Qualisys system, yellow), the proposed (M_E, blue) and standard version (M_O, orange) of the data fusion process. The visual fixation trigger state S_t is presented in purple. During visual fixation phases (S_t, purple) the proposed eye tracking supported version M_E does drift less compared to the standard implementation M_O. In dynamic motion phases both filter versions do accumulate the same amount of drift.

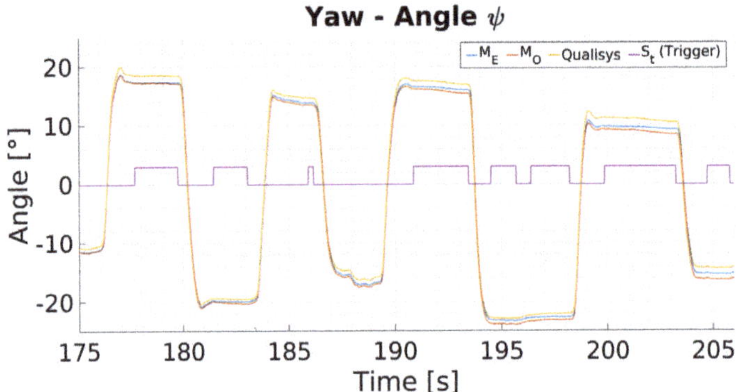

Figure 7. Typical yaw angle measurements for a motion sequence from the Qualisys system (yellow), the proposed (M_E, blue) and standard version (M_O, orange) of the data fusion. During stationary motion, the trigger signal S_t (purple) is set to high and indicates zero orientation change.

The performance of the proposed approach for the heading estimation is presented as total Euler angle error (degrees see Figure 8) as well as mean error reduction rate (e_ψ, unitless see Table 1). The total Euler angle errors are calculated as the difference between the ground truth of absolute orientation from the Qualisys system and the orientation estimation of the proposed (M_E) and standard version (M_O) of the data fusion process (see Figure 7). The dashed black line indicates the reference for zero heading error. Figure 8 presents two sets of Euler angle errors in degrees over time for the complete 900 s duration of two different trials. At the start of the trial the eye tracking supported version of the filter (M_E, blue) as well as the standard GDA filter (M_O, orange) perform identical. During dynamic phases both filters accumulate the same amount of error due to uncorrected gyroscope bias. However, when stationary phases are indicated and the trigger signal S_t is enabled, the eye tracker supported GDA filter version accumulates less gyroscope drift in contrast to the standard implementation, see Figure 8. This effect covers the entire duration of the trials. The orientation estimation errors rise significantly over time for both solution. The GDA based approach with eye tracking based zero orientation change update results in nearly 50% less absolute orientation, see Figure 8.

The mean error reduction rate e_ψ and its standard deviation is calculated based on the absolute error quotient between the proposed (M_E) and standard version (M_O) of the fusion process at 900 s. Table 1 presents total Euler angle errors from six different trials for the proposed (M_E) and standard version (M_O) of the fusion after 900 s and the calculated error reduction rate. On average, the eye tracking supported GDA filter approach accumulates near 50% less orientation error (0.46 ± 0.07) compared to the GDA filter without eye tracking data support, see Table 1.

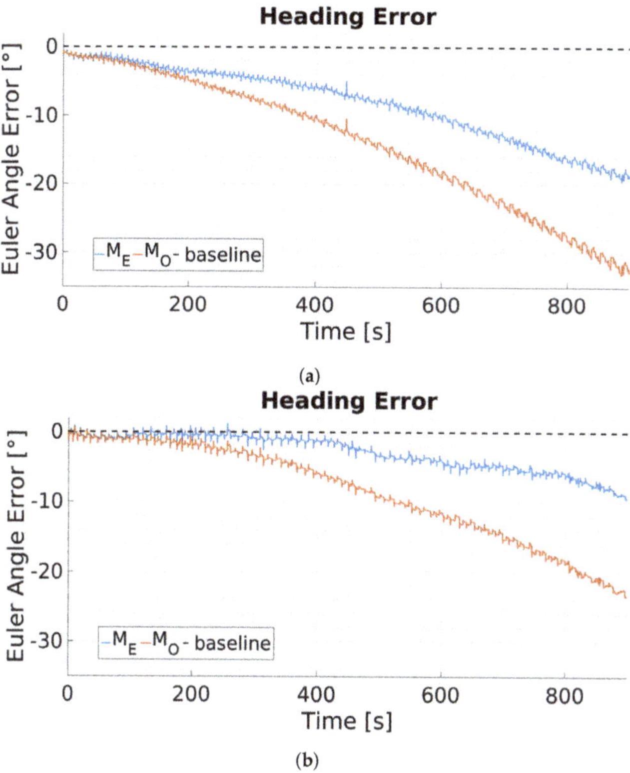

Figure 8. Typical results for absolute heading error referenced to the Qualisys system from the GDA based approach with (M_E, blue) and without eye tracking (M_O, orange) support for two different trials. The eye tracking supported filter results in (**a**) a total of 18.84° error whereas the standard version results in 32.76° and (**b**) in a total of 8.82° error for the eye tracking supported version whereas the standard version results in 22.55°.

Table 1. Absolute error values and error reduction rate (e_ψ) for the GDA based data fusion with (M_E) and without eye tracking (M_O) support after 900 s.

	M_E [°]	M_O [°]	$e_\psi = 1 - \frac{M_E}{M_O}$
1.	-18.84	-32.76	0.42
2.	-8.82	-22.55	0.59
3.	-13.89	-23.17	0.40
4.	-11.33	-19.09	0.40
5.	-12.63	-26.06	0.51
6.	-7.16	-13.52	0.45
average			0.46 ± 0.07

7. Conclusions

Utilizing eye tracking data to support sensor data fusion of MARG-sensor shows improvements of the heading accuracy in magnetically disturbed environments or for IMU sensors that do not feature a heading reference in the order of 50%. Because of the physiological coupling between eye and head rotations, eye tracking can deliver an indicator signal for near zero head orientation change. Furthermore, this trigger signal allows for individual and adaptive noise parameterization through gyroscope capturing and could be used in the context of adaptive noise estimation with respect to head motion while a sufficient amount of data is captured. The proposed method can be used with any mobile eye tracking devices that either feature a build-in IMU or MARG-sensor or are expanded by a custom or third-party sensor. The presented approach does not need a world camera and is therefore mostly independent of surrounding environmental lightning conditions. In addition, the proposed use of interchangeable north direction vector substitutes enables switching between full MARG and IMU-mode, without the need for an additional set of equations in a given filter. This guarantees a continuous quaternion computation and convergence of the filter.

Limitations

The magnitude of error compensation does scale with respect to total fixation duration and amount of stationary motion phases. However, the solution does not reduce the effect during dynamic motion phases, since it does not directly estimate and correct the dc-bias term of the raw gyroscope signal. This is due to various other noise effects that are present in the raw gyroscope signal. Main noise terms among other that influence the in-run dc-bias estimation are ac-noise, oscillating head motion, output rate limitations and possible desynchronization between timestamps of both systems.

When estimating in-run dc-bias the presence of these noise terms can lead to a wrong estimation. Since the dc-bias is subtracted from the raw gyroscope signal at every time step, it effects the complete measurement from that point forward and might result in a worse heading estimate. However, if a sufficient amount of sensor data has been gathered, a low-pass filtered dc-bias estimation might be used to reduce the drift at a smaller scale since the data are only captured during near stationary motion phases and therefore restrain heavy amplitude changes.

The proposed solution can be affected by very slow motion acceleration triggering the visual fixation detection plugin and falsely labeling a static phases. This effect however only appears if the resulting angular rate of the head motion is smaller than the angular rate constraint derived from the dispersion and time threshold of the fixation detection plugin and stays below the median angular rate threshold that is sampled throughout the trial. In this work the angular rate constraint from the fixation detection plugin that might lead to a wrong classification during fixation and simulatanious head motion is $0.95 \frac{°}{s}$ for a 220 ms measurement window. This angular rate results in the maximum dispersion of $0.21°$. This would result in a fixation detection which would in turn trigger the zero rotation update mechanism for one cycle. After this the dispersion threshold is exceeded, setting the trigger to false which in turn resets the online fixation detection sliding window.

8. Future Work

Future research will focus on adaptive gyroscope noise parameter estimation based on the proposed visual fixation trigger for head motion detection. The gyroscope noise parameter estimation can be used to reduce the heading errors even further and without the visual fixation trigger being active. While a sufficient amount of samples is gathered during visual fixations an adapted noise parameter can be estimated and used to identify no motion phases just as the visual fixation trigger. A second instance of the filter running in parallel could be used to compute orientation that includes the estimated gyroscope noise and compare it to the first instance of the filter in real-time. Based on the deviation between both solutions, the estimated bias could be used or discarded from that point on which in turn will lead to improved heading accuracy. Advanced parameter specification of the proposed fusion method will be explored by a broader set of experiments, including experiments in real use cases, multiple age varying paticipants as well as the influence of gyroscope noise parameter estimation on the proposed method.

Author Contributions: Conceptualization, L.W.; methodology, L.W.; software, L.W.; validation, L.W., and M.G.; formal analysis, L.W..; investigation, L.W. and M.G.; resources, M.G.; data curation, L.W.; writing–original draft preparation, L.W.; writing–review and editing, L.W. and M.G.; visualization, L.W.; supervision, M.G.; project administration, M.G.; funding acquisition, M.G. All authors have read and agreed to the published version of the manuscript.

Funding: This work was funded by the Federal Ministry of Education and Research of Germany grant number 13FH011IX6. We acknowledge support by the Open Access Publication Fund of the Westfälische Hochschule, University of Applied Sciences.

Conflicts of Interest: The authors declare no conflict of interest.

Abbreviations

The following abbreviations are used in this manuscript:

MARG	MagneticAngularate Gravity sensor
IMU	Inertial Measurement Unit
GDA	Gradient Descent Algorithm
MEMS	Micro-Electro-Mechanical Systems
API	Application Programming Interface
NED	North East Down
MCU	Microcontroller Unit
dc-bias	direct current bias

References

1. Rudigkeit, N.; Gebhard, M. AMiCUS—A Head Motion-Based Interface for Control of an Assistive Robot. *Sensors* **2019**, *19*, 2836. [CrossRef] [PubMed]

2. Jackowski, A.; Gebhard, M.; Thietje, R. Head Motion and Head Gesture-Based Robot Control: A Usability Study. *IEEE Trans. Neural Syst. Rehabil. Eng.* **2018**, *26*, 161–170. [CrossRef] [PubMed]

3. Alsharif, S.; Kuzmicheva, O.; Gräser, A. Gaze Gesture-Based Human Robot Interface. *Technische Unterstützungssysteme, die die Menschen Wirklich Wollen* **2016**, *12*, 339.

4. Wendel, J. *Integrierte Navigationssysteme: Sensordatenfusion, GPS und Inertiale Navigation*; Walter de Gruyter: Berlin, Germany, 2011.

5. Shiau, J.K.; Huang, C.X.; Chang, M.Y. Noise characteristics of MEMS gyro's null drift and temperature compensation. *J. Appl. Sci. Eng.* **2012**, *15*, 239–246.

6. Valenti, R.G.; Dryanovski, I.; Xiao, J. A linear Kalman filter for MARG orientation estimation using the algebraic quaternion algorithm. *IEEE Trans. Instrum. Meas.* **2016**, *65*, 467–481. [CrossRef]

7. Gebre-Egziabher, D.; Elkaim, G.; Powell, J.D.; Parkinson, B. A non-linear, two-step estimation algorithm for calibrating solid-state strapdown magnetometers. In Proceedings of the 8th St. Petersburg International Conference on Integrated Navigation Systems (IEEE/AIAA), St. Petersburg, Russia, 28–30 May 2001.

8. Wöhle, L.; Gebhard, M. A robust quaternion based Kalman filter using a gradient descent algorithm for orientation measurement. In Proceedings of the 2018 IEEE International Instrumentation and Measurement Technology Conference (I2MTC), Houston, TX, USA, 14–17 May 2018; pp. 1–6. [CrossRef]

9. Sabatini, A.M. Quaternion-based extended Kalman filter for determining orientation by inertial and magnetic sensing. *IEEE Trans. Biomed. Eng.* **2006**, *53*, 1346–1356. [CrossRef] [PubMed]

10. Roetenberg, D.; Luinge, H.; Slycke, P. *Xsens MVN: Full 6DOF Human Motion Tracking Using Miniature Inertial Sensors*; Technical Report; Xsens Motion Technologies BV: Enschede, The Netherlands, 8 April 2009; pp. 1–9.

11. Madgwick, S.O.; Harrison, A.J.; Vaidyanathan, R. Estimation of IMU and MARG orientation using a gradient descent algorithm. In Proceedings of the IEEE International Conference on Rehabilitation Robotics (ICORR), Zurich, Switzerland, 29 June–1 July 2011; pp. 1–7.

12. Robert-Lachaine, X.; Mecheri, H.; Larue, C.; Plamondon, A. Effect of local magnetic field disturbances on inertial measurement units accuracy. *Appl. Ergon.* **2017**, *63*, 123–132. [CrossRef] [PubMed]

13. Wu, Z.; Sun, Z.; Zhang, W.; Chen, Q. Attitude and gyro bias estimation by the rotation of an inertial measurement unit. *Meas. Sci. Technol.* **2015**, *26*, 125102. [CrossRef]

14. Wu, Y.; Zou, D.; Liu, P.; Yu, W. Dynamic Magnetometer Calibration and Alignment to Inertial Sensors by Kalman Filtering. *IEEE Trans. Control Syst. Technol.* **2018**, *26*, 716–723. [CrossRef]

15. Aqel, M.O.; Marhaban, M.H.; Saripan, M.I.; Ismail, N.B. Review of visual odometry: Types, approaches, challenges, and applications. *SpringerPlus* **2016**, *5*, 1897. [CrossRef] [PubMed]

16. Valenti, R.; Sebe, N.; Gevers, T. What are you looking at? *Int. J. Comput. Vis.* **2012**, *98*, 324–334. [CrossRef]

17. Kassner, M.; Patera, W.; Bulling, A. Pupil: An Open Source Platform for Pervasive Eye Tracking and Mobile Gaze-based Interaction. In *Adjunct Proceedings of the 2014 ACM International Joint Conference on Pervasive and Ubiquitous Computing*; ACM: New York, NY, USA, 2014; pp. 1151–1160.

18. Wöhle, L.; Miller, S.; Gerken, J.; Gebhard, M. A Robust Interface for Head Motion based Control of a Robot Arm using MARG and Visual Sensors. In Proceedings of the 2018 IEEE International Symposium on Medical Measurements and Applications (MeMeA), Rome, Italy, 11–13 June 2018; pp. 1–6. [CrossRef]

19. Pupil Labs GmbH. Pupil Core. Open Source Eye Tracking Platform Home Page. Available online: https://pupil-labs.com/products/core/ (accessed on 12 April 2020).

20. Larsson, L.; Schwaller, A.; Holmqvist, K.; Nyström, M.; Stridh, M. Compensation of head movements in mobile eye-tracking data using an inertial measurement unit. In Proceedings of the 2014 ACM International Joint Conference on Pervasive and Ubiquitous Computing: Adjunct Publication, New York, NY, USA, 13–17 September 2014; pp. 1161–1167. [CrossRef]

21. Salvucci, D.D.; Goldberg, J.H. Identifying fixations and saccades in eye-tracking protocols. In Proceedings of the 2000 Symposium on Eye Tracking Research & Applications, Palm Beach Gardens, FL, USA, 6–8 November 2000; pp. 71–78. [CrossRef]

22. Duchowski, A.T.; Jörg, S. Eye animation. In *Handbook of Human Motion*; Springer: Berlin/Heidelberg, Germany, 2016; pp. 1–19.

23. Diebel, J. Representing attitude: Euler angles, unit quaternions, and rotation vectors. *Matrix* **2006**, *58*, 1–35.

24. Real Time Engineers Ltd. FreeRTOS. Real-Time Operating System for Microcontrollers. Available online: https://www.freertos.org/Documentation/RTOS_book.html (accessed on 12 April 2020).

25. STMicroelectronics. Parameters and Calibration of a Low-g 3-Axis Accelerometer, Application Note 4508. Available online: https://www.st.com/resource/en/application$_$note/dm00119044-parameters-and-calibration-of-a-lowg-3axis-accelerometer-stmicroelectronics.pdf (accessed on 12 April 2020).

26. Gebre-Egziabher, D.; Elkaim, G.H.; David Powell, J.; Parkinson, B.W. Calibration of strapdown magnetometers in magnetic field domain. *J. Aerosp. Eng.* **2006**, *19*, 87–102. [CrossRef]

Article

The Validity and Reliability of the Microsoft Kinect for Measuring Trunk Compensation during Reaching

Matthew H. Foreman *,† and Jack R. Engsberg

Program in Occupational Therapy, Washington University in St. Louis, 4444 Forest Park Ave.,
St. Louis, MO 63108, USA; engsbergj@wusm.wustl.edu
* Correspondence: mforeman@methodist.edu; Tel.: +1-(910)-482-5426
† Current address: Doctor of Occupational Therapy Program, Methodist University, Fayetteville, NC 28311, USA.

Received: 12 November 2020; Accepted: 4 December 2020; Published: 10 December 2020

Abstract: Compensatory movements at the trunk are commonly utilized during reaching by persons with motor impairments due to neurological injury such as stroke. Recent low-cost motion sensors may be able to measure trunk compensation, but their validity and reliability for this application are unknown. The purpose of this study was to compare the first (K1) and second (K2) generations of the Microsoft Kinect to a video motion capture system (VMC) for measuring trunk compensation during reaching. Healthy participants (n = 5) performed reaching movements designed to simulate trunk compensation in three different directions and on two different days while being measured by all three sensors simultaneously. Kinematic variables related to reaching range of motion (ROM), planar reach distance, trunk flexion and lateral flexion, shoulder flexion and lateral flexion, and elbow flexion were calculated. Validity and reliability were analyzed using repeated-measures ANOVA, paired *t*-tests, Pearson's correlations, and Bland-Altman limits of agreement. Results show that the K2 was closer in magnitude to the VMC, more valid, and more reliable for measuring trunk flexion and lateral flexion during extended reaches than the K1. Both sensors were highly valid and reliable for reaching ROM, planar reach distance, and elbow flexion for all conditions. Results for shoulder flexion and abduction were mixed. The K2 was more valid and reliable for measuring trunk compensation during reaching and therefore might be prioritized for future development applications. Future analyses should include a more heterogeneous clinical population such as persons with chronic hemiparetic stroke.

Keywords: trunk; upper extremity; compensation; reaching; Kinect; video motion capture; validity; reliability

---

1. Introduction

Upper extremity (UE) motor impairments are highly prevalent in many clinical populations such as stroke [1]. Impaired UE movement is frequently accompanied by compensatory strategies that help a person adapt to limitations in motor function but may impact recovery and cause negative effects if used long term [2–4]. There are numerous well-researched, standardized assessments that measure UE abilities according to factors such as speed, strength, range of motion (ROM), and movement quality, but few that directly measure the amount of compensation utilized during task performance [5–7]. Without objective measurement and subsequent intervention, continued compensatory movements can reduce the amount of task-driven neuroplastic change achieved following neurologic injury and ultimately contribute to maladaptive plasticity, learned disuse or non-use, and chronic pain or injury [2–4]. Objective assessment of targeted and compensatory UE movements often relies on video motion capture cameras (VMC) or electromagnetic sensors that, while extremely accurate, are typically expensive and not feasible for application in a clinical setting. Because the amount of motor recovery achieved, and inversely the amount of compensation used, is highly predictive of participation and quality of life in persons living with long-term UE impairments, a clinically feasible, affordable, accurate, and objective measure of movement compensation may be an important innovation in rehabilitation science [8].

The Microsoft Kinect (Microsoft Corp., Redmond, WA, USA) is a low-cost, off-the-shelf motion sensor originally designed for video games that can be adapted for quantitative assessment of UE clinical movements [9–12]. The measurement abilities of the first-generation Kinect (K1) have been established for UE movements, spatiotemporal gait variables, standing balance, postural control, and even static foot posture [9,10,13–15]. The abilities of the second-generation Kinect (K2) are not as robustly established, but have been investigated for some UE, gait, and postural movements [11,16,17]. A recent study within our laboratory found both sensors to be valid relative to the gold standard of a VMC system when measuring reaching (forward and side) and angular shoulder movements (frontal, transverse, sagittal) [12]. Both sensors have also been frequently used within our laboratory for virtual reality (VR)-based motor rehabilitation aimed at improving UE motor abilities of persons with various impairments [18–21]. The Kinect sensors have some advantages over widely used optical and inertial sensor systems, namely significantly lower cost, higher portability, easier deployment in a lab or clinic, wider accessibility, and marker-less motion tracking with simpler throughput for the control of video games and VR applications. Conversely, the Kinects typically produce significantly lower resolution and less reliable data compared to gold-standard motion capture systems such as VMC and wearable inertial sensors [9–11,16].

Reaching is one of the most rigorously researched UE movements due to its involvement in many activities of daily living (ADLs). The kinematics of reaching in populations such as chronic stroke have been investigated in many different studies that often rely on VMC systems [22,23]. Not only do persons with stroke reach less accurately, slower, and with less motor control, they also utilize trunk flexion earlier and to a greater degree compared to the healthy population [22]. While differences in symmetry and joint coordination exist between healthy and impaired reaching, placing objects beyond the arm's length of healthy participants has been found to elicit trunk movement similar to that used by hemiparetic stroke patients reaching to objects within arm's length [23]. Few previous studies have examined the abilities of both generations of the Kinect sensor for measuring trunk compensation during reaching [24], and only one existing study has compared the measurement abilities of both sensors to simultaneous video motion capture [12]. The current study aims to go beyond previous work performed in our laboratory [12] to include a larger sample size of participants and movement trials with a focus on trunk kinematics during reaches that require trunk compensation. The purpose of this investigation was to establish the validity and reliability of two versions of the Microsoft Kinect for measuring UE and trunk kinematics during different reaching conditions.

2. Materials and Methods

2.1. Participants

A convenience sample of five healthy participants (3 women and 2 men, mean age 24.8 years) were recruited to participate in this study. A small sample size was considered due to the large sample of reaches (240 repetitions) performed by each participant and the overall focus of this study being the comparison of repeatable reaching motions across sensors and testing days. All participants gave informed written consent and the study protocol was approved by the university's Institutional Review Board.

2.2. Hardware

Both the K1 and K2 combine standard red-green-blue (RGB) video and an infrared (IR) depth sensor with advanced pattern recognition algorithms to provide full-body, three-dimensional (3D) skeletal motion capture without the use of wearable trackers. Both sensors provide data at approximately 30 frames per second (fps), but the K2 generally boasts improved hardware compared to the K1 (Table A1) [25]. For example, the K2 collects high definition RGB images (1920 × 1080 pixels) while the K1 collects standard definition RGB (640 × 480 pixels) that fails to compete with most modern webcams [25]. The RGB and IR cameras in the K2 also have wider fields of view and, when combined with updated tracking algorithms, can track greater numbers of skeletal landmarks and overall users [25]. Most importantly, the K2 utilizes a time-of-flight algorithm for motion tracking that is more robust, less noisy, and more reliable than the structured light algorithm used by the K1 [25]. The VMC system was considered the gold standard for comparison in this case and consisted of eight IR motion capture cameras (MAC Eagle Digital Cameras, Motion Analysis Corp., Santa Rosa, CA, USA) measuring at 60 fps with a 3D resolution accurate to within one millimeter.

2.3. Experimental Procedure

Participants performed a set of targeted reaching movements similar to a previously developed reaching performance task [12,26] while simultaneously being measured by the K1, the K2, and an 8-camera VMC system. Each participant was seated on a stool in the center of the VMC capture volume with the K1 and K2 positioned at a midline distance of approximately 2.0 m and a height of 1.2 m [12]. Each movement set involved reaching towards a target in the sagittal (forward), scaption (45 degree angle), or frontal (lateral) planes at either a non-extended or extended distance. The non-extended distance was defined relative to each participant's anthropometrics as shoulder height and arm's length, while the extended distance was moved 20 cm beyond arm's length (Figure 1). This extended reach required a healthy participant to flex the trunk and displace the shoulder to meet the target, similar to compensatory movements employed for reaching by persons with hemiparetic stroke [23]. Participants were provided verbal instruction but, given that they were healthy participants performing a relatively simple targeted reaching movement, no formal training was provided. On two different testing days, five repetitions were performed within each of four sets for the three directions and two conditions, resulting in a total of 240 repetitions for each of five participants. Given the large number of movements, participants were consistently asked for signs of fatigue and pain. None of the healthy participants reported any pain or fatigue in the UE. Participants were also given short breaks between movement sets (approx. 3–5 min) to mitigate fatigue. These breaks allowed researchers to code and save data files, check for data errors, and double check or adjust experimental setup and procedures.

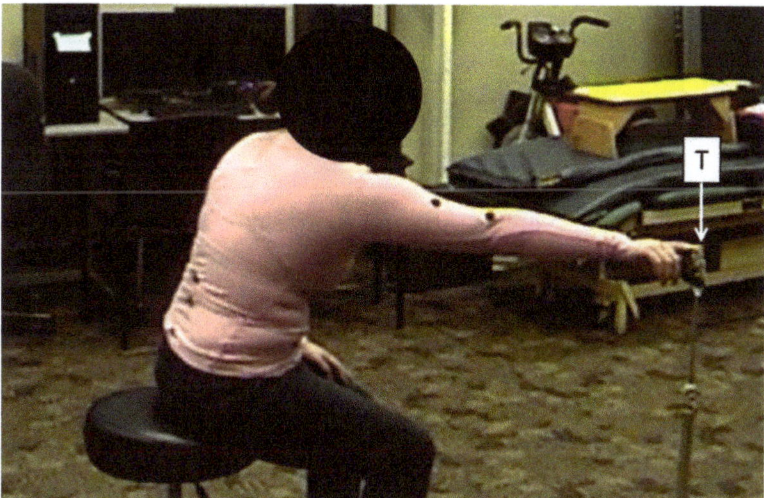

Figure 1. An example of a participant reaching towards the target (T) during an extended scaption reach while wearing retroreflective markers.

2.4. Data Collection

Kinematic data were collected for the K1 and K2 using the Microsoft Kinect for Windows Software Development Kit (SDK v1.8 and v2.0) [27], a virtual reality peripheral network (VRPN) server [28], and custom software designed in MATLAB (r2012a, Mathworks Inc., Natick, MA, USA). The 3D positions of 11 upper body landmarks for the K1 and K2 were measured relative to each sensor's origin (Figure 2). Common landmarks were head, neck, shoulders, elbows, wrists, and hands. The K1 defined torso as the body centroid, while the K2 defined the torso as a mid-spine landmark. Similar data were simultaneously collected for the VMC system using Motion Analysis software (Cortex, Motion Analysis Corp., Santa Rosa, CA, USA) to measure the positions of 25 retroreflective markers placed on bony landmarks on the participant's upper body. Markers were placed on the top of the head (vertex); C7, T10, L5, and S4 vertebrae; sternal notch; xiphoid process; acromion processes; medial and lateral epicondyles; ulnar and radial styloids; anterior superior iliac spines; dorsal hands; and index fingers. Two redundant markers were placed on the humerus and forearm.

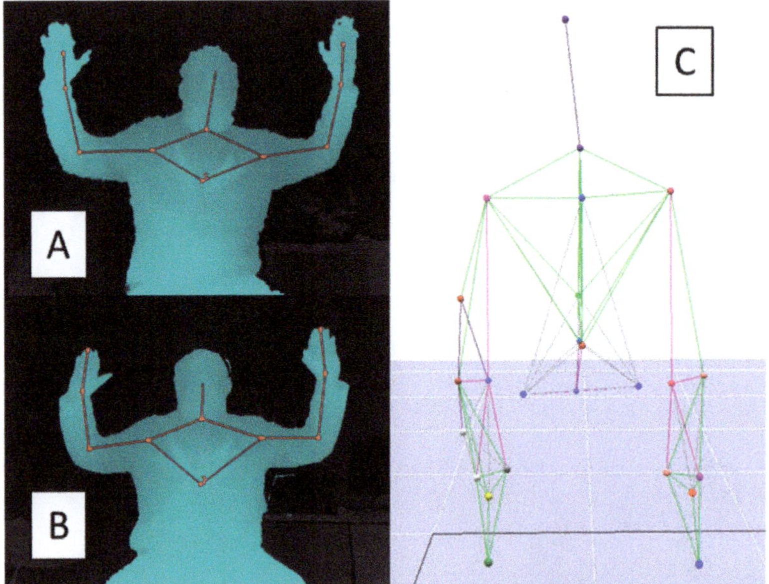

Figure 2. Examples of the kinematic body landmarks measured by the K1 (**A**), K2 (**B**), and VMC (**C**). The K1 and K2 measured 11 body landmarks. The VMC measured the position of 25 body landmarks.

2.5. Analysis Procedure

Once collected, Kinect data were filtered (6th order, 6 Hz Butterworth) and used to create body segment vectors including spine (torso-neck), humerus (shoulder-elbow), and forearm (elbow-wrist/hand). VMC data were similarly filtered (6th order, 6 Hz Butterworth), imported into MATLAB, and used to create analogous body segments using marker midpoints and biomechanical conventions [29]. Clinically relevant variables were calculated including reaching ROM, planar reaching distance (sagittal and frontal), shoulder flexion and abduction, trunk flexion and lateral flexion, and elbow flexion. Reaching ROM was defined as the Euclidean distance between the shoulder and the hand, while planar reaching distance was defined as the distance traveled by the hand in the sagittal or frontal plane. Shoulder flexion and abduction were defined as the angle between the humerus and spine in the sagittal and frontal planes, respectively. Trunk flexion and lateral flexion were similarly defined as the angle between the spine and the vertical coordinate axis in the sagittal and frontal planes, respectively. Finally, elbow flexion was defined as the angle between the forearm and the humerus.

2.6. Statistical Approach

A peak detection algorithm was used to determine the start and stop of each reach in terms of the maximum and minimum distance of the hand from the target. The target's position was not inherently available from the Kinect data, therefore an estimation was calculated as the average hand position at its maximum Euclidean distance from neutral. The first repetition of each trial was disregarded due to variable starting positions of the arm and hand. A three standard deviation algorithm was used to identify and remove outliers due to motion tracking errors. Validity was investigated using data from the first testing day (D1) to calculate magnitude differences, Pearson's correlations (r), Bland-Altman 95% limits of agreement (LOA), and a repeated measures analysis of variance (ANOVA) with Bonferroni corrections across sensors. Reliability was investigated using averages within each testing day to calculate magnitude differences, intra-class correlations (ICC), Pearson's correlations (r), Bland-Altman 95% LOA, and paired *t*-tests between days [30,31]. Estimates of correlations in terms of r and ICC were evaluated as excellent (0.75–1), modest (0.4–0.74), or poor (0–0.39) [31]. Bland-Altman analyses for validity (Table A2) and reliability (Table A4) as well as Pearson's correlations for reliability (Table A3) are presented in the Appendix A.

3. Results

3.1. Trunk Compensation

For trunk flexion and trunk lateral flexion, the K2 was closer in magnitude to the VMC than the K1 in all directions and for both non-extended and extended reaches (Table 1). For trunk flexion, when considering Bland-Altman LOA for all movements, the K2 was within −3.5°–6.6° and the K1 was within −2.7°–14.2° of the VMC (Table A2). Similarly for trunk lateral flexion, the K2 was within −5.9–7.9° and the K1 was within −9.0–13.4° of the VMC. Significant differences were found between K2 and VMC for trunk flexion during extended forward reaching and lateral flexion during extended scaption reaching (Table 1). Significant differences were found between K1 and VMC for trunk flexion during all extended reaches and lateral flexion in all conditions but extended lateral reaching.

The K2 was more valid than the K1 for measuring trunk movements during extended reaches (Table 2). The K2 showed excellent agreement with the VMC for measuring trunk flexion (r = 0.77–0.88) and lateral flexion (r = 0.77–0.89) during extended reaches. The K1 showed moderate–excellent agreement with the VMC for trunk flexion (r = 0.52–0.78) and moderate agreement for lateral flexion (r = 0.50–0.60) during extended reaches. For non-extended reaches, the K2 showed only moderate agreement (r = 0.43) for measuring trunk flexion during lateral reaching. All other correlations were poor for both the K1 and K2. Bland-Altman analyses show that mean biases for trunk flexion and lateral flexion were smaller and with narrower LOA for the K2 than the K1 when compared to VMC (Table A2).

Reliability results were mixed for all three sensors when measuring the trunk (Table 3). The K2 showed excellent reliability for measuring trunk flexion during lateral reaching (ICC = 0.91), but poor–modest reliability for trunk flexion in all other reach directions (ICC = −0.53–0.69). The K2 also showed excellent reliability for lateral flexion in the scaption (ICC = 0.75), lateral (ICC = 0.82), and extended forward (ICC = 0.84) directions, but poor–modest reliability in all other directions (ICC = 0.12–0.66). The K1 showed modest–excellent reliability (ICC = 0.62–0.88) for trunk measurements during reaches in all directions except forward (ICC = 0.28–0.34). The VMC showed mixed results similar to K2, with poor–excellent reliability in the forward direction (ICC = −0.42–0.93), poor–excellent reliability in the scaption direction (ICC = 0.08–0.89), and modest–excellent reliability in the lateral direction (ICC = 0.66–0.82) for both trunk flexion and lateral flexion. Pearson's correlations between testing days mirror these results (Table A3). Bland-Altman LOA analyses show small mean biases for trunk flexion and lateral flexion between testing days for the K1 (bias = −1.4°–0.8°), K2 (bias = −3.2°–1.4°), and VMC (bias = −3.0°–1.8°) (Table A4).

Table 1. Mean (± SD) magnitudes for the K1, K2, and VMC for all kinematic variables and all movements on the two different testing days D1 and D2. Each of five participants performed four sets of five reaches (N = 100) for each direction and condition. This sample was repeated on two separate testing days (D1 and D2).

	D1			D2		
	K1	**K2**	**VMC**	**K1**	**K2**	**VMC**
Forward (N = 100)						
Reaching ROM (cm)	43.9 ± 11.6 *	49.2 ± 15.7 *	32.7 ± 14.0	36.3 ± 13.2	42.1 ± 19.6	25.4 ± 17.3
Sagittal reach distance (cm)	49.2 ± 4.4	54.0 ± 12.2	42.0 ± 6.1	45.4 ± 6.5	49.5 ± 14.9	41.1 ± 6.6
Shoulder flexion (deg)	77.7 ± 7.5 *	78.0 ± 6.5 *	60.8 ± 6.1	83.0 ± 10.7	74.7 ± 9.0	62.9 ± 5.3
Trunk flexion (deg)	−2.2 ± 0.9	−0.4 ± 0.8	0.4 ± 1.3	−2.7 ± 1.2	−0.2 ± 0.5	0.2 ± 1.5
Trunk lateral flexion (deg)	0.6 ± 0.5 *	0.0 ± 0.4	−1.1 ± 0.7	0.6 ± 0.5	−0.1 ± 0.4	−3.1 ± 11.3
Elbow flexion (deg)	110.1 ± 43.6	104.4 ± 38.1	86.9 ± 29.5	87.5 ± 50.9	80.1 ± 48.3	69.0 ± 33.0
Forward Extend (N = 100)						
Reaching ROM (cm)	37.4 ± 18.3	52.2 ± 19.4 *	30.3 ± 13.7	34.5 ± 16.5	51.5 ± 16.8	26.7 ± 15.4
Sagittal reach distance (cm)	58.6 ± 12.8	72.7 ± 14.2	62.6 ± 6.8	57.7 ± 12.3	72.2 ± 15.9	61.0 ± 9.4
Shoulder flexion (deg)	88.5 ± 9.5	103.0 ± 8.4 *	81.7 ± 7.2	87.8 ± 15.4	95.6 ± 16.3	78.8 ± 9.0
Trunk flexion (deg)	10.3 ± 2.7 *	15.0 ± 3.5 *	18.7 ± 3.0	11.7 ± 1.6	17.9 ± 2.9	21.7 ± 3.5
Trunk lateral flexion (deg)	0.9 ± 1.8 *	−0.9 ± 1.4	−3.7 ± 2.5	1.0 ± 2.1	−0.9 ± 2.3	−4.2 ± 5.7
Elbow flexion (deg)	109.3 ± 44.7	111.3 ± 45.2	87.9 ± 29.3	99.4 ± 42.0	97.2 ± 43.0	73.9 ± 32.1
Scaption (N = 100)						
Reaching ROM (cm)	39.1 ± 14.4	37.3 ± 16.7	33.3 ± 15.5	34.5 ± 13.4	30.6 ± 17.8	27.8 ± 17.3
Sagittal reach distance (cm)	25.0 ± 5.9	26.8 ± 11.6	24.0 ± 6.0	25.9 ± 5.1	23.4 ± 11.6	24.0 ± 5.0
Frontal reach distance (cm)	42.9 ± 7.5 *	45.2 ± 10.4	37.8 ± 6.5	37.9 ± 7.5	39.2 ± 10.2	34.3 ± 5.5
Shoulder flexion (deg)	65.9 ± 12.3 *	57.7 ± 9.9 *	41.4 ± 11.6	67.4 ± 7.9	61.1 ± 6.5	46.3 ± 5.4
Shoulder abduction (deg)	52.8 ± 17.2 *	55.2 ± 13.5 *	36.1 ± 11.4	57.7 ± 11.1	60.1 ± 9.9	37.0 ± 6.8
Trunk flexion (deg)	−3.4 ± 1.0 *	−0.2 ± 0.7	0.0 ± 0.8	−3.7 ± 1.2	−0.2 ± 0.6	0.3 ± 1.1
Trunk lateral flexion (deg)	−7.3 ± 1.5 *	−0.1 ± 0.5	−0.4 ± 0.8	−6.6 ± 1.8	−0.1 ± 0.5	−0.3 ± 0.9
Elbow flexion (deg)	112.1 ± 44.7	101.1 ± 42.4	88.2 ± 33.3	88.5 ± 51.5	82.9 ± 41.6	74.4 ± 34.0
Scaption Extend (N = 100)						
Reaching ROM (cm)	36.6 ± 14.6	40.5 ± 17.8 *	31.8 ± 14.5	28.3 ± 16.8	31.5 ± 19.3	24.7 ± 17.8
Sagittal reach distance (cm)	30.4 ± 6.3	37.3 ± 11.2	37.8 ± 7.3	31.3 ± 6.5	34.3 ± 11.6	38.8 ± 5.3
Frontal reach distance (cm)	47.4 ± 14.2	54.6 ± 14.9	51.2 ± 9.3	42.7 ± 11.0	48.6 ± 13.5	46.1 ± 7.4
Shoulder flexion (deg)	72.0 ± 8.7	88.9 ± 7.6 *	62.1 ± 7.5	72.2 ± 8.1	87.8 ± 5.0	65.1 ± 6.9
Shoulder abduction (deg)	66.7 ± 11.0 *	86.3 ± 8.9 *	58.5 ± 7.3	67.6 ± 8.6	87.6 ± 7.4	57.5 ± 7.2
Trunk flexion (deg)	5.5 ± 1.4 *	12.5 ± 2.5	15.0 ± 2.8	6.4 ± 1.9	13.1 ± 2.5	15.1 ± 4.9
Trunk lateral flexion (deg)	10.2 ± 1.7 *	13.4 ± 2.4 *	16.0 ± 3.5	10.8 ± 3.1	13.3 ± 3.5	15.9 ± 3.9
Elbow flexion (deg)	107.4 ± 44.9	108.8 ± 42.7	88.6 ± 32.6	85.7 ± 48.6	86.4 ± 46.5	72.0 ± 36.6
Lateral (N = 100)						
Reaching ROM (cm)	25.6 ± 16.0	27.7 ± 16.8	29.0 ± 16.7	23.3 ± 14.0	22.6 ± 16.5	27.4 ± 17.9
Frontal hand distance (cm)	49.7 ± 10.8	57.8 ± 12.6	51.6 ± 5.4	44.6 ± 10.2 **	50.9 ± 14.9 **	47.1 ± 7.5
Shoulder abduction (deg)	51.3 ± 12.2	53.1 ± 10.5 *	42.6 ± 10.3	49.3 ± 12.0	49.5 ± 12.7	39.2 ± 9.6
Trunk flexion (deg)	0.3 ± 0.9	0.2 ± 0.7	−0.2 ± 0.7	0.4 ± 0.9	0.0 ± 0.3	0.1 ± 0.6
Trunk lateral flexion (deg)	−7.8 ± 1.3 *	−0.6 ± 0.9	0.0 ± 1.4	−7.7 ± 2.5	−0.5 ± 0.6	−0.5 ± 0.9
Elbow flexion (deg)	91.1 ± 48.4	91.6 ± 44.5	79.2 ± 35.9	84.8 ± 48.8	80.2 ± 42.9	72.9 ± 36.6
Lateral Extend (N = 100)						
Reaching ROM (cm)	13.1 ± 14.8 *	23.7 ± 16.1	25.3 ± 15.4	13.8 ± 17.3	20.5 ± 18.2	24.5 ± 18.1
Frontal hand distance (cm)	55.7 ± 13.1 *	69.9 ± 14.9	69.4 ± 7.6	52.6 ± 15.2	65.2 ± 19.6	65.5 ± 11.8
Shoulder abduction (deg)	77.4 ± 8.5	88.5 ± 9.2 *	72.9 ± 9.4	72.1 ± 10.3	81.0 ± 11.4	67.4 ± 11.5
Trunk flexion (deg)	0.0 ± 2.2 *	3.8 ± 3.0	3.8 ± 3.1	−0.8 ± 1.3	2.5 ± 2.7 **	3.0 ± 2.9
Trunk lateral flexion (deg)	18.9 ± 3.9	21.7 ± 3.4	23.9 ± 4.6	18.4 ± 4.9	20.5 ± 3.8	22.1 ± 6.4
Elbow flexion (deg)	87.5 ± 48.4	93.6 ± 42.6	77.3 ± 32.4	81.3 ± 51.6	86.9 ± 46.6	73.2 ± 36.3

* $p < 0.05$ for Bonferonni-corrected pairwise *t*-test between Kinect and VMC. ** $p < 0.05$ for paired *t*-test between testing days. K1: KinectV1; K2: KinectV2; VMC: video motion capture; D1: day one of testing; D2: day two of testing.

Table 2. Validity measured by Pearson's correlation coefficients (r) between the K1 and VMC and the K2 and VMC on D1.

	Forward		Scaption		Lateral	
	K1	**K2**	**K1**	**K2**	**K1**	**K2**
Non-Extended						
Reaching ROM (cm)	0.93 *	0.95 *	0.94 *	0.94 *	0.94 *	0.94 *
Sagittal reach distance (cm)	0.60 *	0.79 *	0.75 *	0.81 *	-	-
Frontal reach distance (cm)	-	-	0.93 *	0.97 *	0.92 *	0.94 *
Shoulder flexion (deg)	0.19	0.24	0.77 *	0.80 *	-	-
Shoulder abduction (deg)	-	-	0.97 *	0.97 *	0.88 *	0.96 *
Trunk flexion (deg)	−0.19	0.01	−0.44 *	0.22	−0.03	0.17
Trunk lateral flexion (deg)	0.25 *	0.10	−0.36 *	0.20	−0.23 *	0.43 *
Elbow flexion (deg)	0.95 *	0.94 *	0.97 *	0.99 *	0.96 *	0.99 *
Extended						
Reaching ROM (cm)	0.95 *	0.91 *	0.90 *	0.98 *	0.90 *	0.95 *
Sagittal reach distance (cm)	0.91 *	0.82 *	0.67 *	0.84 *	-	-
Frontal reach distance (cm)	-	-	0.97 *	0.96 *	0.94 *	0.95 *
Shoulder flexion (deg)	0.23 *	0.36 *	0.31 *	0.66 *	-	-
Shoulder abduction (deg)	-	-	0.90 *	0.91 *	0.72 *	0.89 *
Trunk flexion (deg)	0.78 *	0.88 *	0.52 *	0.77 *	0.72 *	0.83 *
Trunk lateral flexion (deg)	0.51 *	0.77 *	0.60 *	0.89 *	0.50 *	0.78 *
Elbow flexion (deg)	0.98 *	0.97 *	0.96 *	0.98 *	0.99 *	0.99 *

* $p < 0.05$ for Pearson's correlation between Kinect and VMC. K1: KinectV1; K2: KinectV2; VMC: video motion capture; D1: day one of testing.

Table 3. Reliability measured by intra-class correlation coefficients (ICC) between testing days D1 and D2 for each of the three sensors.

	Forward			Scaption			Lateral		
	K1	**K2**	**VMC**	**K1**	**K2**	**VMC**	**K1**	**K2**	**VMC**
Non-Extended									
Reaching ROM (cm)	0.59	0.86	0.78	0.88	0.82	0.84	0.98	0.96	0.99
Sagittal reach distance (cm)	0.54	0.90	0.82	0.58	0.73	0.74	-	-	-
Frontal reach distance (cm)	-	-	-	0.74	0.77	0.76	0.93	0.94	0.84
Shoulder flexion (deg)	0.37	0.46	−0.08	0.52	−0.22	−0.02	-	-	-
Shoulder abduction (deg)	-	-	-	0.56	0.56	0.84	0.99	0.93	0.96
Trunk flexion (deg)	0.28	0.31	0.93	0.74	0.43	0.89	0.86	0.69	0.66
Trunk lateral flexion (deg)	0.34	0.14	0.84	0.69	0.75	0.36	0.85	0.82	0.81
Elbow flexion (deg)	0.75	0.82	0.74	0.76	0.72	0.80	0.99	0.97	0.99
Extended									
Reaching ROM (cm)	0.95	0.99	0.94	0.85	0.77	0.83	0.97	0.96	0.99
Sagittal reach distance (cm)	0.90	0.98	0.79	0.65	0.67	0.78	-	-	-
Frontal reach distance (cm)	-	-	-	0.92	0.87	0.70	0.97	0.97	0.92
Shoulder flexion (deg)	−0.57	−0.23	−0.61	0.17	−0.69	−1.47	-	-	-
Shoulder abduction (deg)	-	-	-	−0.18	−0.04	0.26	0.52	0.66	0.88
Trunk flexion (deg)	0.74	−0.53	−0.42	0.75	0.66	0.74	0.63	0.91	0.82
Trunk lateral flexion (deg)	0.87	0.84	0.65	0.62	0.20	0.08	0.88	0.66	0.71
Elbow flexion (deg)	0.94	0.92	0.91	0.82	0.75	0.82	0.98	0.98	0.97

K1: Kinect V1; K2: KinectV2; VMC: video motion capture; D1: day one of testing; D2: day two of testing.

3.2. Upper Extremity Movements

The movement traces for the three planar reaching conditions (i.e., sagittal, scaption, frontal) illustrate directional differences between the Kinects and the VMC (Figure 3). Discrepancies in reaching magnitude between the Kinects and the VMC were dependent on the direction of movement. Differences in reaching ROM and planar distance were greatest during forward reaching, reduced during scaption reaching, and least during lateral reaching (Figure 3). Reaching ROM, planar reach distance, and elbow flexion measurements consistently showed excellent validity for the K2 (r = 0.79–0.99) and moderate–excellent validity for the K1 (r = 0.60–0.95) (Table 2). Reliability of these measurements was moderate–excellent for all three sensors (Table 3). Validity and reliability of shoulder flexion and abduction measurements varied from poor to excellent for all three sensors (Tables 2 and 3).

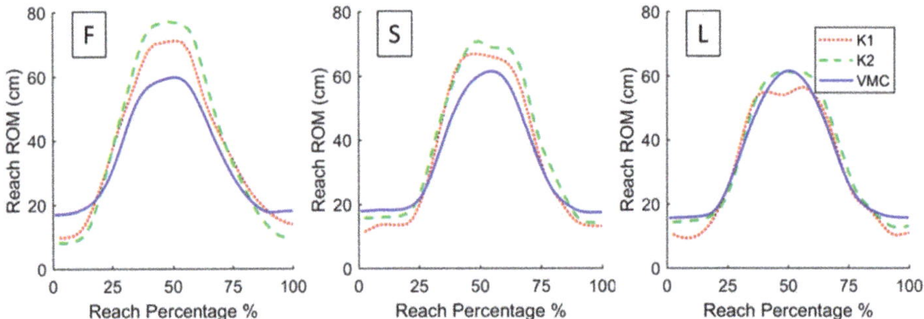

Figure 3. Three sets of curves showing reach ROM from start to stop of a typical reaching movement. The **left** curve (F) represents a forward reach, the **middle** curve (S) represents a scaption reach, and the **right** curve (L) represents a lateral reach. Curves for the K1, K2, and VMC are shown separately (see legend).

4. Discussion

The purpose of this investigation was to establish the validity and reliability of two versions of the Microsoft Kinect for measuring UE and trunk kinematics during various reaching conditions. Specifically, participants were asked to perform both a non-extended and extended reach in each of three directions (forward, scaption, lateral) while their movements were recorded by the K1, K2, and the gold-standard VMC simultaneously. The K2 measured the trunk more similarly to the VMC as shown by smaller average magnitude differences in trunk flexion and lateral flexion. Validity results for trunk measurement were excellent for the K2 and modest–excellent for the K1 during extended reaching conditions intended to simulate movements that might be used by persons with chronic stroke. Reliability for trunk measurement was modest–excellent for extended reaching with the K1, with the exception of the forward direction, but varied from poor to excellent for the K2. Results for both sensors were generally excellent for measuring arm and hand displacement, excellent for measuring elbow flexion, and mixed for shoulder measurement, with reaches in the scaption and lateral directions providing more valid and reliable results than the forward direction.

The results of this study are supported by previous research that examines the validity of the K1 and K2 in terms of other functional movements. Bonnechere and colleagues [9] found similar results when comparing the K1 to VMC during the performance of four functional movements including shoulder abduction (similar to lateral reaching) and elbow flexion (similar to forward reaching). Clark and colleagues [11] found the K2 to have excellent concurrent validity for measuring trunk movements during dynamic balance tasks and anterior–posterior movements, but poor–moderate validity for static tasks and medial–lateral movements. In the current investigation, the K2 similarly shows the greatest validity for measuring trunk flexion during an extended movement in the anterior–posterior direction. Reither et al. [12] found similar results while measuring the K1, K2, and VMC simultaneously with a single participant reaching forward, reaching to the side, and performing shoulder movements in various planes, but did not investigate trunk kinematics during such movements. In summary, Reither et al. [12] similarly found a greater range in single-day correlations between K1 and VMC (r = 0.31–0.96) than between the K2 and VMC (r = 0.45–0.96) with correlation magnitudes dependent on movement plane. The authors also found varied day-to-day reliability results for both K1 and K2 and, in general, a greater direction-dependent underestimation of kinematics displayed by the K1 [12]. The current study goes beyond the methods of Reither et al. [12] by utilizing an increased sample size of participants and movements, the inclusion of extended reaches to elicit trunk compensations, analysis of the trunk along with the UE, and movements in the scaption plane along with sagittal, frontal, and transverse planes.

We found several low and negative reliability (ICC) values (Table 3), particularly for shoulder flexion, shoulder abduction, trunk flexion, and trunk lateral flexion during non-extended reaching in the forward and scaption directions for all sensors including VMC. Negative ICC values are not ideal and can often be attributed to low between-subjects variance in the phenomenon being measured [32]. Accordingly, these results might be due to small between-day variance in the kinematic variables being tested. For example, a negative ICC value (ICC = −0.53) was calculated for the K2 between days for trunk flexion during the extended forward reach, but Bland-Altman analysis shows a small mean bias (bias = −3.0°) and LOA (LOA = −13.2–6.8°). This suggests a relatively small mean difference, and thus satisfactory repeatability, between testing days even in the face of a negative ICC calculation that may be due to small and non-systematic variance. A more heterogeneous clinical population may improve correlation results by increasing variance in the sample. Pearson's correlations (Table A3) and Bland-Altman LOA (Table A4) were included to give a broader picture of absolute and relative reliability for all three sensors. Additional, more advanced analyses may also provide further insight into these discrepancies; for example, dynamic time warping (DTW) is an advanced signal processing technique that could provide a measure of signal match for the time series data collected by the K1 and K2 [33].

The most notable limitation to this work is the use of healthy participants rather than a sample of participants with hemiparesis. As mentioned previously, persons with hemiparesis reach significantly differently than unimpaired persons, namely with slower movement, less accuracy, impaired interjoint coordination, and increased use of compensatory movement at the trunk [22,23]. Targets placed beyond the reach of healthy participants can elicit a similar compensatory response at the trunk, but persons with hemiparesis exhibit less symmetry and earlier trunk recruitment in comparison [23]. Healthy reaching is simply not the same as hemiparetic reaching. However, the purpose of the current study is to validate the measurement capabilities of the K1 and K2 relative to each other and to a gold-standard VMC system. Numerous referenced studies use healthy participants for sensor validation with intentions for future clinical application [9–17]. Healthy participants are more accessible, can perform the large number of required movements without fatigue or pain, and can more readily reproduce movements across trials and testing days for validity and reliability analyses. Given that the ultimate application of this study is implementation for clinical measurement of neurologically impaired populations, the ecological validity of future work would greatly benefit from testing with a more heterogeneous sample of persons with hemiparetic stroke.

The current study provides some insights for the design of such future work; for example, it may be necessary to recruit more individuals and reduce the overall repetitions performed to better capture variability, mitigate fatigue, and enhance the generalizability of results for real-world clinical populations. In addition, the experimental protocol could be adjusted to provide detailed instruction and training for impaired populations to reduce trial variability and enhance the efficiency of data reduction and cleaning. Given that the evidence shows that persons with hemiparetic stroke recruit the trunk earlier and more often than healthy populations [23], it may be necessary to eliminate or reduce the distance of the extended reach to maximize reaching performance and reduce frustration. Finally, given the results of the current study, it may be prudent to focus on the planes of movement best measured by the K1 and K2 due to their hardware constraints (e.g., lateral > scaption > forward).

Other variations in results might be attributed to various study limitations. First, the Kinect SDK uses a tracking algorithm that does not rely on the specific placement of markers on palpable bony landmarks as does the VMC. While this is convenient for users, it has been previously noted as a limitation in the Kinect's ability to accurately measure kinematics of movement due to variable body segment lengths; however, previous studies have developed algorithms through regression that may be able to correct for this during real-time tracking [9]. Second, it was clear through both observation and the relatively high standard deviations attributed to each movement (Table 1) that different strategies were used for reaching by individual participants. No neutral starting point was defined a priori, and some participants returned their arm to their lap between repetitions while others remained in a flexed position. This resulted in large variations in range of motion, namely with elbow flexion. Finally, reliability results varied inconsistently for all three sensors, and it should be noted that, on top of statistical limitations, there are intra-individual differences across trials and across days in each participant's reaching kinematics. Participants were given similar instructions for each trial and testing day, but differences in the repeatability of human movement yet exist and may be attributable to the slight variance in between-day correlation and significance testing. Participants were provided verbal instruction but no formal training at the simple reaching movements, so movement may have differed between movement sets and even testing days due to subtle learning effects. It is also possible that the placement of motion capture markers varied slightly between days, resulting in reliability differences. Increasing the overall sample size in the future could mitigate these intra- and inter-individual differences in repeatable movement.

This study shows that the K1 and K2 may serve as useful tools for objectively measuring UE and trunk kinematics, but application may depend on the body segment, joint, and movement plane of interest. Few studies have investigated their relative measurement properties, but both sensors are widely employed as the basis for VR-based interventions for persons with motor impairments including stroke and cerebral palsy [19,21]. Use of such interventions continues to grow along with client interest, professional knowledge, and technological accessibility [34]. The current investigation may inform future VR development, namely the inclusion of real-time measurement of trunk compensation using the K2.

5. Conclusions

In conclusion, the K1 and K2 have been shown to be valid and reliable for measuring some aspects of UE and trunk kinematics during reaching. In particular, the K2 exhibited slightly better characteristics for measuring the trunk during standard and extended reaching in different directions, and may be recommended over the K1 in future development for purposes of measuring trunk compensation in clinical populations.

Author Contributions: M.H.F. contributed in research design, data collection, data analysis, manuscript development, and manuscript submission. J.R.E. contributed as the research mentor in research supervision and manuscript editing. All authors have read and agreed to the published version of the manuscript.

Funding: This research received no external funding.

Acknowledgments: This research was supported by the Program in Occupational Therapy at the Washington University School of Medicine. All study participants provided informed consent, and this study was approved by the Institutional Review Board at the Washington University School of Medicine.

Conflicts of Interest: The authors have no conflict of interest to declare.

Appendix A

Table A1. Comparison of the first-generation Microsoft Kinect V1 (K1) and the second-generation Microsoft Kinect V2 (K2). The K2 boasts improved motion sensing hardware, particularly in resolution, field of view, and sensing algorithms. This is adapted from Pagliari and Pinto (2015) [25].

	Kinect V1	Kinect V2
RGB camera (pixels)	640 × 480	1920 × 1080
Depth camera (pixels)	640 × 480	512 × 424
Max depth distance (m)	4.0	4.5
Min depth distance (m)	0.8	0.5
Horizontal field of view (deg)	57	70
Vertical field of view (deg)	43	60
Skeletal markers	20	26
Possible skeletons tracked	2	6
USB capability	2.0	3.0

RGB: red-green-blue; USB: universal serial bus.

Table A2. Validity measured by mean bias (Bias) and Bland-Altman limits of agreement, lower bound (LB) and upper bound (UB), for K1 and K2 in comparison to VMC on D1.

	K1			K2		
	Bias	LB	UB	Bias	LB	UB
Forward						
Reaching ROM (cm)	−11.2	−21.4	−0.9	−17.5	−27.0	−8.0
Sagittal reach distance (cm)	−7.2	−16.9	2.5	−12.6	−28.7	3.5
Frontal reach distance (cm)	-	-	-	-	-	-
Shoulder flexion (deg)	−16.9	−34.0	0.2	−16.9	−31.2	−2.5
Shoulder abduction (deg)	-	-	-	-	-	-
Trunk flexion (deg)	2.7	−0.8	6.1	0.8	−2.0	3.5
Trunk lateral flexion (deg)	−1.7	−3.2	−0.2	−1.1	−2.7	0.6
Elbow flexion (deg)	23.3	−11.9	58.4	19.9	−7.9	47.7
Forward Extend						
Reaching ROM (cm)	−7.2	−20.7	6.3	−22.5	−38.8	−6.2
Sagittal reach distance (cm)	4.0	−10.1	18.1	−10.7	−29.2	7.8
Frontal reach distance (cm)	-	-	-	-	-	-
Shoulder flexion (deg)	−6.8	−27.5	13.9	−21.3	−37.5	−5.1
Shoulder abduction (deg)	-	-	-	-	-	-
Trunk flexion (deg)	8.4	4.6	12.2	3.3	0.0	6.6
Trunk lateral flexion (deg)	−4.6	−9.0	−0.2	−2.6	−5.9	0.8
Elbow flexion (deg)	20.5	−13.8	54.8	23.9	−9.0	56.9
Scaption						
Reaching ROM (cm)	−5.8	−16.2	4.5	−4.0	−14.8	6.8
Sagittal reach distance (cm)	−0.9	−9.1	7.2	−2.8	−17.5	12.0
Frontal reach distance (cm)	5.2	−0.4	10.7	7.5	−1.2	16.1
Shoulder flexion (deg)	−24.6	−38.9	−10.3	−16.2	−29.2	−3.3
Shoulder abduction (deg)	−16.7	−30.1	−3.3	−19.1	−26.4	−11.7
Trunk flexion (deg)	3.4	0.3	6.5	0.2	−1.6	2.0
Trunk lateral flexion (deg)	6.9	3.1	10.7	−0.2	−1.9	1.4
Elbow flexion (deg)	23.9	−5.1	52.9	12.9	−8.4	34.2
Scaption Extend						
Reaching ROM (cm)	−4.8	−17.3	7.7	−8.7	−18.0	0.7
Sagittal reach distance (cm)	7.3	−3.6	18.2	0.5	−12.2	13.1
Frontal reach distance (cm)	−3.8	−14.8	7.2	3.4	−9.2	16.0
Shoulder flexion (deg)	−9.8	−26.9	7.4	−26.7	−37.4	−16.0
Shoulder abduction (deg)	−8.5	−18.1	1.0	−28.0	−33.9	−22.1
Trunk flexion (deg)	9.5	4.8	14.2	2.5	−1.1	6.1
Trunk lateral flexion (deg)	5.8	0.3	11.3	2.7	−0.7	6.0
Elbow flexion (deg)	18.8	−12.7	50.3	20.2	−4.3	44.7

Table A2. *Cont.*

	K1			K2		
	Bias	**LB**	**UB**	**Bias**	**LB**	**UB**
Lateral						
Reaching ROM (cm)	3.3	−7.7	14.3	1.3	−9.7	12.2
Sagittal reach distance (cm)	-	-	-	-	-	-
Frontal reach distance (cm)	−1.8	−13.9	10.2	6.2	−8.9	21.2
Shoulder flexion (deg)	-	-	-	-	-	-
Shoulder abduction (deg)	−8.7	−20.1	2.7	−10.6	−16.3	−4.9
Trunk flexion (deg)	−0.5	−2.7	1.6	−0.4	−1.8	1.1
Trunk lateral flexion (deg)	7.8	3.6	11.9	0.5	−2.0	3.0
Elbow flexion (deg)	11.9	−21.5	45.4	12.4	−7.4	32.2
Lateral Extend						
Reaching ROM (cm)	12.2	−0.9	25.2	1.6	−8.7	11.9
Sagittal reach distance (cm)	-	-	-	-	-	-
Frontal reach distance (cm)	−13.7	−26.5	−0.8	0.5	−15.3	16.4
Shoulder flexion (deg)	-	-	-	-	-	-
Shoulder abduction (deg)	−4.5	−17.7	8.7	−15.7	−24.2	−7.1
Trunk flexion (deg)	3.9	−0.4	8.1	0.0	−3.5	3.5
Trunk lateral flexion (deg)	5.0	−3.4	13.4	2.2	−3.5	7.9
Elbow flexion (deg)	10.2	−23.7	44.1	16.3	−6.4	39.1

Bias: mean bias of VMC-Kinect for each sensor; LB: lower bound for Bland-Altman limits of agreement; UB: Upper bound for Bland-Altman limits of agreement; K1: KinectV1; K2: KinectV2; VMC: video motion capture; D1: day one of testing.

Table A3. Reliability measured by Pearson's correlation coefficients (r) between testing days D1 and D2 for each of the three sensors.

	Forward			Scaption			Lateral		
	K1	**K2**	**VMC**	**K1**	**K2**	**VMC**	**K1**	**K2**	**VMC**
Non-Extended									
Reaching ROM (cm)	0.48 *	0.83 *	0.64 *	0.76 *	0.65 *	0.68 *	0.95 *	0.93 *	0.97 *
Sagittal reach distance (cm)	0.30 *	0.83 *	0.62 *	0.23 *	0.51 *	0.52 *	-	-	-
Frontal reach distance (cm)	-	-	-	0.64 *	0.65 *	0.66 *	0.84 *	0.93 *	0.81 *
Shoulder flexion (deg)	0.19	0.08	−0.04	0.29 *	−0.11	−0.01	-	-	-
Shoulder abduction (deg)	-	-	-	0.39 *	0.32 *	0.68 *	0.92 *	0.82 *	0.93 *
Trunk flexion (deg)	0.05	0.06	0.70 *	0.30 *	0.01	0.70 *	0.42 *	0.04	0.31 *
Trunk lateral flexion (deg)	−0.06	0.11	0.45 *	0.56 *	0.36 *	0.13	0.68 *	0.41 *	0.63 *
Elbow flexion (deg)	0.64 *	0.79 *	0.60 *	0.67 *	0.55 *	0.65 *	0.97 *	0.94 *	0.96 *
Extended									
Reaching ROM (cm)	0.82 *	0.91 *	0.82 *	0.74 *	0.63 *	0.72 *	0.90 *	0.92 *	0.97 *
Sagittal reach distance (cm)	0.74 *	0.90 *	0.60 *	0.34 *	0.45 *	0.56 *	-	-	-
Frontal reach distance (cm)	-	-	-	0.84 *	0.77 *	0.56 *	0.89 *	0.96 *	0.93 *
Shoulder flexion (deg)	−0.25 *	−0.16	−0.25 *	0.08	−0.28 *	−0.44 *	-	-	-
Shoulder abduction (deg)	-	-	-	−0.11	−0.02	0.10	0.36 *	0.59 *	0.82 *
Trunk flexion (deg)	0.59 *	−0.32 *	−0.24 *	0.36 *	0.33 *	0.56 *	0.48 *	0.68 *	0.53 *
Trunk lateral flexion (deg)	0.52 *	0.60 *	0.54 *	0.38 *	0.11	0.04	0.64 *	0.37 *	0.47 *
Elbow flexion (deg)	0.83 *	0.87 *	0.83 *	0.69 *	0.61 *	0.71 *	0.94 *	0.96 *	0.93 *

* $p < 0.05$ for Pearson's correlation between D1 and D2. K1: Kinect V1; K2: KinectV2; VMC: video motion capture; D1: day one of testing; D2: day two of testing.

Table A4. Reliability analysis using mean bias (Bias) and Bland-Altman limits of agreement, lower bound (LB) and upper bound (LB), to compare all three sensors across testing days D1 and D2.

	K1			K2			VMC		
	Bias	LB	UB	Bias	LB	UB	Bias	LB	UB
Forward									
Reaching ROM (cm)	8.0	−16.9	32.8	6.8	−15.6	29.3	7.4	−19.2	33.9
Sagittal reach distance (cm)	3.9	−9.1	17.0	4.0	−13.2	21.2	0.9	−9.9	11.8
Frontal reach distance (cm)	-	-	-	-	-	-	-	-	-
Shoulder flexion (deg)	−5.0	−28.3	18.2	2.0	−16.8	20.8	−2.1	−18.2	14.1
Shoulder abduction (deg)	-	-	-	-	-	-	-	-	-
Trunk flexion (deg)	0.4	−2.6	3.4	−0.1	−1.7	1.4	0.2	−1.9	2.4
Trunk lateral flexion (deg)	0.0	−1.5	1.4	0.1	−1.0	1.2	0.2	−1.6	2.1
Elbow flexion (deg)	−18.7	−91.3	53.9	−16.1	−71.0	38.8	−17.9	−72.9	37.2
Forward Extend									
Reaching ROM (cm)	3.0	−17.7	23.7	−0.5	−15.8	14.8	3.6	−13.7	20.9
Sagittal reach distance (cm)	0.9	−17.0	18.8	−1.2	−15.5	13.1	1.6	−13.3	16.5
Frontal reach distance (cm)	-	-	-	-	-	-	-	-	-
Shoulder flexion (deg)	0.7	−38.4	39.8	7.1	−30.9	45.1	3.0	−22.4	28.3
Shoulder abduction (deg)	-	-	-	-	-	-	-	-	-
Trunk flexion (deg)	−1.4	−5.7	3.0	−3.2	−13.2	6.8	−3.0	−13.2	7.1
Trunk lateral flexion (deg)	−0.1	−3.9	3.7	0.0	−3.4	3.5	0.5	−9.0	10.0
Elbow flexion (deg)	−9.8	−59.7	40.1	−11.4	−57.3	34.4	−14.2	−49.5	21.1
Scaption									
Reaching ROM (cm)	4.9	−13.9	23.8	6.7	−21.6	34.9	5.5	−20.5	31.5
Sagittal reach distance (cm)	−0.6	−13.8	12.7	3.4	−18.9	25.8	0.0	−10.7	10.7
Frontal reach distance (cm)	−4.5	−16.8	7.7	−6.1	−22.9	10.7	−3.5	−13.4	6.4
Shoulder flexion (deg)	0.1	−22.1	22.2	−3.0	−26.5	20.5	−4.5	−28.6	19.5
Shoulder abduction (deg)	−3.3	−34.0	27.5	−5.0	−32.3	22.4	−0.8	−17.4	15.7
Trunk flexion (deg)	0.3	−2.3	2.8	−0.1	−1.8	1.6	−0.3	−1.9	1.3
Trunk lateral flexion (deg)	−0.7	−3.8	2.3	−0.1	−1.1	1.0	−0.1	−2.4	2.1
Elbow flexion (deg)	−16.0	−89.3	57.3	−18.2	−96.5	60.1	−13.9	−69.3	41.6
Scaption Extend									
Reaching ROM (cm)	8.2	−14.2	30.7	9.0	−22.5	40.5	7.1	−17.1	31.3
Sagittal reach distance (cm)	−0.8	−15.2	13.5	3.0	−20.5	26.5	−1.0	−13.2	11.1
Frontal reach distance (cm)	−4.7	−19.7	10.4	−5.9	−24.9	13.0	−5.1	−20.8	10.6
Shoulder flexion (deg)	−0.1	−22.5	22.2	1.2	−18.8	21.1	−2.1	−24.2	20.0
Shoulder abduction (deg)	−0.9	−22.7	21.0	−0.1	−20.4	20.1	1.5	−16.4	19.4
Trunk flexion (deg)	−0.7	−4.1	2.7	−0.6	−6.2	5.0	−0.1	−8.1	8.0
Trunk lateral flexion (deg)	−0.6	−6.3	5.2	0.0	−7.8	7.9	0.1	−9.9	10.1
Elbow flexion (deg)	−21.6	−93.6	50.3	−22.4	−100.2	55.4	−16.6	−68.8	35.7
Lateral									
Reaching ROM (cm)	2.3	−8.2	12.7	5.1	−7.5	17.7	1.6	−6.4	9.6
Sagittal reach distance (cm)	-	-	-	-	-	-	-	-	-
Frontal reach distance (cm)	−5.2	−17.0	6.7	−6.9	−17.8	4.1	−4.5	−13.1	4.2
Shoulder flexion (deg)	-	-	-	-	-	-	-	-	-
Shoulder abduction (deg)	2.0	−7.4	11.4	3.6	−10.4	17.7	3.4	−4.1	10.9
Trunk flexion (deg)	−0.1	−2.0	1.8	0.1	−1.1	1.2	−0.3	−1.7	1.2
Trunk lateral flexion (deg)	−0.1	−3.6	3.5	−0.1	−1.8	1.5	0.5	−1.6	2.6
Elbow flexion (deg)	−6.3	−30.1	17.5	−11.4	−40.7	17.9	−6.3	−27.0	14.4
Lateral Extend									
Reaching ROM (cm)	−0.7	−15.8	14.4	3.2	−11.1	17.5	0.8	−8.9	10.4
Sagittal reach distance (cm)	-	-	-	-	-	-	-	-	-
Frontal reach distance (cm)	−3.1	−16.6	10.5	−4.7	−17.9	8.5	−3.9	−14.5	6.7
Shoulder flexion (deg)	-	-	-	-	-	-	-	-	-
Shoulder abduction (deg)	5.3	−15.7	26.3	7.6	−11.1	26.2	5.5	−7.5	18.4
Trunk flexion (deg)	0.8	−3.2	4.7	1.4	−3.2	5.9	0.9	−4.7	6.6
Trunk lateral flexion (deg)	0.4	−7.0	7.9	1.1	−6.8	9.0	1.8	−9.8	13.3
Elbow flexion (deg)	−6.2	−41.4	29.0	−6.7	−32.1	18.7	−4.1	−30.5	22.3

Bias: mean bias of D1-D2 for each sensor; LB: lower bound for Bland-Altman limits of agreement; UB: Upper bound for Bland-Altman limits of agreement; K1: KinectV1; K2: KinectV2; VMC: video motion capture.

References

1. Olsen, T.S. Arm and leg paresis as outcome predictors in stroke rehabilitation. *Stroke* **1990**, *21*, 247–251. [CrossRef] [PubMed]
2. Alaverdashvili, M.; Foroud, A.; Lim, D.H.; Whishaw, I.Q. "Learned baduse" limits recovery of skilled reaching for food after forelimb motor cortex stroke in rats: A new analysis of the effect of gestures on success. *Behav. Brain Res.* **2008**, *188*, 281–290. [CrossRef]
3. Levin, M.F.; Kleim, J.A.; Wolf, S.L. What do motor "recovery" and "compensation" mean in patients following stroke? *Neurorehabil. Neural Repair* **2009**, *23*, 313–319. [CrossRef] [PubMed]
4. Roby-Brami, A.; Feydy, A.; Combeaud, M.; Biryukova, E.; Bussel, B.; Levin, M. Motor compensation and recovery for reaching in stroke patients. *Acta Neurol. Scand.* **2003**, *107*, 369–381. [CrossRef]
5. Fugl-Meyer, A.R.; Jääskö, L.; Leyman, I.; Olsson, S.; Steglind, S. The post-stroke hemiplegic patient. 1. a method for evaluation of physical performance. *Scand. J. Rehabil. Med.* **1975**, *7*, 13–31.
6. Lyle, R.C. A performance test for assessment of upper limb function in physical rehabilitation treatment and research. *Int. J. Rehabil. Res.* **1981**, *4*, 483–492. [CrossRef]
7. Wolf, S.L.; Catlin, P.A.; Ellis, M.; Archer, A.L.; Morgan, B.; Piacentino, A. Assessing Wolf motor function test as outcome measure for research in patients after stroke. *Stroke* **2001**, *32*, 1635–1639. [CrossRef] [PubMed]
8. Desrosiers, J.; Noreau, L.; Rochette, A.; Bourbonnais, D.; Bravo, G.; Bourget, A. Predictors of long-term participation after stroke. *Disabil. Rehabil.* **2006**, *28*, 221–230. [CrossRef] [PubMed]
9. Bonnechere, B.; Jansen, B.; Salvia, P.; Bouzahouene, H.; Omelina, L.; Moiseev, F.; Sholukha, V.; Cornelis, J.; Rooze, M.; Jan, S.V.S. Validity and reliability of the Kinect within functional assessment activities: Comparison with standard stereophotogrammetry. *Gait Posture* **2014**, *39*, 593–598. [CrossRef] [PubMed]
10. Clark, R.A.; Pua, Y.-H.; Fortin, K.; Ritchie, C.; Webster, K.E.; Denehy, L.; Bryant, A.L. Validity of the Microsoft Kinect for assessment of postural control. *Gait Posture* **2012**, *36*, 372–377. [CrossRef]
11. Clark, R.A.; Pua, Y.-H.; Oliveira, C.C.; Bower, K.J.; Thilarajah, S.; McGaw, R.; Hasanki, K.; Mentiplay, B.F. Reliability and concurrent validity of the Microsoft Xbox One Kinect for assessment of standing balance and postural control. *Gait Posture* **2015**, *42*, 210–213. [CrossRef]
12. Reither, L.R.; Foreman, M.H.; Migotsky, N.; Haddix, C.; Engsberg, J.R. Upper extremity movement reliability and validity of the Kinect version 2. *Disabil. Rehabil. Assist. Technol.* **2017**, 1–9. [CrossRef]
13. Huber, M.; Seitz, A.; Leeser, M.; Sternad, D. Validity and reliability of Kinect skeleton for measuring shoulder joint angles: A feasibility study. *Physiotherapy* **2015**, *101*, 389–393. [CrossRef] [PubMed]
14. Mentiplay, B.F.; Clark, R.A.; Mullins, A.; Bryant, A.L.; Bartold, S.; Paterson, K. Reliability and validity of the Microsoft Kinect for evaluating static foot posture. *J. Foot Ankle Res.* **2013**, *6*, 14. [CrossRef] [PubMed]
15. Yeung, L.; Cheng, K.C.; Fong, C.; Lee, W.C.; Tong, K.-Y. Evaluation of the Microsoft Kinect as a clinical assessment tool of body sway. *Gait Posture* **2014**, *40*, 532–538. [CrossRef] [PubMed]
16. Dehbandi, B.; Barachant, A.; Smeragliuolo, A.H.; Long, J.D.; Bumanlag, S.J.; He, V.; Lampe, A.; Putrino, D. Using data from the Microsoft Kinect 2 to determine postural stability in healthy subjects: A feasibility trial. *PLoS ONE* **2017**, *12*, e0170890. [CrossRef]
17. Kuster, R.P.; Heinlein, B.; Bauer, C.M.; Graf, E.S. Accuracy of KinectOne to quantify kinematics of the upper body. *Gait Posture* **2016**, *47*, 80–85. [CrossRef]
18. Behar, C.; Lustick, M.; Foreman, M.H.; Webb, J.; Engsberg, J.R. Personalized virtual reality for upper extremity rehabilitation: Moving from the clinic to a home exercise program. *J. Intellect. Disabil. Diagn. Treat.* **2016**, *4*, 160–169.
19. Lauterbach, S.A.; Foreman, M.H.; Engsberg, J.R. Computer games as therapy for persons with stroke. *Games Health Res. Dev. Clin. Appl.* **2013**, *2*, 24–28. [CrossRef]
20. Mraz, K.; Eisenberg, G.; Diener, P.; Amadio, G.; Foreman, M.H.; Engsberg, J.R. The effects of virtual reality on the upper extremity skills of girls with rett syndrome: A single case study. *J. Intell. Disabil. Diagn. Treat.* **2016**, *4*, 152–159. [CrossRef]
21. Sevick, M.; Eklund, E.; Mensch, A.; Foreman, M.; Standeven, J.; Engsberg, J. Using free internet videogames in upper extremity motor training for children with cerebral palsy. *Behav. Sci.* **2016**, *6*, 10. [CrossRef] [PubMed]
22. Cirstea, M.; Levin, M.F. Compensatory strategies for reaching in stroke. *Brain* **2000**, *123*, 940–953. [CrossRef]
23. Levin, M.F.; Michaelsen, S.M.; Cirstea, C.M.; Roby-Brami, A. Use of the trunk for reaching targets placed within and beyond the reach in adult hemiparesis. *Exp. Brain Res.* **2002**, *143*, 171–180. [CrossRef] [PubMed]

24. Valdés, B.A.; Glegg, S.M.; Van der Loos, H.M. Trunk compensation during bimanual reaching at different heights by healthy and hemiparetic adults. *J. Motor Behav.* **2017**, *49*, 580–592. [CrossRef] [PubMed]
25. Pagliari, D.; Pinto, L. Calibration of kinect for xbox one and comparison between the two generations of microsoft sensors. *Sensors* **2015**, *15*, 7569. [CrossRef] [PubMed]
26. Wagner, J.M.; Rhodes, J.A.; Patten, C. Reproducibility and minimal detectable change of three-dimensional kinematic analysis of reaching tasks in people with hemiparesis after stroke. *Physic. Therap.* **2008**, *88*, 652–663. [CrossRef] [PubMed]
27. Microsoft Corporation. The Microsoft Kinect Software Development Kit SDK v2.0, Redmond, WA, USA, 2015. Available online: http://www.microsoft.com/en-us/download/details.aspx?id=44561 (accessed on 1 November 2020).
28. Suma, E.A.; Krum, D.M.; Lange, B.; Koenig, S.; Rizzo, A.; Bolas, M. Adapting user interfaces for gestural interaction with the flexible action and articulated skeleton toolkit. *Comput. Graph.* **2013**, *37*, 193–201. [CrossRef]
29. Wu, G.; Van der Helm, F.C.; Veeger, H.D.; Makhsous, M.; Van Roy, P.; Anglin, C.; Nagels, J.; Karduna, A.R.; McQuade, K.; Wang, X. ISB recommendation on definitions of joint coordinate systems of various joints for the reporting of human joint motion—Part II: Shoulder, elbow, wrist and hand. *J. Biomech.* **2005**, *38*, 981–992. [CrossRef]
30. Berchtold, A. Test–retest: Agreement or reliability? *Method. Innov.* **2016**, *9*. [CrossRef]
31. Bland, J.M.; Altman, D. Statistical methods for assessing agreement between two methods of clinical measurement. *Lancet* **1986**, *327*, 307–310. [CrossRef]
32. McGraw, K.O.; Wong, S.P. Forming inferences about some intraclass correlation coefficients. *Psych. Methods* **1996**, *1*, 30. [CrossRef]
33. Hage, R.; Detrembleur, C.; Dierick, F.; Pitance, L.; Jojczyk, L.; Estievenart, W.; Buisseret, F. DYSKIMOT: An ultra-low-cost inertial sensor to assess head's rotational kinematics in adults during the Didren-Laser Test. *Sensors* **2020**, *20*, 833. [CrossRef] [PubMed]
34. Laver, K.E.; Lange, B.; George, S.; Deutsch, J.E.; Saposnik, G.; Crotty, M. Virtual reality for stroke rehabilitation. *Cochrane Database Syst. Rev.* **2017**, *11*. [CrossRef] [PubMed]

Publisher's Note: MDPI stays neutral with regard to jurisdictional claims in published maps and institutional affiliations.

Review

Upper Limb Physical Rehabilitation Using Serious Videogames and Motion Capture Systems: A Systematic Review

Andrea Catherine Alarcón-Aldana [1,*], **Mauro Callejas-Cuervo** [2] **and Antonio Padilha Lanari Bo** [3]

1   Software Research Group, Universidad Pedagógica y Tecnológica de Colombia, Tunja 150002, Colombia
2   School of Computer Science, Universidad Pedagógica y Tecnológica de Colombia, Tunja 150002, Colombia; mauro.callejas@uptc.edu.co
3   School of Information Technology and Electrical Engineering, University of Queensland, Brisbane 4072, Australia; antonio.plb@uq.edu.au
*   Correspondence: andrea.alarconaldana@uptc.edu.co; Tel.: +57-300-217-6098

Received: 15 September 2020; Accepted: 20 October 2020; Published: 22 October 2020

Abstract: The use of videogames and motion capture systems in rehabilitation contributes to the recovery of the patient. This systematic review aimed to explore the works related to these technologies. The PRISMA method (Preferred Reporting Items for Systematic reviews and Meta-Analyses) was used to search the databases Scopus, PubMed, IEEE Xplore, and Web of Science, taking into consideration four aspects: physical rehabilitation, the use of videogames, motion capture technologies, and upper limb rehabilitation. The literature selection was limited to open access works published between 2015 and 2020, obtaining 19 articles that met the inclusion criteria. The works reported the use of inertial measurement units (37%), a Kinect sensor (48%), and other technologies (15%). It was identified that 26% used commercial products, while 74% were developed independently. Another finding was that 47% of the works focus on post-stroke motor recovery. Finally, diverse studies sought to support physical rehabilitation using motion capture systems incorporating inertial units, which offer precision and accessibility at a low cost. There is a clear need to continue generating proposals that confront the challenges of rehabilitation with technologies which offer precision and healthcare coverage, and which, additionally, integrate elements that foster the patient's motivation and participation.

Keywords: serious videogames; motion capture; upper limbs; physical rehabilitation; telerehabilitation; inertial sensors; inertial measurement unit (IMU); state of the art

---

1. Introduction

One of the sustainable development objectives suggested by the United Nations (UN) is oriented toward the universal and integral coverage of health services, and the reduction of its inequalities, in order for everyone to be in good health [1]. In accordance with the above, it is taken into account that inequalities contribute to millions of people with disabilities facing difficulties in carrying out their basic daily activities. This is more pronounced among people from communities with fewer opportunities and resources, which are generally geographically located in areas that are distant from the services required for rehabilitation processes [2].

Of the different types of disabilities, motor disability is considered to be one of the main limitations to human beings carrying out their basic activities, affecting the quality of life of the individual, as well as that of those around them [3]. In the last few years, telemedicine and telerehabilitation have been strengthened with the implementation of diverse technologies that support rehabilitation processes, oriented toward providing patients with the services required, reducing the number of journeys to main cities, where, in general, specialists, hospitals, clinics, and centers equipped with the technology for the therapies are located. The benefits of telemedicine are more evident in cases associated with traveling and the mobility of the patient, costs, or other factors, for instance, in a situation of isolation or confinement such as that experienced worldwide due to COVID 19, which does not allow people to travel somewhere that is adapted for the necessary therapy session for the patients' recovery [4].

Although in the last few years there have been many technological proposals that support physical rehabilitation, there are still difficulties and gaps in the area which represent an opportunity to contribute to improvements in biomechanical data capture accuracy, the coverage and affordability of health services, and the flexibility and motivation offered to the patients.

With the purpose of identifying the advances and the options available, in order to contribute to the improvement of motor rehabilitation processes, this review includes works published between 2015 and June 2020, oriented toward the support of upper limb physical rehabilitation, which use videogames and a motion capture system. These publications mainly show the use of the Kinect sensor and inertial sensors as motion capture systems. At the same time, it is identified that the works included mainly support motor rehabilitation in people who have suffered a stroke, and another aspect that stands out is the use of commercial systems on the market, which offer different videogames for motor rehabilitation. The objective of this systematic review is to determine the main contributions to this type of rehabilitation in order to identify the opportunities and challenges that should be taken into consideration in future proposals, focused on the improvement in quality of life of people with motor disabilities.

2. Materials and Methods

This section provides a description of the process and criteria taken into account to conduct the article selection included in this documental research, according to aspects of the PRISMA method (Preferred Reporting Items for Systematic reviews and Meta-Analyses) [5]. This allowed the authors to critically identify, select, and evaluate the relevant research, as well as compile and analyze the data from the studies included in the review.

2.1. Eligibility Criteria

The eligibility criteria taken into consideration for inclusion of the studies in this review were (i) that they were published in English, (ii) that they were published within the last 5 years, in the period 2015–June 2020, (iii) that the full text was open access, and (iv) that the type of document was an article, systematic review, state-of-the-art review, or journal.

Concerning the second aspect, the period mentioned was selected, given that as, from 2010, when Kinect was created, and until 2015, its use became popular in different contexts. After 2015, it is noticeable that there was an upsurge of companies and projects using other motion capture systems and integrating serious videogames, in addition to the Kinect sensor, in the field of rehabilitation, which is the main interest of the present study. Another relevant element in this review is that the studies included had therapeutic purposes of rehabilitation or telerehabilitation of the upper limb using videogames and some motion capture system, regardless of the gender and age of the population which participated in the validation of the proposals described.

2.2. Search Strategy

The search of the publications was carried out in four academic databases: Scopus, PubMed, IEEE Xplore, and Web of Science. The following search terms, classified into four groups, were used: (i) medical aspect: rehabilitation, health, "physical therapy", musculoskeletal, telerehabilitation, "tele-rehabilitation", "tele rehabilitation"; (ii) use of videogames: videogames, "video games", video-games, "serious videogames", "serious games", "serious video games", exergames, exergaming, "active videogames"; (iii) motion capture system technology: "inertial sensor", "motion capture", mocap, "motion capture system", wearable; iv) segment or part of the body the rehabilitation is focused on: "upper limb", elbow, shoulder, arm, wrist, humerus. In the search parameters used in the databases (see Table 1), in each group, the operator OR was included between the different terms considered to be synonyms, and, to separate the groups, the operator AND was used, thereby enabling the search to include at least one relevant term from each group in the data consultation.

Table 1. Search parameters in the different databases.

Database	Search Parameters
Scopus	TITLE-ABS-KEY (((rehabilitation OR health OR "physical therapy" OR "musculoskeletal") AND (videogames OR "video games" OR "video-games" OR "serious videogames" OR "serious games" OR "serious video games" OR "exergames" OR "exergaming" OR "active videogames") AND ("upper limb" OR "elbow" OR "shoulder" OR "arm" OR "wrist" OR "humerus") AND ("inertial sensor" OR "motion capture" OR "motion capture system" OR mocap OR wearable))) AND (LIMIT-TO (PUBYEAR, 2020) OR LIMIT-TO (PUBYEAR, 2019) OR LIMIT-TO (PUBYEAR, 2018) OR LIMIT-TO (PUBYEAR, 2017) OR LIMIT-TO (PUBYEAR, 2016) OR LIMIT-TO (PUBYEAR, 2015))
PubMed	((rehabilitation OR health OR "physical therapy" OR "musculoskeletal") AND (videogames OR "video games" OR "video-games" OR "serious videogames" OR "serious games" OR "serious video games" OR "exergames" OR "exergaming" OR "active videogames") AND ("upper limb" OR "elbow" OR "shoulder" OR "arm" OR "wrist" OR "humerus") AND ("inertial sensor" OR "motion capture" OR "motion capture system" OR mocap OR wearable))
IEEE Xplore and Web of Science	((rehabilitation OR health OR "physical AND therapy" OR musculoskeletal) AND (videogames OR "video AND games" OR video-games OR "serious AND videogames" OR "serious AND games" OR "serious AND video AND games" OR exergames OR exergaming OR "active AND videogames") AND ("upper AND limb" OR "elbow" OR "shoulder" OR "arm" OR "wrist" OR "humerus") AND ("inertial AND sensor" OR "motion AND capture" OR "mocap" OR "motion AND capture AND system" OR wearable))

The terminology used to refer to motion capture technology often changes between scientific domains. For instance, in clinical studies, it may be possible that focus was given to the manufacturer name. In other papers, alternative terms may have been used, such as simply "accelerometers" or "motion sensing". We recognize this is a limitation of the methodology adopted in this paper, which may have prevented some papers from being listed in the first stage.

2.3. Description of the Selection Process of the Study

The selection process of the works related to the review topic included four phases: firstly, the identification of the studies, in which all the records that respond to the search parameters in each database were taken into account; secondly, the application of a filter, using the eligibility criteria, in order to select the works related to the purpose of the review, which are available and can be accessed; thirdly, a "screening" phase, which filtered out works, eliminating those that did not adjust to the focus of the investigation and/or those which appeared in multiple databases; finally, an inclusion phase, allowing for the identification of documents to be part of the detailed analysis of the systematic review.

3. Results

This section shows the findings of the selection process of the study, as well as the characteristics of the works included in the analysis and the individual results presented in those publications.

3.1. Selection of the Study

Figure 1 presents the systematic process for the selection of peer-reviewed articles, in which it was identified that a total of 122 documents, published between 2015 and June 2020 and which included the search terms, were found on the databases. In essence, they are studies that focused on the support of upper limb physical rehabilitation, with the use of videogames and motion capture systems. From the total, after applying the eligibility criteria described, 31 works were left; afterward, 11 were eliminated as they were duplicated, and one referred to a book of abstracts from a conference [6]. Thus, the number was reduced to 19 documents, which were directly related to the topic of this review.

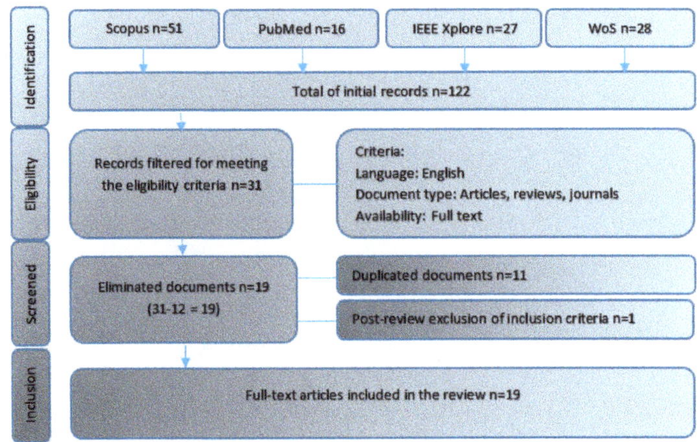

Figure 1. Systematic process used in the selection of articles, based on PRISMA (Preferred Reporting Items for Systematic reviews and Meta-Analyses).

3.2. General Characteristics of the Study

The main characteristics of the 19 works included in the review could be classified into four groups: (i) according to the motion capture system used; (ii) according to the diagnoses or clinical condition the investigation focuses on; (iii) population included in the validation process; (iv) availability (affordability) of the technology used (motion capture system, videogame, technological platform) in the investigation.

3.2.1. Motion Capture Systems Reported in the Studies

Regarding the motion capture systems reported to be used in the 19 studies, nine (48%) used Microsoft Kinect, seven (37%) used inertial measurement units (IMUs), one (5%) used a passive orthosis (which integrates inertial sensors, which would add up to 42% for the use of inertial sensors in these works), one (5%) of the studies used Microsoft HoloLens, and the remaining 5% corresponded to a study which reported a systematic review in a period different from that established and, therefore, it was not considered in the review (see Figure 2).

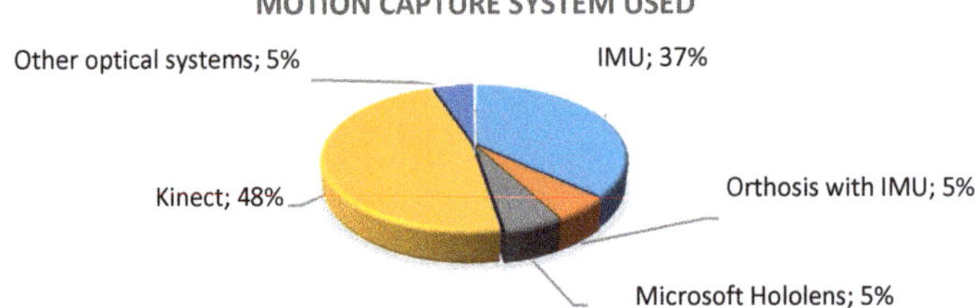

Figure 2. Percentage of the use of motion capture systems.

3.2.2. Diagnosis or Clinical Condition on Which the Technology Described in the Works Was Focused

In the studies analyzed, it was identified that 47% of the investigations focused on the treatment of people who suffered a stroke, 11% addressed situations related to the range of movement (ROM), another 11% contributed to the treatment of any injury in the upper limb, 5% oriented their investigation toward people with Friedreich's ataxia, 5% focused on the treatment of children with cerebral palsy, 5% analyzed energy expenditure in the execution of physical activity, and 16% did not focus on an illness or clinical condition in particular, but on the analysis of technology; thus, they were classified as "not applicable" (N/A), as observed in Figure 3.

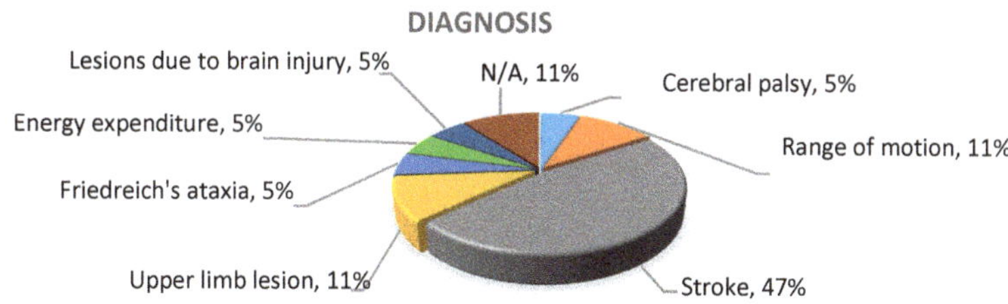

Figure 3. Diagnoses in the works analyzed.

3.2.3. Population Involved in the Validation of the Results

In the validation process (Figure 4), 47% of the studies involved patients (with a 21.66 median and a 19.77 standard deviation), 32% validated their proposal only with healthy participants, 11% made a correlational validation between patients and healthy participants, and the remaining 11% did not validate their proposal with a specific population, given that it had a technical focus (drift correction or systematic review of the literature, mainly).

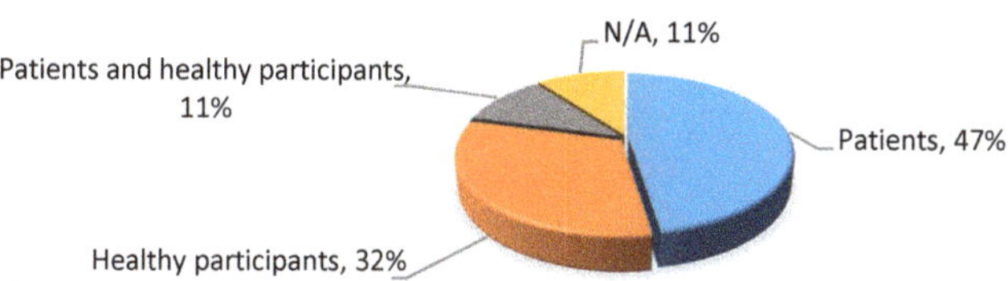

Figure 4. Percentage of the population involved in the study.

3.2.4. Affordability of the Technology Used

With regard to technology availability and affordability, 21% of the works included in this review used commercial products focused on physical rehabilitation. Another 21% proposed systems, referring to the development of technology in academic and/or investigative environments. Most of the studies (58%) were classified as "mixed", given that the technology used involved a combination of commercial products and some personalized development (mainly videogames), as observed in Figure 5.

Figure 5. Distribution of the technology used.

An additional aspect in the global analysis of the literature is that, although, within the search parameters, the upper limbs were included, it was identified that there was diversity regarding the part of the body being focused on in the works, as presented in Figure 6. It can be observed that 47%

referred, in a general way, to the upper limb, 16% referred specifically to the wrist and the hand, 5% analyzed the range of movement of the shoulder joint, 5% included both upper limbs and lower limbs, 11% oriented the treatment toward the upper part of the body, another 11% focused on the movement of the whole human body, and the remaining 11% had a different focus; thus, they did not analyze any part of the human body.

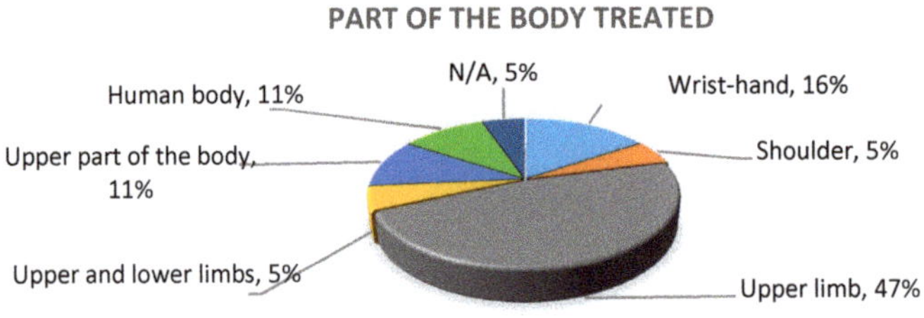

Figure 6. Distribution of the part of the body treated in the study.

3.3. Technologies as Support in the Physical Rehabilitation of the Upper Limb

Table 2 presents the main characteristics of each of the 19 studies analyzed and identifies how they supported physical rehabilitation using videogames and motion capture systems.

Table 2. Search parameters in the different databases. IMU, inertial measurement unit; MS, Microsoft; ROM, range of motion; N/A, not applicable.

No.	Mocap System	Clinical Condition	Population (Sample) *	Technology Used **	Part of the Body Rehabilitated	Reference
1	IMU	Cerebral palsy	19 P	Mixed: Myo bracelet, adapted commercial videogame (Dashy Square and personalized software development)	Hand and wrist	[7]
2	MS HoloLens	ROM	25 H	Mixed: MS HoloLens and developed videogame	Shoulder	[8]
3	IMU	Stroke	8 H	Proposed system: an environment of games and software for the therapist	Upper and lower limbs	[9]
4	MS Kinect	Upper limb lesions	10 P	Mixed: MS Kinect V2, videogame development, and web application	Arm	[10]
5	IMU	N/A	11 H	Proposed system	Arm	[11]
6	IMU	N/A	N/A	Commercial: ArmeoSenso	N/A	[12]
7	IMU	Upper limb lesions	10 P	Mixed: Myo bracelet and a developed videogame	Arm	[13]
8	MS Kinect	Stroke	30 H	Commercial: MS Kinect V2 and Mystic Isle (videogame integrated to Kinect)	Upper part of the human body	[14]
9	MS Kinect	Stroke	11 P	Mixed: MS Kinect and a developed videogame	Arm	[15]
10	MS Kinect	Stroke	24 P	Mixed: MS Kinect and Recovery Rapids ™ (personalized videogame)	Arm	[16]
11	MS Kinect	ROM	10 H	Mixed: MS Kinect and development of a personalized system	Arm	[17]
12	MS Kinect	Friedreich's ataxia	27 P, 43 H	Mixed: MS Kinect and development of a videogame.	Arm	[18]
13	IMU	Stroke	29 P	Commercial: Bimeo	Arm	[19]
14	IMU	Stroke	11 P	Commercial: ArmeoSenso.	Arm	[20]
15	MS Kinect	Stroke	74 P	Commercial: JRS Wave	Human body	[21]
16	MS Kinect	Stroke	18 P, 12 H	Proposed system	Upper part of the human body	[22]
17	MS Kinect	Energy expenditure	19 H	Mixed: MS Kinect and development of a system	Human body	[23]
18	Other optical systems	Lesions due to brain injury	N/A	Mixed	Hand	[24]
19	Orthosis with IMU	Stroke	7 P	Proposed system	Wrist and hand	[25]

* Population: P = patients; H = healthy participants. ** Technology used: commercial and/or developed.

The terms rehabilitation and habilitation, according to the World Health Organization [26], are two processes which "enable persons with disabilities to attain and maintain their maximum independence, full physical, mental, social, and vocational ability, and full inclusion and participation in all aspects of life". Rehabilitation is defined as the group of methods geared toward the recuperation of an activity or function lost or diminished by a trauma or illness, and it covers a wide variety of activities, including medical care rehabilitation, physiotherapy, psychotherapy, language therapy, occupational therapy, and support services. In this sense, physical rehabilitation is oriented to the recovery of the patient's motor function by the physical medicine and rehabilitation team.

For the autonomous development of different basic activities, the movement of various parts of the body is required, especially the upper limbs, which allow the realization of diverse complex manual activities [27]. In this sense, in the literature, diverse proposals were found oriented toward processes of upper limb physical rehabilitation, denoting marked trends concerning the use of motion capture systems and videogames.

3.3.1. Use of Motion Capture Systems in Upper Limb Physical Rehabilitation

Motion capture (MOCAP) is the process of acquiring motion by combining software and hardware [28] and is understood as a technique for recording motion and its corresponding transformation into a digital model. It is commonly used in areas such as entertainment, robotics, medicine, and physical rehabilitation, among others [29,30]. Specifically in the field of physical rehabilitation, it is used to identify the effectiveness of appropriate therapy plans [31,32], which when integrated with information and communication technologies in this field, provides therapeutic assistance to patients under the modality of telemedicine. Biomechanical motion capture systems can mainly be optical and non-optical, as shown in Figure 7.

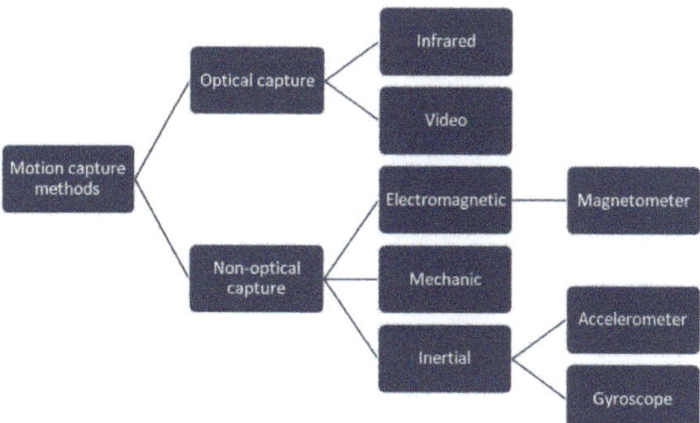

Figure 7. Main motion capture system methods [33].

Optical Systems Used

Optical systems that use infrared light require the location of markers at specific points on the individual's body. Then, using a configuration with multiple cameras, properly placed around the capture space, the position of the reflective markers is recorded [34].

The measurement of human movement with optoelectronic systems offers precision due to the position of the retroreflective markers, and that depends, to a great extent, on the optical characteristics of the camera system and the algorithms implemented in the monitoring software [35]. Microsoft Kinect is an example of an optical system for motion capture without markers. This system can detect 25 joints of the human body of six people at the same time and provides precise information on depth data or corresponding original red/green/blue (RGB) data [36].

In this review, most of the works described how they involved a Kinect sensor as an optical system for motion capture in the development of the research. Some used the sensor, and, in addition, they proposed new products to support rehabilitation. For example, [10] evaluated the usability and performance of the KineActiv platform developed in Unity Engine and incorporating Microsoft Kinect V2. Its purpose was to encourage patients to do the rehabilitation exercises prescribed by the specialist, who could control the patient's performance and correct errors in their execution along the way. In addition, this work included a web platform allowing the physiotherapist to monitor the results of the session, control the patient's health, and adjust the rehabilitation routines. At the same time, [17] proposed a system denominated GoNet V2, which was associated with the Microsoft Kinect V2 game controller. It was aimed toward physical and rehabilitation specialists, and, through the recording, storage, and management of information, it supported the treatment and evaluation of the range of movement of the joint. In [18], the Kinect sensor and a game developed in a previous project (ICT4Rehab) were used in order to corroborate whether serious videogames could be used as an evaluation tool for the functioning of the upper limbs in the treatment of motor deterioration in patients with Friedreich's Ataxia, even with a patient sitting in a wheelchair.

Furthermore, [14] determined the spatial precision and the validity of the measurement of Microsoft Kinect V2, using the videogame Mystic Isle, developed as a rehabilitation game. In this case, they compared the results of the sensor with a motion capture system using standard markers, Vicon, which is another optical motion capture system incorporating markers, which uses infrared cameras to track the three-dimensional location of the reflective markers placed on the body. This work presented satisfactory results in the improvement of the motor function and the performance of daily activities in people with a chronic cerebrovascular accident. Regarding the results of the visual comparative analysis with Vicon, for the case of the hand and the elbow, Kinect V1 showed good precision in the calculation of the movement trajectory, but its validity was limited in terms of the movement of the shoulder. For its part, [15] presented five experiments, three of which were application cases, using devices part of the research project called REHABITATION. In one of these cases, a videogame was proposed in addition to the use of Kinect, which fostered the rehabilitation of the upper limbs in stroke patients. In this case, the purpose was the evaluation of the usability perceived. In this aspect, it was ranked as "excellent" on the scale of usability system (SUS) and as "good" on the modified scale of usability system (mSUS).

Other investigations did not focus on the development of new products, but rather on the validation of different attributes in the use of technologies in motor rehabilitation. In [22], the authors evaluated Kinect's capacity to find movement performance indices through a reliability analysis between sessions and tests. Specifically, reliability was analyzed using eight performance indices: medium velocity, normalized medium velocity, peaks of normalized velocity, logarithm of dimensionless jerk, curvature, spectral arc length, shoulder angle, and elbow angle. In the results of the study, acceptable reliability and sensitivity were mentioned in all the sessions for medium velocity, logarithm of dimensionless jerk, and curvature measured by Kinect for healthy individuals and stroke patients.

In the same way, in [21], the feasibility, efficiency, and safety of the JRS Wave commercial system were evaluated. This software is part of the rehabilitation system called Jintronix (JRS) which was launched by the company Jintronix [37] and uses Microsoft Kinect as its motion capture system. JRS Wave has tasks already set up regarding the upper limb and balance, standing, and walking, and it was used in the rehabilitation of patients hospitalized due to stroke. At the same time, it has a telemedicine system allowing doctors to manage the information of the patients and monitor the physical rehabilitation tasks. The main result referred to the efficiency in the differences of activity levels of the use of rehabilitation technology in comparison to regular rehabilitation. At the same time, in [15], five experiments were described, three of which were cases of application, using devices proposed in the framework of the investigation project developed by the authors. One of these cases, in particular, was related to the aspects addressed in this document, in which the authors proposed a videogame and, along with Microsoft Kinect, fomented the rehabilitation of upper limbs in post-stroke patients.

Another approach identified in the works was that of proposals to optimize the data capture by the Kinect sensor. For example, in [16], a methodology was proposed to extract and evaluate the therapeutic movements of the game-based rehabilitation, executed in environments which were not controlled or supervised. This methodology was oriented toward isolating the relevant movements and eliminating strange movements from the data captured by Kinect, involving the development of computer models that can efficiently process large volumes of data for their later kinematic analysis. Using the Kinect sensor and Microsoft SDK, in [23], three predictive algorithmic models were applied: a Gaussian process regression (GPR), a locally weighted k-nearest-neighbor regression, and linear regression (LR), in order to calculate the mechanical work carried out by the human body and subsequent metabolic energy. The determination of the body segment properties, such as segment mass, length, center-of-mass position, and radius of gyration, were calculated from the Zatsiorsky–Seluyanov's equations of de Leva, with adjustments made for posture cost. The results showed that the Gaussian process regression slightly outperformed the other two techniques and that it was possible to determine the physical activity energy expenditure during exercise, using the Kinect sensor. Therefore, the estimates for high-energy activities, such as jumps, could be made with accuracy, but not for activities which require low energy such as squats and other activities with stationary positions.

With regard to the use of optical systems, in addition to Kinect, the use of a glove called the 5DT Data Glove Ultra from the company 5DT [38] was presented in the research. This glove was initially designed for computer animation, but it has since been used in other fields. It is fabricated with an elastic material and uses fiber-optic sensors in each of the five fingers to detect changes in the global position of the finger [39]. In the review presented in [24], two documents were included that used this glove as a motion capture system. The first presented the development of a videogame platform with virtual reality that integrated a 5DT Data Glove Ultra and a PlayStation 3 videogame console, for the rehabilitation of adolescents affected by cerebral palsy. This had the purpose of contributing to improving hand movement and the consistency of the bones in the forearm. The other document presented a rehabilitation plan involving videogames, using a PlayStation 3 console and the 5DT Data Glove Ultra for the rehabilitation of the hand of pediatric patients with hemiplegia. In this review, a third document was presented that used an infrared transmitter fixed with a Velcro strap to the hand of the patient and an infrared camera (Nintendo Wiimote) as a motion capture system, which captured the infrared transmission in order to generate an image of the patient in the virtual environment.

On the other hand, in [8], the potential offered by Microsoft HoloLens was explored, i.e., an optical device placed on the head which does not require markers or sensors for the following of the arm or the hand. An application was developed with augmented reality, using the engine from the game Unity and the Microsoft HoloToolkit, for the improvement of the range of movement of the shoulder, allowing a perfect remote interaction with the personal doctor. The work, using the Likert questionnaire, identified good levels of motivation and ergonomics in the proposed technology, from the perspective of a group of patients, as well as that of rehabilitation specialists.

Non-Optical Systems Used

Non-optical systems are based on small inertial sensors with built-in accelerometers, gyroscopes, and magnetometers, which allow the recording of data associated with movement in an integrated storage device; these systems are characterized by their low cost, accuracy, and ease of use in ambulatory environments [33]. Portable systems with IMUs are ergonomic, portable, and sensitive, and they can obtain relevant data quickly and accurately in order to make correct decisions related to the intervention of the patient [40].

Different works implemented IMUs due to their potential, such as the case presented in [9], in which a rehabilitation system that integrates videogames and portable technology was proposed, allowing exercises to be realized at home, in order to help the remote recuperation of stroke patients presenting a disability in the upper limbs. The system developed had two principal components: the game engine environment and the software of the therapist to remotely track and follow the progress and achievements of the patients. With regard to the hardware proposed for motion capture, a server in a Raspberry Pi connected wirelessly to a development platform and an MPU6050 sensor was implemented, with a flexible sensor for the detection of flexion and resistance of the fingers and a pulse sensor in order to control the cardiac frequency. Through a survey, the authors identified a great potential for the developed system to facilitate the rehabilitation process of patients from the comfort of their homes and under the remote supervision of the therapist.

Understanding the advantages of the use of inertial sensors, some works focused on an improvement in their efficiency and, taking into account that one of the limitations that IMUs have shown is the problem of drift, in [12], a drift correction method was proposed on the basis of a rest pose magnetometer (RPMC), for the measurement of combined inertia and the following of the arm in real time with a magnetometer. This method corrected drift while the user was relaxing, involving a precalibrated direction of the magnetic field. The commercial system ArmeoSenso was used and a videogame was developed to validate a method following arm movement, resulting in precise monitoring, low latency, and good rhythm, including in environments with proximity to ferromagnetic materials, such as in the home. In the same way, another work optimizing the data generated by IMUs was presented in [11], whose authors began from the premise that classifying a large number of arm movements with IMU-based systems is a difficult task. Therefore, they built a single wrist-mounted device with an inertial sensor and a temperature sensor, to explore the possibility of increasing the classification accuracy of IMU-based systems. The data obtained were pre-processed, and the secondary characteristics were calculated using principal component analysis (PCA) for dimensionality reduction; then, several automated learning models were applied to select the optimal model for speed and accuracy. The results showed that adding a thermal sensor to the IMU-based system significantly increased the classification accuracy in 24 arm movements in healthy participants from 75% to 93.55%.

Another aspect of the research was the accuracy offered by IMUs, which is why they were compared with systems recognized as gold standards in motion analysis [41–45], obtaining acceptable and trustworthy results in different fields, including that of medicine. Moreover, one of the five experiments described in [14] approached the comparison of IMUs to the Vicon Motion Analysis system configured with seven Vicon Bonita infrared cameras. The experiment was conducted on the measurement of the range of movement of the shoulder joint, specifically with three movements: shoulder abduction, external rotation of the shoulder joint, and horizontal adduction. The papers showed the high utility of IMUs in simple monitoring activities, thanks to their ease of connection and handling. Furthermore, the type of system used and the expected period of use influenced motion detection and its characteristics.

Furthermore, [25] took advantage of IMUs in the implementation of a passive orthosis in order to detect movements of the elbow and hand through a classification mechanism, in order to evaluate the progress, or its opposite, in the motor recovery of post-stroke patients, implementing a system that can be used at the patient's home, demonstrating that the mechanism of adaptation was effective in 78.6% of the sessions, making it appropriate as a self-adjusting tool for machine-based exercise.

Other works used commercial systems involving IMUs, which presented stability in their operation and offerred reliability in the data they provide, allowing the evaluation of their contribution to rehabilitation processes. On the one hand, Bimeo is a sensor-based rehabilitation device aimed at stroke patients and other neurological patients. This device offers a motivating virtual reality environment, which aims to make the therapy effective and motivating for patients and also offers therapists a support tool to monitor and control the Bimeo process [46]. This system was used in [19] to evaluate the short-term effects of competitive and collaborative games in arm rehabilitation. The participants' subjective experience was quantified using the "intrinsic motivation inventory" questionnaire after each game, and they also used a final questionnaire on game preferences. Exercise intensity was quantified using the Bimeo system, according to wearable inertial sensors that measured hand speed in each game. The results of the work indicated that both competition and cooperation could increase patient motivation to play and that exercise intensity increased when the play partner was a family member or friend.

The ArmeoSenso system [47] involves virtual reality and is based on IMUs for the training of the function of the upper limbs. It also includes therapy software with videogames and automatically evaluates the arm movement [48]. In [20], a feasibility study was carried out on the development of unsupervised arm therapies in self-directed rehabilitation processes carried out in patients' homes. In this study, after the training given by the specialist, patients with arm hemiparesis used the system in their homes for six weeks with an average duration of 137 min per week, identifying that home therapy is safe and contributes to guiding the rehabilitation process.

The Myo bracelet, developed by Thalmic laboratories, is a portable movement and gesture control device. The newest version of the system, which consists of eight EMG sensors and IMUs, allows the user to control events from a computer (or other device) via a Bluetooth connection, which has been widely used in research environments because of its accessibility [49]. The Myo bracelet is an electromyographic detection device, i.e., the sensors can detect biometric changes in the user's arm muscles as they move, determining the user's intentions [50], offering high precision, depending on the location and orientation in which it is used [51]. The bracelet has been used in different contexts, and one of the concerns regarding home interventions is low adherence; thus, [7] evaluated the feasibility of a new intervention that combines a gaming technology integrating evidence-based biofeedback and training strategies. In this case, the purpose was to use the bracelet and videogames in the experiment to identify the recruitment rate of 8–18 year old patients with cerebral palsy and their continuity in home therapy for one month. The Myo bracelet was also used by children with upper limb disabilities in [13] to evaluate a game developed and adapted to be controlled with the bracelet. According to the results, they identified that the participants felt comfortable and were able to interact with the game and, therefore, there was high acceptance due to the fun that was experienced. In this way, the authors reported that the Myo bracelet made it possible to improve accessibility to videogames and improve the exercise of the upper limbs.

3.3.2. Use of Videogames in Upper Limb Physical Rehabilitation

Another aspect of interest in this review was to identify how the use of videogames is being addressed in upper limb physical rehabilitation. It was found that, although some studies involved the use of commercial products, most of them developed new videogames adapted to the needs of the target population of the investigation.

In [7], the commercial videogame "Dashy Square" was used [52], which was launched in 2016 by KasSanity and was adapted so that the participants in the investigation executed therapeutic gestures with their hands to control the actions of the game on the screen; this was used as a motivational environment involving goals to tackle muscle weakness and selective motor control. It was determined that the training focusing on the solution, proposed in this work, in combination with videogames that provide biofeedback, had a positive influence on the activities that require enrolment of participants and practice at home, and that there was more retention of patients during a monthly intervention, which were the parameters defined. Another commercial product used was the videogame therapy software included in the ArmeoSenso system, which has been previously mentioned, which is oriented toward the recovery of the function of the arm. In [12], a therapy game called "Meteors" was implemented in ArmeoSenso. This game involves a virtual robotic arm that coincides with the movement of the arm of the player and is used to catch meteors which fall on a planet. At the same time, in [20], in addition to the videogame "Meteors", the game "Slingshot" was used, with the purpose of training arm coordination and improving precision in the movements for aiming and extending the arm. In this game, the patient exercises the flexion/extension of the elbow. To this end, the patient holds a virtual slingshot with which they have to shoot stones to set targets which may be stationary or in motion, while the size and velocity may vary. In these games, the score is calculated according to performance. The level of difficulty can be dynamically adjusted in order to maintain motivation and commitment during the recovery of the patient.

In [21], the JRS Wave software was used, which was designed in collaboration with occupational and physical therapists, using criteria of motor relearning. In this system, amusing and attractive videogames were programmed in order to exercise the upper limbs, practice balance, and walking. In this work, although the authors did not provide details about the videogames used, they claimed that they could be adjusted to different levels of complexity and speed, and they determined that these tools increase adherence and joy for exercising, thereby increasing the amount of repetitive exercise carried out by people with limited mobility.

Furthermore, works were identified in which, regardless of whether the motion capture system was commercial or an independent proposal, specific videogames for the development of the investigation were presented. In this group of works, the development of videogames that used the platform Unity was noticeable [53]. For example, in [8], with the purpose of treating any deficit of the upper limb which deteriorates the range of motion, a videogame was designed and developed from traditional rehabilitation exercises with the Rolyan range-of-motion shoulder arc. The game presents a curved tube, with mobile colored rings around it. They have to be moved from one side of the tube to the other, achieving a complete range of motion of the upper limb. This improves motor planning abilities and visual monitoring. The videogame can also be used with the HoloLens glasses using augmented reality. In this case, the user, with the movement of the hand, controls a virtual cursor throughout a predefined virtual trajectory. Furthermore, in [10] using Unity, active videogames were created to be executed with Kinect in the platform KineActiv, through which patients interact with a gamified user interface that implements a personalized game environment for each type of exercise.

Using Unity 3D and the C# programming language, [13] developed and evaluated a videogame involving a jigsaw puzzle with three levels of difficulty, adapted to be used with the Myo bracelet. In the game, the gestures perceived by electromyography such as double touch, shake the hand to the right and the left, close fist, and separate the fingers, were the commands to interact with the videogame and put together the jigsaw puzzle, contributing to motor recovery, as well as to cognitive aspects of the patient. The evaluation tool was based on an evaluation questionnaire of educational games and showed that the videogame was stimulating and attractive, and that it fulfilled the expectations of the patients (5–15 years old children with disabilities). It was also identified that the genres preferred by the children were those of adventure, reasoning, and creativity. In the same way, [14] developed "Mystic Isle" with Unity 3D, a multiplane, full-body rehabilitation videogame which uses Kinect V2 as an input device. This game, depending on the therapeutic treatment, can be used in either a sitting or a standing position, and different movements can be followed: gross motor movements (steps, jumps, squats) or fine motor movements (shake a hand, turn the palm face up, open and close the hand). With this system, the player is tracked in a three-dimensional space and, afterward, the data are registered in real time by the associated software, showing good results related to motor function and the execution of basic daily activities in chronic post-stroke patients.

In one of the cases in [15], a rehabilitation videogame was presented which implements C# and UnityScript. This game is oriented toward the physical recovery of upper limbs in neurological patients, through the interaction with two scenarios related to their daily lives. The first scenario presents a bookshelf, and the player has to avoid the books in it falling. The second scenario simulates a kitchen where the player, in a given time, has to pick up an object requested in a written text. Both games are controlled by the movement of the hands, which is detected by Kinect V2. For that reason, the first time that the game was used, a calibration was carried out in order to guarantee that the patient could reach all corners of the screen with their virtual hands.

The use of Kinect and its utilities was also noticeable in [16], through "Recovery Rapids", a personalized videogame to be used with Kinect, in which data were captured to evaluate the relevant movements of the continual therapeutic game. In this work, a methodology was presented to isolate the relevant movements and eliminate strange movements from the data obtained through Kinect during the therapeutic game, incorporating the implementation of computational models to efficiently process great volumes of motion capture data compiled in noncontrolled environments.

In [18], a serious videogame called "WipeOut" was used (developed by the authors for a previous project), and, in conjunction with the use of the Kinect sensor, a functional evaluation of the upper limbs was carried out. With the movement of the arm and the position of the hand, the player must wipe the screen to discover an image. The evaluation was carried out contrasting the performance (time and precision) in the execution of the activity in two groups: one of patients with Friedreich's ataxia and another of healthy individuals.

Using C# and Microsoft XNA game studio, in [22], a game was designed to be used with Kinect, the purpose of which was to be able to monitor the movement of the hands of the participants. For this, the player had to move their hands to intercept and catch several colored balls which went in the direction of the person, according to the guide given by the system. The program registers the positions of the joints of the upper part of the body (hand, wrist, elbow, shoulder, center of the shoulder, position of the head, and waist) to be able to carry out the respective analysis. The results of the study indicated acceptable reliability in the session and between sessions when the Kinect sensor and the proposed game were utilized, which as comparable to the data calculated with robotic systems or clinical evaluation scales.

Another example of videogame development for physical rehabilitation was included in the SCRIPT system [25], designed for wrist and hand rehabilitation, through the execution of daily exercises mediated by three interactive videogame options. In one of these games, the patient must open and close their hand, with this movement controlling a seashell that opens and closes to catch fish. They also used the videogame "Crocco", in which the subject must move a crocodile on the screen. This game is available in four variants. In the simplest one, the player flexes and extends the wrist to avoid obstacles; in the second case, lateral arm movements are added to move the crocodile laterally on the screen; the third variant of the game requires the grip movement to simulate the crocodile eating fruit, and the fourth variant includes all the previous movements. The third game included in the system described is called "Labyrinth", in which the patient moves the cursor through a maze, in three variations of the game. In the simplest one, the cursor can be moved up and down with anteroposterior hand movements, as well as to the left and right, encouraging the flexion–extension of the wrist. In the second variation, prone supination of the hand is used to open and close the doors in the labyrinth. The third variation includes a gripping gesture to take a key before opening doors. In this research, although the participants did not practice as much as initially advised, such that great variability in the duration of the sessions was obtained, relative ease in the movement of the wrist compared to the movement of the hand was identified, as well as that the proposed system can be used as an adaptive regulator of the difficulty of the exercise, depending on the performance of each subject. Likewise, in [10], a rehabilitation environment was presented, in which one of its modules corresponded to a videogame engine, which included three games designed specifically for rehabilitation purposes located in the patient's home. However, in this case, the system allowed the recovery of both the upper and the lower limb, through the use of designed games.

Another approach to the use of developed videogames, specifically for motor recovery, was the comparison of competitive, cooperative, or individual videogames. In this regard, [19] designed four videogames for arm rehabilitation. One of those videogames was competitive, in which the patient plays against another person (a friend, relative, or therapist). There were also two cooperative games, in which the patient and another player play together against the computer; lastly, there was an individual game in which the patient plays alone against the computer. It was identified that competitive games contributed, to a greater extent, to the functional recovery and improvement in the quality of life of the patients in comparison to conventional rehabilitation exercises.

3.3.3. Diagnosis and Treatments Supported by Technology

The publications analyzed in this review were oriented toward the support of physical rehabilitation using technology. Most of these works fostered the recovery of patients who suffered from a particular clinical condition. Below, there is a description of the contribution made by each of them.

Technological Support in Post-Stroke Motor Recovery

A cerebrovascular accident, also known as a stroke, is an acute event, caused primarily by a blockage (accumulation of fatty deposits on the inner walls of the blood vessels), which prevents blood from flowing to the brain. They can also be caused by bleeding from a blood vessel in the brain or by blood clots [54]. One of the main effects of a stroke, in patients and their families, is the limitation in carrying out basic daily activities; thus, one of the main purposes of rehabilitation therapies is to improve the movements of the arm and promote the recovery of lost function through rehabilitation therapy [55].

In the review, it was identified that most of the works were geared toward the support of motor rehabilitation treatments in post-stroke patients, as is the case of [9,14,16,19–22,25], and the five cases presented in [15]. These works described different experiments that involved videogames and motion capture systems in the motor recovery of neurological patients to contribute to the improvement in their quality of life and facilitate the work of the medical staff involved.

Technological Support in the Recovery from Other Diagnoses

Another diagnosis mentioned was cerebral palsy, which is the most frequent cause of motor disability in children and the third cause of disorders in neurological development. It is, in essence, a group of nonprogressive disorders which occur during the development of the brain, in the fetal phase or the first years of life. These disorders affect mobility and postural development; therefore, they also make carrying out different activities more difficult [56]. In [7], the feasibility of using technology involving games was identified, integrating biofeedback from the evidence and training strategies, focusing on the solution proposed, in order to support the execution of the therapy efficiently in the home of young people with cerebral palsy. In [24], a systematic review was presented which analyzed in depth three works associated with lesions due to brain injury: cerebral palsy, stroke, and children with hemiparesis; concern with developing works contributing to the motor recovery of neurological patients was noted.

Additionally, in [18], the contribution of serious games was validated, which were developed for rehabilitation, and it was determined how they can be used as an evaluation tool for the function of the upper limb in patients with advanced Friedreich's ataxia. This is a hereditary disease of the central nervous system and the peripheral nervous system which causes gait ataxia, dysmetria, dysarthria, dysphagia, severe proprioceptive and superficial sensory loss, weakness, limb atrophy, and loss of muscle tone or spasticity or a combination of both, among other complications related with the senses [57].

Another aspect of interest related to human health is that of energy expenditure in the execution of physical activity, given that energy expenditure and the metabolism substrate are important elements when considering physical activity, and, from their characteristics, it is possible to establish treatments to improve a person's quality of life [58]. Hence, in [23], through the use of Kinect and Microsoft SDK, an estimation was made of the mechanical work carried out by the body and, thus, it was possible to calculate the metabolic energy using predictive algorithmic models.

Concerning the evaluation and analysis of the range of movement, and the lesions in the upper limbs derived from different clinical conditions, in [7,9,12,16], and in one of the experiments presented in [15], it was identified that the contribution of these technologies to a patient's motor recovery is positive, as it helps to overcome the limitation of traditional rehabilitation methods.

In conclusion, the findings of the analysis of the articles which met the inclusion criteria of the review can be classified as is shown in the Venn diagram in Figure 8.

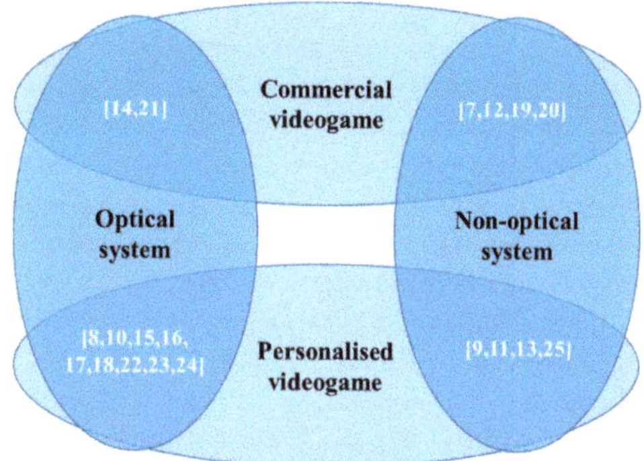

Figure 8. Paper classification according to the technologies used.

It is necessary to mention that, while optical motion capture systems (11 papers) presented problems with sensor occlusion, in the systems using IMUs or non-optical systems (eight papers), the investigations were left open in order for future research to obtain better precision and to correct the drift generated by the magnetometer. In addition, for the years 2019 and 2020, the literature consulted registered an increase in the use of non-optical motion caption systems (five papers), as opposed to optical systems (two papers).

Furthermore, regarding the use of videogames in physical rehabilitation, there was a clear trend toward the development of personalized videogames (13 papers) and, on fewer occasions, commercial videogames were used (six papers).

Taking into account that the objective of the investigation included the integration of motion capture systems and videogames in upper limb physical rehabilitation, it is appropriate to mention that future investigations could focus on the development of technological tools involving IMUs and the independent development of videogames for the support of said processes.

4. Discussion

According to the importance of physical and functional rehabilitation in the quality of life of patients and the people around them, in this review, the technological contributions developed in the past few years in this field were identified, mainly regarding the inclusion of videogames and motion capture systems as support in the motor recovery of the upper limb. In the literature, a wide use of Kinect was identified as the motion capture system, although there were some limits regarding the movements carried out in the depth and occlusal planes of the limbs, i.e., the visual interruption between the camera and some of the body segments, as well as the capture of data in some specific positions (for example, sitting). Furthermore, aspects related to precision were considered in [59–61], with greater emphasis when it comes to physical rehabilitation, where precision can be a determining factor in the process. Even so, this sensor was used as a complement in the motor recovery therapies or in works focused on the validation of different attributes such as the usability of the technologies proposed or the verification of motion evaluation methods [10,14,16–18,21–23]. Among commercial products, not only Microsoft Kinect was used; the use of Nintendo Wii with its Balance Board and the Myo bracelet was reported, allowing validations in the medical field thanks to the fact that they have a lesser cost in comparison with clinical systems, such as Vicon, OptiTrack, and Qualisys, among others.

In this sense, comparisons were made of different motion capture systems with respect to Vicon, OptiTrack, or Qualisys considered to be the gold standard, against which those systems using inertial measurement units have shown a comparable performance [41–45,62] denoting the reliability, accessibility, accuracy, and portability offered by IMUs. In this way, inertial sensors become a good option to be used in the medical field to support motor and functional recovery processes, which require precise measurements with an accessible cost in order to be mass-produced.

Currently, novel motion capture technology involving video alone is available. Using tools from machine learning, researchers have demonstrated that tracking joints of multiple human figures may be achieved [63]. The potential of this approach is enormous, since it would enable implementing games for rehabilitation using hardware available in most dwellings. Nevertheless, for real-time operation, these methods still require powerful graphics hardware, which limits their availability at the moment.

On the other hand, the use of serious videogames has increased due to the lack of motivation of patients when they are in the process of motor recovery. In the face of this, individual, cooperative, and competitive video games have been used. Commercial video games were used in [7,12,19–21], which, despite encouraging the execution of physical activity and supporting the player's motivation, were not adapted to the particular characteristics of physical rehabilitation. For this reason, most of the studies proposed active video games specifically for rehabilitation, to increase motivation and adherence to therapies [8,11,13,15–18,22,23,25], in some cases associated with a configuration module allowing the health professional to adjust the characteristics of the game according to the diagnosis and progress of the patient in treatment [9,10,14,24]. In the particular case of commercial rehabilitation products, such as ArmeoSenso, Bimeo, or JRS Wave, they respond adequately to such requirements in the area of physical rehabilitation, although the additional costs involved must be taken into account.

When referring to the use of commercial products, i.e., videogames and motion capture systems, it should be noted that they are an important contribution to the field of rehabilitation. However, they are not certified as medical products [15] and, therefore, to include them in a clinical routine, it is recommended that a thorough preliminary study be carried out or, if possible, a design and development procedure, guided by health professionals, to obtain products that respond to the specific needs of the rehabilitation process. Among the particular characteristics of a videogame for rehabilitation, it is worth mentioning that it should have simple visual backgrounds, clinical diagrams in accordance with the patient's situation, and configurability in terms of range of movement, speed, and recovery time, among other aspects of the process [18].

Although this review included works that used videogames and motion capture systems in physical rehabilitation, not all the works analyzed integrate these components into a single product or system, i.e., the information generated by these technologies was disconnected, making complete and timely analysis difficult in motor recovery therapy.

One of the fundamental aspects in order to achieve the objectives of a physical rehabilitation process is that it is adequately monitored and controlled, and that it is adjustable in a timely manner regardless of whether the patient and the health professional are in the same geographical location or not. For this reason, an optimal system to support physical rehabilitation should integrate various functionalities and technologies, including an accurate and portable motion capture system, as well as a customized active video game module to encourage patient motivation and guide them properly in the execution of therapy. It is also important that the system has a management and monitoring module of the rehabilitation plan assigned to each patient in real time, making it possible to manage the electronic medical record of rehabilitation processes.

In this sense, out of the works included in this review only five presented home rehabilitation systems that allow the therapist to remotely adjust and monitor the configuration of the game according to the patient's rehabilitation objectives, incorporating the recording of information in an associated computer system. Out of these, in [8], IMUs are used, in [9,13,20] the Kinect sensor was used, and two of the three works analyzed in [23] used the 5DT Data Glove Ultra and the Nintendo Wiimote. Moreover, in four of these five works, videogames developed specifically for rehabilitation were proposed. In this sense, it was identified that this type of system offers a significant contribution to the processes of motor recovery and that it is important that the information gained from the therapies carried out by a patient in a location is convenient for them and registered correctly, such that the process is evaluated in a timely and reliable manner. Thus, telerehabilitation involving a system with these components offers proper support for the management of the process, benefitting patients, their caregivers, and the medical team involved.

5. Conclusions

The ability to carry out basic daily activities autonomously is an aspect related to an individual's quality of life. People can lose their mobility and their capacity to execute daily activities for different reasons, as in the case of neurologic diseases or other clinical conditions. In order to recover functionality, physical rehabilitation systems are implemented that require, in addition to knowledge and orientation from professionals in the area, tools and technologies which provide precision and optimize the process. Motivation and commitment of the patient are also required, as reported in the works analyzed in this research.

This review included studies which support the physical rehabilitation of the upper limb with the use of videogames and motion capture systems, and it identified 19 documents which met the criteria of eligibility defined for this investigation. In the documents analyzed, it was found that, concerning motion capture systems, the use of Microsoft Kinect is prominent, due to its affordability and ease of use. There was also a strong trend regarding the implementation of IMUs given their precision and portability.

Concerning the affordability of the technologies used, it can be stated that most of the works used commercial systems and complemented them with the development of components allowing the adjustment of the technology to rehabilitation processes. Development mainly involved personalized and configurable videogames that respond to some requirements of the motor rehabilitation process, especially attending to the need to foment, increase, and maintain the motivation of the patient in the execution of the therapy. In general, the works showed the advantages provided by the use of active videogames in the recovery of patients, as long as they are designed and developed with the accompaniment of physical and functional rehabilitation professionals, and that they can be used in the patient's environment.

The studies analyzed included videogames, as well as motion capture systems, although only 26% of these works integrated the different components into one sole product and complemented them with a system that manages the data of the patients for respective monitoring throughout therapy. Thus, in general, this review identified that an optimal system to support physical rehabilitation should include a motion capture system that offers precision and portability, a module of active videogames that are configurable to the particular needs of each patient's recovery, which permit motivation and proper guidance in the execution of the therapies and, lastly, a computer system which allows the management and monitoring of the rehabilitation plan assigned to each patient, attending to the fundamental aspects of telerehabilitation.

Author Contributions: Conceptualization, A.C.A.-A. and M.C.-C.; methodology, A.C.A.-A. and M.C.-C.; software, A.C.A.-A.; validation, A.C.A.-A., M.C.-C., and A.P.L.B.; formal analysis, A.C.A.-A.; investigation, A.C.A.-A., M.C.-C., and A.P.L.B.; resources, M.C.-C.; data curation, A.C.A.-A., M.C.-C., and A.P.L.B.; writing—original draft preparation, A.C.A.-A.; writing—review and editing: A.C.A.-A., M.C.-C., and A.P.L.B.; visualization, A.C.A.-A.; supervision, M.C.-C. and A.P.L.B.; project administration, M.C.-C.; funding acquisition, M.C.-C. All authors have read and agreed to the published version of the manuscript.

Funding: This study was funded by Universidad Pedagógica y Tecnológica de Colombia (project number SGI 2947) and the APC was funded by the same institution.

Acknowledgments: This work was supported by the Software Research Group GIS from the School of Computer Science, Engineering Department, Universidad Pedagógica y Tecnológica de Colombia (UPTC).

Conflicts of Interest: The authors declare no conflict of interest.

References

1. PNUD Objetivos de Desarrollo Sostenible | PNUD. Available online: https://www.undp.org/content/undp/es/home/sustainable-development-goals.html (accessed on 8 November 2019).
2. World Health Organization; The World Bank. *World Report on Disability*; WHO Press: Geneva, Switzerland, 2011.

3. International Disability Alliance | International Disability Alliance. Available online: http://www. internationaldisabilityalliance.org/ (accessed on 3 February 2020).

4. Turolla, A.; Rossettini, G.; Viceconti, A.; Palese, A.; Geri, T. Musculoskeletal Physical Therapy During the COVID-19 Pandemic: Is Telerehabilitation the Answer? *Phys. Ther.* **2020**, *100*, 1260–1264. [CrossRef]

5. Moher, D.; Liberati, A.; Tetzlaff, J.; Altman, D.G.; Altman, D.; Antes, G.; Atkins, D.; Barbour, V.; Barrowman, N.; Berlin, J.A.; et al. Preferred reporting items for systematic reviews and meta-analyses: The PRISMA statement. *PLoS Med.* **2009**, *6*. [CrossRef]

6. Bionics, B.T.O. ISPO 17th World Congress Abstract Book. *Prosthet. Orthot. Int.* **2019**, *43*, 1–600. [CrossRef]

7. MacIntosh, A.; Desailly, E.; Vignais, N.; Vigneron, V.; Biddiss, E. A biofeedback-enhanced therapeutic exercise video game intervention for young people with cerebral palsy: A randomized single-case experimental design feasibility study. *PLoS ONE* **2020**, *15*, e0234767. [CrossRef] [PubMed]

8. Condino, S.; Turini, G.; Viglialoro, R.; Gesi, M.; Ferrari, V. Wearable Augmented Reality Application for Shoulder Rehabilitation. *Electronics* **2019**, *8*, 1178. [CrossRef]

9. Agyeman, M.O.; Al-Mahmood, A.; Hoxha, I. A Home Rehabilitation System Motivating Stroke Patients with Upper and/or Lower Limb Disability. *ACM Int. Conf. Proceeding Ser.* **2019**. [CrossRef]

10. Fuertes, G.; Mollineda, R.; Gallardo, J.; Pla, F. A RGBD-Based Interactive System for Gaming-Driven Rehabilitation of Upper Limbs. *Sensors* **2019**, *19*, 3478. [CrossRef] [PubMed]

11. Lui, J.; Menon, C. Would a thermal sensor improve arm motion classification accuracy of a single wrist—Mounted inertial device? *Biomed. Eng. Online* **2019**, *18*, 1–17. [CrossRef]

12. Wittmann, F.; Lambercy, O.; Gassert, R. Magnetometer-Based Drift Correction During Rest in IMU Arm Motion Tracking. *Sensors* **2019**, *19*, 1312. [CrossRef]

13. Fernandes, F.G.; Cardoso, A.; Lopes, R.D.A. Games applied to children with motor impairment using the myo wearable device. *An. Acad. Bras. Cienc.* **2020**, *92*, 1–14. [CrossRef]

14. Ma, M.; Proffitt, R.; Skubic, M. Validation of a Kinect V2 based rehabilitation game. *PLoS ONE* **2018**, *13*, 1–15. [CrossRef] [PubMed]

15. David, V.; Forjan, M.; Paštěka, R.; Scherer, M.; Hofstätter, O. Development of a multi-purpose easy-to-use set of tools for home based rehabilitation: Use cases and applications developed during the rehabitation project. *ACM Int. Conf. Proc. Ser.* **2018**, 323–330. [CrossRef]

16. Yang, Z.; Rafiei, M.H.; Hall, A.; Thomas, C.; Midtlien, H.A.; Hasselbach, A.; Adeli, H.; Gauthier, L.V. A Novel Methodology for Extracting and Evaluating Therapeutic Movements in Game-Based Motion Capture Rehabilitation Systems. *J. Med. Syst.* **2018**, *42*. [CrossRef] [PubMed]

17. Da, J.S.; Neto, C.; Filho, P.P.R.; Ferreira, G.P.; Silva, D.A.; Da, N.B.; Olegario, C.; Duarte, B.F.; Albuquerque, V.H.C.D.E.; Ao, J.O. Dynamic Evaluation and Treatment of the Movement Amplitude Using Kinect Sensor. *IEEE Access* **2018**, *6*, 17292–17305. [CrossRef]

18. Bonnechère, B.; Jansen, B.; Haack, I.; Omelina, L.; Feipel, V.; Jan, S.V.S.; Pandolfo, M. Automated functional upper limb evaluation of patients with Friedreich ataxia using serious games rehabilitation exercises. *J. Neuroeng. Rehabil.* **2018**, *15*, 1–9. [CrossRef]

19. Goršič, M.; Cikajlo, I.; Novak, D. Competitive and cooperative arm rehabilitation games played by a patient and unimpaired person: Effects on motivation and exercise intensity. *J. Neuroeng. Rehabil.* **2017**, *14*, 1–18. [CrossRef] [PubMed]

20. Wittmann, F.; Held, J.P.; Lambercy, O.; Starkey, M.L.; Curt, A.; Höver, R.; Gassert, R.; Luft, A.R.; Gonzenbach, R.R. Self-directed arm therapy at home after stroke with a sensor-based virtual reality training system. *J. Neuroeng. Rehabil.* **2016**, *13*, 1–10. [CrossRef] [PubMed]

21. Bird, M.L.; Cannell, J.; Callisaya, M.L.; Moles, E.; Rathjen, A.; Lane, K.; Tyson, A.; Smith, S. "FIND Technology": Investigating the feasibility, efficacy and safety of controller- free interactive digital rehabilitation technology in an inpatient stroke population: Study protocol for a randomized controlled trial. *Trials* **2016**, *17*, 1–6. [CrossRef]

22. Mobini, A.; Behzadipour, S.; Saadat, M. Test–retest reliability of Kinect's measurements for the evaluation of upper body recovery of stroke patients. *Biomed. Eng. OnLine* **2015**, *14*, 1–13. [CrossRef]

23. Nathan, D.; Huynh, D.Q.; Rubenson, J.; Rosenberg, M. Estimating Physical Activity Energy Expenditure with the Kinect Sensor in an Exergaming Environment. *PLoS ONE* **2015**, *10*, 1–22. [CrossRef]

24. Callejas-cuervo, M.; Díaz, G.M.; Ruíz-olaya, A.F. Integration of emerging motion capture technologies and videogames for human upper-limb telerehabilitation: A systematic review. *Rev. DYNA* **2015**, *82*, 68–75. [CrossRef]
25. Basteris, A.; Nijenhuis, S.M.; Buurke, J.H.; Prange, G.B. Lag—Lead based assessment and adaptation of exercise speed for stroke survivors. *Rob. Auton. Syst.* **2015**, *73*, 144–154. [CrossRef]
26. OMS Atención Médica y Rehabilitación. Available online: https://www.who.int/disabilities/care/es/ (accessed on 8 November 2019).
27. Guzik-Kopyto, A.; Michnik, R.; Wodarski, P.; Chuchnowska, I. Determination of Loads in the Joints of the Upper Limb during Activities of Daily Living. In *Proceedings of the Advances in Intelligent Systems and Computing*; Piętka, E., Badura, P., Kawa, J., Wieclauek, W., Eds.; Springer: Kamień Śląski, Poland, 2016; Volume 472, pp. 99–108.
28. Tobón, R. *The Mocap Book: A Practical Guide to the Art of Motion Capture*; Foris Force: Orlando, FL, USA, 2010.
29. Srinivasan, S.R.; Sridhar, S.; Balasubramanian, G.; Vasu, K. Complex Animal Movement Capture and Live Transmission (CAMCALT). In Proceedings of the TENCON 2018—2018 IEEE Region 10 Conference, Jeju, Korea, 28–31 October 2018; pp. 0978–0981.
30. Lammatha, K.A.; Chinnusamy, K.; Malgireddy, N.R.C. Marker less motion capture (Mocap) information compressor & facts mining IRC monocular object. *J. Int. Pharm. Res.* **2019**, *46*, 117–124. [CrossRef]
31. Liao, Y.; Vakanski, A.; Xian, M.; Paul, D.; Baker, R. A review of computational approaches for evaluation of rehabilitation exercises. *Comput. Biol. Med.* **2020**, *119*, 103687. [CrossRef]
32. Mandery, C.; Terlemez, Ö.; Do, M.; Vahrenkamp, N.; Asfour, T. Unifying Representations and Large-Scale Whole-Body Motion Databases for Studying Human Motion. *IEEE Trans. Robot.* **2016**, *32*, 796–809. [CrossRef]
33. Callejas-Cuervo, M.; Ruiz, O.A.; Gutiérrez, R. Métodos de captura de movimiento biomecánico enfocados en telefisioterapia. *Pan Am. Health Care Exch. PAHCE* **2013**, 1–6. [CrossRef]
34. Solberg, R.T.; Jensenius, A.R. Optical or inertial? Evaluation of two motion capture systems for studies of dancing to electronic dance music. In Proceedings of the SMC 2016—13th Sound Music Computing Conference Proceedings, Hamburg, Germany, 31 August–3 September 2016; pp. 469–474.
35. Topley, M.; Richards, J.G. A comparison of currently available optoelectronic motion capture systems. *J. Biomech.* **2020**, *106*, 109820. [CrossRef]
36. Wang, X.; Ge, W.; Li, J. Hand tracking and fingertip detection based on Kinect. In *Proceedings of the IOP Conference Series: Materials Science and Engineering*; IOP Publishing: Xi'an, China, 2020; Volume 740.
37. Jintronix Sense Your Progress—Jintronix. Available online: http://www.jintronix.com/ (accessed on 19 February 2020).
38. 5DT Technologies Home—5DT. Available online: https://5dt.com/ (accessed on 29 July 2020).
39. Golomb, M.R.; McDonald, B.C.; Warden, S.J.; Yonkman, J.; Saykin, A.J.; Shirley, B.; Huber, M.; Rabin, B.; AbdelBaky, M.; Nwosu, M.E.; et al. In-Home Virtual Reality Videogame Telerehabilitation in Adolescents With Hemiplegic Cerebral Palsy. *Arch. Phys. Med. Rehabil.* **2010**, *91*, 1–8. [CrossRef] [PubMed]
40. Pollind, M.; Soangra, R. Development and Validation of Wearable Inertial Sensor System for Postural Sway Analysis. *Meas. J. Int. Meas. Confed.* **2020**, *165*, 108101. [CrossRef]
41. Sers, R.; Forrester, S.; Moss, E.; Ward, S.; Ma, J.; Zecca, M. Validity of the Perception Neuron inertial motion capture system for upper body motion analysis. *Meas. J. Int. Meas. Confed.* **2020**, *149*, 107024. [CrossRef]
42. Feuvrier, F.; Sijobert, B.; Azevedo, C.; Griffiths, K.; Alonso, S.; Dupeyron, A.; Laffont, I.; Froger, J. Inertial measurement unit compared to an optical motion capturing system in post-stroke individuals with foot-drop syndrome. *Ann. Phys. Rehabil. Med.* **2019**, *63*, 195–201. [CrossRef] [PubMed]
43. Khobkhun, F.; Hollands, M.A.; Richards, J.; Ajjimaporn, A. Can we accurately measure axial segment coordination during turning using inertial measurement units (IMUs)? *Sensors* **2020**, *20*, 2518. [CrossRef] [PubMed]
44. Beange, K.H.E.; Chan, A.D.C.; Beaudette, S.M.; Graham, R.B. Concurrent validity of a wearable IMU for objective assessments of functional movement quality and control of the lumbar spine. *J. Biomech.* **2019**, *97*, 109356. [CrossRef] [PubMed]
45. Hafer, J.F.; Provenzano, S.G.; Kern, K.L.; Agresta, C.E.; Grant, J.A.; Zernicke, R.F. Measuring markers of aging and knee osteoarthritis gait using inertial measurement units. *J. Biomech.* **2020**, *99*, 109567. [CrossRef]

Sensors **2020**, *20*, 5989

46. Kinestica Bimeo PRO Neurological Rehabilitation/Stroke—Kinestica—Motivating Neurological Rehabilitation. Available online: http://www.kinestica.com/bimeo-pro.html (accessed on 19 February 2020).

47. Hocoma Armeo®Senso—Hocoma. Available online: https://www.hocoma.com/solutions/armeo-senso/ (accessed on 19 February 2020).

48. Wittmann, F.; Lambercy, O.; Gonzenbach, R.R.; Van Raai, M.A.; Hover, R.; Held, J.; Starkey, M.L.; Curt, A.; Luft, A.; Gassert, R. Assessment-driven arm therapy at home using an IMU-based virtual reality system. *IEEE Int. Conf. Rehabil. Robot.* **2015**, 707–712. [CrossRef]

49. Boyali, A.; Hashimoto, N.; Matsumoto, O. Hand posture and gesture recognition using MYO armband and spectral collaborative representation based classification. In Proceedings of the 2015 IEEE 4th Global Conference on Consumer Electronics (GCCE), Osaka, Japan, 27–30 October 2015; pp. 200–201.

50. Fan, Y.; Yang, C.; Wu, X. Improved teleoperation of an industrial robot arm system using leap motion and MYO armband. In Proceedings of the 2019 IEEE International Conference on Robotics and Biomimetics (ROBIO), Dali, China, 6–8 December 2019.

51. Popovici, I. Experimental results on the accuracy of the Myo Armband for short-range pointing tasks. In Proceedings of the 15th International Conference on Development and Application Systems, DAS 2020—Proceedings, Suceava, Romania, 21–23 May 2020; pp. 185–188.

52. Valve Corporation Dashy Square en Steam. Available online: https://store.steampowered.com/app/461230/Dashy_Square/ (accessed on 10 August 2020).

53. Unity Technologies Unity. Available online: https://unity.com (accessed on 12 August 2020).

54. World Health Organization Cardiovascular Diseases. Available online: https://www.who.int/health-topics/cardiovascular-diseases/#tab=tab_1 (accessed on 2 July 2020).

55. Ramlee, M.H.; Gan, K.B. Function and Biomechanics of Upper Limb in Post-Stroke Patients—A Systematic review. *J. Mech. Med. Biol.* **2017**, *17*, 1–19. [CrossRef]

56. Martínez de Zabarte Fernández, J.M.; Ros Arnal, I.; Peña Segura, J.L.; García Romero, R.; Rodríguez Martínez, G. Bone health impairment in patients with cerebral palsy. *Arch. Osteoporos.* **2020**, *15*, 3–9. [CrossRef]

57. Koeppen, A.H.; Mazurkiewicz, J.E. Friedreich ataxia: Neuropathology revised. *J. Neuropathol. Exp. Neurol.* **2013**, *72*, 78–90. [CrossRef]

58. Li, S.; Xue, J.J.; Hong, P.; Song, C.; He, Z.H. Comparison of energy expenditure and substrate metabolism during overground and motorized treadmill running in Chinese middle-aged women. *Sci. Rep.* **2020**, *10*, 1–7. [CrossRef]

59. Naeemabadi, M.R.; Dinesen, B.; Andersen, O.K.; Najafi, S.; Hansen, J. Evaluating accuracy and usability of microsoft kinect sensors and wearable sensor for tele knee rehabilitation after knee operation. In Proceedings of the 11th International Joint Conference on Biomedical Engineering Systems and Technologies, Madeira, Portugal, 19–21 January 2018; pp. 128–135.

60. Schlagenhauf, F.; Sreeram, S.; Singhose, W. Comparison of Kinect and Vicon Motion Capture of Upper-Body Joint Angle Tracking. In Proceedings of the 2018 IEEE 14th International Conference on Control and Automation (ICCA), Anchorage, AK, USA, 12–15 June 2018; pp. 674–679.

61. Napoli, A.; Glass, S.; Ward, C.; Tucker, C.; Obeid, I. Performance analysis of a generalized motion capture system using microsoft kinect 2.0. *Biomed. Signal Process. Control* **2017**, *38*, 265–280. [CrossRef]

62. Callejas-cuervo, M.; Gutierrez, R.M.; Hernandez, A.I. Joint amplitude MEMS based measurement platform for low cost and high accessibility telerehabilitation: Elbow case study. *J. Bodyw. Mov. Ther.* **2016**, *21*, 574–581. [CrossRef] [PubMed]

63. Cao, Z.; Hidalgo Martinez, G.; Simon, T.; Wei, S.-E.; Sheikh, Y.A. OpenPose: Realtime Multi-Person 2D Pose Estimation using Part Affinity Fields. *IEEE Trans. Pattern Anal. Mach. Intell.* **2019**, 1–14. [CrossRef]

Publisher's Note: MDPI stays neutral with regard to jurisdictional claims in published maps and institutional affiliations.

Article

DataSpoon: Validation of an Instrumented Spoon for Assessment of Self-Feeding

Tal Krasovsky [1,2,*], **Patrice L. Weiss** [3], **Oren Zuckerman** [4], **Avihay Bar** [4,5], **Tal Keren-Capelovitch** [3] **and Jason Friedman** [6]

[1] Department of Physical Therapy, University of Haifa, Haifa 3498838, Israel
[2] Pediatric Rehabilitation Department, Sheba Medical Center, Tel Hashomer 52621, Israel
[3] Department of Occupational Therapy, University of Haifa, Haifa 3498838, Israel;
    plweiss@gmail.com (P.L.W.); talkrn@gmail.com (T.K.-C.)
[4] Media Innovation Lab, The Interdisciplinary Center (IDC) Herzliya, Herzliya 4610101, Israel;
    orenz@idc.ac.il (O.Z.); avihaybar@gmail.com (A.B.)
[5] Efi Arazi School of Computer Science, The Interdisciplinary Center (IDC) Herzliya, Herzliya 4610101, Israel
[6] Department of Physical Therapy, Sackler Faculty of Medicine, Tel Aviv University, Tel Aviv 6997801, Israel;
    jason@tau.ac.il
*   Correspondence: tkrasovsk@univ.haifa.ac.il

Received: 9 February 2020; Accepted: 7 April 2020; Published: 9 April 2020

Abstract: Clinically feasible assessment of self-feeding is important for adults and children with motor impairments such as stroke or cerebral palsy. However, no validated assessment tool for self-feeding kinematics exists. This work presents an initial validation of an instrumented spoon (DataSpoon) developed as an evaluation tool for self-feeding kinematics. Ten young, healthy adults (three male; age 27.2 ± 6.6 years) used DataSpoon at three movement speeds (slow, comfortable, fast) and with three different grips: "natural", power and rotated power grip. Movement kinematics were recorded concurrently using DataSpoon and a magnetic motion capture system (trakSTAR). Eating events were automatically identified for both systems and kinematic measures were extracted from yaw, pitch and roll (YPR) data as well as from acceleration and tangential velocity profiles. Two-way, mixed model Intraclass correlation coefficients (ICC) and 95% limits of agreement (LOA) were computed to determine agreement between the systems for each kinematic variable. Most variables demonstrated fair to excellent agreement. Agreement for measures of duration, pitch and roll exceeded 0.8 (excellent agreement) for >80% of speed and grip conditions, whereas lower agreement (ICC < 0.46) was measured for tangential velocity and acceleration. A bias of 0.01–0.07 s (95% LOA [−0.54, 0.53] to [−0.63, 0.48]) was calculated for measures of duration. DataSpoon enables automatic detection of self-feeding using simple, affordable movement sensors. Using movement kinematics, variables associated with self-feeding can be identified and aid clinical reasoning for adults and children with motor impairments.

Keywords: kinematics; concurrent validity; outcome assessment; feasibility; rehabilitation

---

1. Introduction

Recent technological advances have enabled the development of lightweight, wearable inertial motion sensors, which are showing promise as rehabilitation tools [1]. Inertial sensors can monitor movement quality and present valuable information to clinicians during on-site or tele-rehabilitation sessions using affordable equipment [2].

Sensor-based assessment of movement kinematics is currently used for gait analysis in healthy individuals [3] as well as clinical populations such as Parkinson's disease [4], stroke and Huntington's disease [5] and children with cerebral palsy [6]. Upper limb kinematics have been recorded using wearable sensors in healthy individuals [7] as well as individuals after stroke [8], in order to objectively quantify movement patterns. Additionally, inertial sensors are able to detect performance of functional tasks such as drinking or brushing hair, in both healthy and clinical populations [9]. In future applications, information derived from low-cost inertial sensors may be able to be used to provide feedback (e.g., auditory, visual, tactile), and affect motor performance as is currently the case with more high-end motion capture systems [10].

Due to their age or physical condition, the use of body-mounted sensors is problematic for some populations who may be uncomfortable with or encumbered by the use of external measurement devices. To circumvent this problem, sensor-based technology can be embedded within everyday objects thereby creating clinically-feasible tools for the measurement of movement quality during functional movements such as eating. Existing applications of instrumented tools for eating include forks [11] and chopsticks [12] which help assess and promote fine motor skills and healthy eating habits in children.

In this study, we present the initial validation of an instrumented spoon (DataSpoon) [13,14], developed as an assessment tool for clinicians that provides quantitative information regarding self-feeding in children and adults with motor impairments such as cerebral palsy (CP) or stroke. Self-feeding is one of several self-care activities that are critical for the well-being of a child [15], hence it is an important skill to train, develop and monitor in children with motor disorders [13]. Furthermore, self-feeding kinematics is altered in people with neurological conditions, such as Parkinson's disease [16], stroke [17], or Multiple Sclerosis [18], and in children [19] and adults [20] with cerebral palsy. Specifically, both spatial and temporal patterns of reaching with a utensil to the mouth may be altered and movements are slower, more curved and less smooth. Furthermore, due to a reduced ability to individually control the fingers, people with neurological conditions may opt for an alternative grip strategy (e.g., "power grip") which is typical for young children [21] and leads to further changes in kinematics and force production throughout the movement [16,17,22]. Such alterations in kinematics support the need to evaluate self-feeding in people with motor impairments using a clinically-feasible measurement system. The DataSpoon system includes an instrumented spoon wirelessly paired with an Android smartphone application which presents information regarding eating patterns to a clinician. Monitoring self-feeding kinematics was demonstrated to be feasible among children of different ages and a small sample of children with CP [23]. However, before measures of self-feeding kinematics can be used to detect between-group differences in children or adults with or without motor impairments, it is essential that the psychometric properties of measuring self-feeding kinematics be established. Thus, the current work is a preliminary validation of sensor-based information from the spoon vis-à-vis a "gold standard" kinematic measurement, during self-feeding in healthy young adults. This was accomplished by: (1) describing the automated detection of feeding events from an affordable inertial sensor embedded within a teaspoon (DataSpoon) and (2) determining the validity of kinematic measures extracted from DataSpoon when compared with a "gold standard" motion capture system. We chose to evaluate kinematic measures which are associated with linear velocity and acceleration and are considered "gold standard" [24] as well as measures based on angular velocity and acceleration.

2. Materials and Methods

2.1. Participants

Ten young adults (3 male, 27.2 ± 6.6 years old) were recruited from local university students and staff. They were included in this study if they: (1) were 18–40 years of age, (2) were right-hand dominant and (3) did not have any orthopedic or neurological problems affecting arm movement kinematics or causing pain in arm movement. Ethics approval was received from the Tel-Aviv University Human ethics committee (authorization no. 11152802), and the participants signed an informed consent form before participating in the study.

2.2. Instruments

DataSpoon is an instrumented spoon (size: 19*2*1.5 cm, mass: 38 g; Figure 1a) which enables measurement of 6 Degrees of Freedom kinematic data and real-time presentation of movement via a smartphone app. A small, low cost wireless 3D accelerometer with gyroscope and magnetometer (red amber from GemSense, Haifa, Israel, see Table 1) is mounted at the proximal base of the spoon's handle, and a single CR2032 replaceable battery is mounted at the distal end of the spoon's handle. The raw 3D acceleration data and fused absolute orientation angles (in quaternions) were sampled at 50 Hz and transmitted via Bluetooth to a dedicated android smartphone and transformed to Excel files for offline processing. DataSpoon signals were compared to a "gold-standard" magnetic motion capture system (trakSTAR, Ascension Technology Corp., Shelburne, VT, USA, see Table 1). A trakSTAR sensor (Figure 1a) sampling at 200 Hz was attached to the center of the DataSpoon handle such that concurrent data was collected from both systems. The sensor's lightweight cable was attached with medical tape to the subject's forearm, allowing free movement of the spoon while minimizing the forces applied by the cable on the spoon.

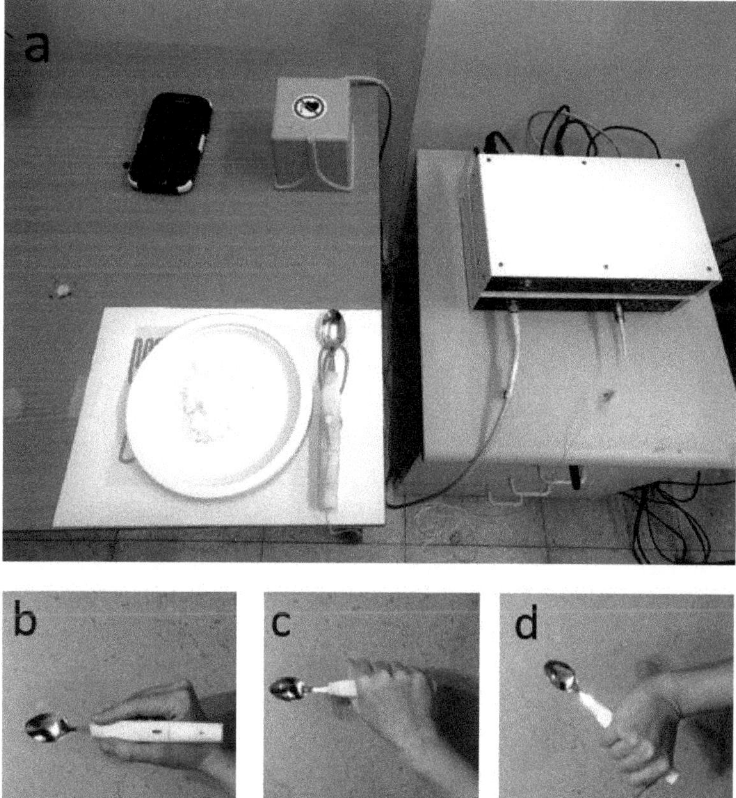

Figure 1. (**a**) Experimental setup. DataSpoon was placed on a placemat pointing towards the distal end of the table. A smartphone captured real-time spoon movement. A trakSTAR sensor was located at the center of the spoon and connected to the trakSTAR box via a lightweight cable. (**b**) Natural grip. (**c**) Power grip. (**d**) Rotated power grip.

Table 1. Comparison of motion capture devices used in the study.

	GemSense Red Amber (Including Battery Extension)	Ascension trakSTAR System with Model 180 Sensor
Size	24 mm diameter	2 mm diameter, 9.9 mm length (not including cable)
Mass	25 g	<5 g (not including cable)
Accuracy	Not available	Position: 1.4 mm RMS, angle: 0.5° RMS
Range	Dependent on Bluetooth (approx. 10 m)	58 cm at highest accuracy level
Approximate cost	USD 40	USD 4000 (for a one-sensor setup)
Sample rate	50 Hz	200 Hz (maximum is 255 Hz)

2.3. Procedures

Participants were seated by a table such that both feet were flat on the floor with hips and knees flexed at 90 degrees. A plate was placed on the table such that the center of the plate aligned with the midline of the participant. A mark on the table to the right of the plate identified the initial and final position of the spoon (Figure 1a). Participants were required to eat small amounts of yoghurt/soft cheese/fruit puree using the DataSpoon at three speeds of movement (slow, comfortable, fast) and with three different grips of the spoon: "natural" grip, power grip and rotated power grip (Figure 1b–d). The power grip is a common grip used among typically developing young children [25,26] as well as children [19] and adults [16,17] with motor impairments due to neurological conditions. Due to limited range of motion in the wrist in the frontal plane (radial/ulnar deviation), this grip type allows for a smaller variety of movements [27]. The rotated power grip was intended to provide an awkward eating posture for participants in order to facilitate variable movement kinematics among healthy individuals, which may be closer to the increased variability of movement kinematics observed in people with motor impairments such as cerebral palsy [22,28]. The inclusion of varied grip positions was intended to provide a variable constraint on hand posture which may translate to variable self-feeding kinematics, and thus challenge the detection of feeding events (such as spoon in mouth) and allow for more accurate computation of validity scores. The instruction to participants was to "hold the spoon as you normally would hold a spoon" for the "natural" grip, which was typically a precision grip, to "keep the thumb below the handle and close to the spoon itself" for the power grip and to "keep the thumb below the handle and oriented towards the distal end of the spoon" for the rotated power grip. Participants performed three repetitions in each condition, such that the total number of eating cycles was ~27.

2.4. Data Analysis

Yaw, Pitch and Roll angles (Figure 2) were obtained directly from the trakSTAR and computed from quaternions for DataSpoon. The angles from the red amber in the DataSpoon are calculated on the device from the raw data using a proprietary algorithm. A filtered derivative for Yaw was calculated (2nd order Butterworth low-pass filter, 1 Hz cutoff) and used to detect eating events using a similar algorithm for both the trakSTAR and DataSpoon signals (Figure 3): (1) Movement onset event: the first point where both the yaw and yaw velocity signals exceed 5% of their respective peaks; (2) Spoon in mouth event: the highest peak in the yaw signal, removing adjacent peaks if inter-peak distance was under 2 s; (3) Spoon down event: time of the first zero crossing in the yaw velocity signal after each "in mouth" event. For visualization purposes, trakSTAR and DataSpoon signals were synchronized by performing a fast rotation (pitch) movement of the spoon prior to each recording. The timing of the peak in pitch was synchronized between the signals automatically using code. However, outcome measures were calculated separately for each device.

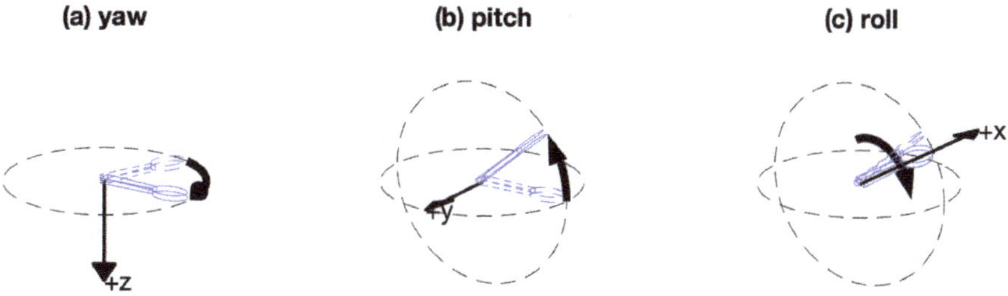

Figure 2. Yaw, Pitch and Roll angles (Tait–Bryan angles). The final orientation consists of three rotations in order: (**a**) yaw is the rotation about the z (up-down) axis; (**b**) pitch is the rotation about the rotated horizontal (y) axis; (c) roll is the rotation about the long axis (rotated x axis) of the spoon.

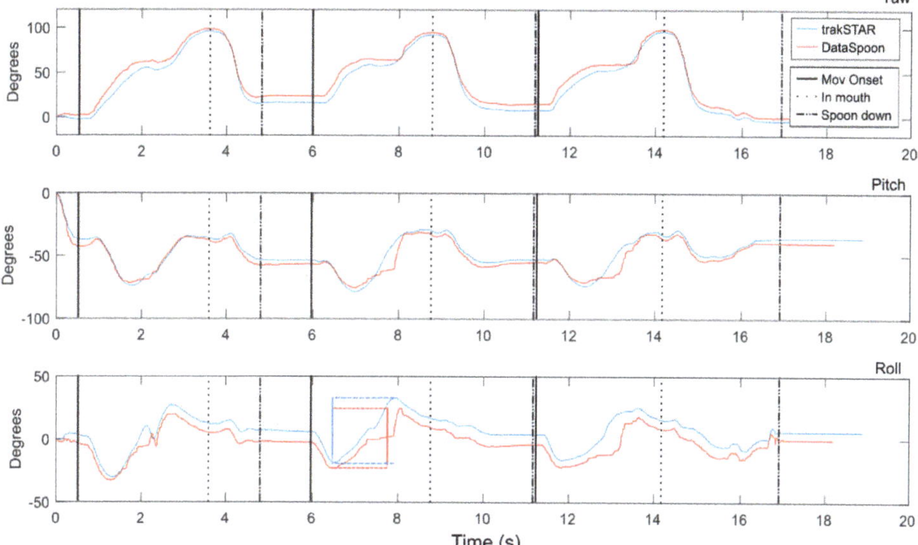

Figure 3. Yaw, Pitch and Roll angles for 3 consecutive eating cycles at natural spoon position and comfortable speed. trakSTAR (blue) and DataSpoon (red) signals were synchronized by a common movement of pitch at onset of recording. Black vertical lines indicate timing of eating cycle events identified for trakSTAR signals. Blue and red vertical lines (bottom panel) demonstrate the calculation of range (in this case - of roll) for one movement part.

The duration of the eating phases (to- and from the mouth) and the range of pitch and roll motion were calculated from Yaw, Pitch and Roll angles (Tait–Bryan angles, Figure 2). Additional measures were extracted from the acceleration signal: in order to obtain tangential velocity, the following procedure was performed: a filtered acceleration signal (2nd order Butterworth low-pass filter, 1 Hz cutoff) was multiplied by the rotation matrix obtained from the spoon. The baseline acceleration signal was subtracted in order to eliminate the effect of gravity, and the acceleration signal was low-pass filtered (4th order Butterworth filter, 3 Hz cutoff low-pass), integrated and high-pass filtered (4th order Butterworth filter, 0.35 Hz cutoff) before calculating the square root of the sum of squares to obtain tangential velocity [29]. The peak tangential velocity was computed for each part of the movement—up (onset to in mouth) and down (in mouth to spoon down). As a measure of movement fluency (i.e., smoothness), the number of zero crossings in the acceleration profile was calculated for each movement axis and summed over the three axes (Figure 4). This number represents the number of peaks in the tangential velocity profile, which is a measure of smoothness (more peaks indicate a jerkier movement) [30].

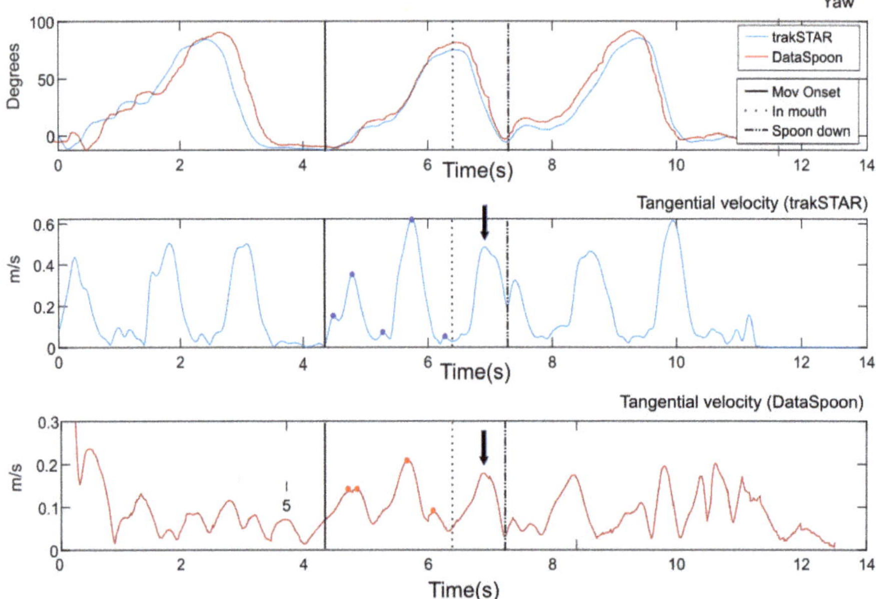

Figure 4. Tangential velocity profiles from trakSTAR (middle panel) and DataSpoon (bottom panel). Yaw for both systems is depicted in the top panel for comparison. One movement duration is marked for both devices. The number of peaks in the tangential velocity profile (i.e., zero crossings in the acceleration profile) is marked for the first part of movement ("to mouth"), and the peak velocity is marked for the second part ("from mouth").

2.5. Statistical Analysis

Concurrent validity was provided using two-way, mixed model Intraclass correlation coefficients (ICCs; single measures) which were computed separately for each movement condition (model ICC (3,2)) [31]; ICC values smaller than 0.4 were defined as poor, $0.41 < ICC < 0.6$ as fair, $0.61 < ICC < 0.8$ as good, and $0.81 < ICC < 1.0$ as excellent agreements. In addition, 95% limits of agreement were calculated by averaging the measurements for each participant under each condition, subtracting the DataSpoon measurement from the trakSTAR measurement and computing mean ±1.96 standard deviations of the difference.

3. Results

Out of a total of 257 movements which were recorded, eight movements were unavailable due to technical issues associated with the spoon (communication lags and disconnections) and 27 movements were removed when events could not be identified reliably by either device (for example when pause between movements was too small). Thus, 222 movements were analyzed in total.

Results of ICCs are detailed in Table 2. ICCs were fair to excellent for measures of duration and range of motion, and poor to fair for measures of peak velocity and movement fluency. Mean differences and 95% Limits of Agreement are detailed in Table 3. Although the mean bias was smaller than 100 ms for temporal measures, and smaller than 1.5 degrees for angular measures (range of roll and of pitch movement in the first stage of eating), 95% limits of agreement exceeded 22 degrees for angular measures and were ~±0.5 s for temporal measures. Tangential velocity and acceleration measures showed some bias, which may have resulted from integration error for the DataSpoon.

Table 2. Intraclass correlation coefficients (ICCs) depicting agreement between trakSTAR and DataSpoon, with 95% confidence interval (square brackets) and significance level below. ICC values higher than 0.8 (excellent agreement) are in bold.

Measure	Natural Grip			Power Grip			Rotated Power Grip		
	Slow	Comfortable	Fast	Slow	Comfortable	Fast	Slow	Comfortable	Fast
Duration of Movement to Mouth	0.99 [0.99, 1.00] <0.01	0.99 [0.97, 0.99] <0.01	0.86 [0.66, 0.94] <0.01	0.95 [0.90, 0.98] <0.01	0.85 [0.66, 0.93] <0.01	0.86 [0.66, 0.94] <0.01	0.97 [0.92, 0.98] <0.01	0.98 [0.95, 0.99] <0.01	0.88 [0.77, 0.94] <0.01
Duration of Movement from Mouth	0.87 [0.74, 0.94] <0.01	0.83 [0.89, 0.98] <0.01	0.55 [0.16, 0.79] <0.01	0.90 [0.79, 0.95] <0.01	0.88 [0.74, 0.94] <0.01	0.50 [0.06, 0.77] 0.02	0.95 [0.89, 0.98] <0.01	0.83 [0.67, 0.92] <0.01	0.89 [0.77, 0.95] <0.01
Duration of Movement (total)	0.97 [0.93, 0.98] <0.01	0.99 [0.97, 0.99] <0.01	0.91 [0.80, 0.96] <0.01	0.94 [0.86, 0.97] <0.01	0.94 [0.87, 0.97] <0.01	0.87 [0.69, 0.95] <0.01	0.96 [0.89, 0.98] <0.01	0.94 [0.87, 0.97] <0.01	0.97 [0.93, 0.98] <0.01
Range of Pitch	0.86 [0.70, 0.93] <0.01	0.81 [0.62, 0.91] <0.01	0.62 [0.27, 0.83] <0.01	0.75 [0.52, 0.88] <0.01	0.64 [0.32, 0.83] <0.01	0.90 [0.75, 0.96] <0.01	0.87 [0.73, 0.94] <0.01	0.92 [0.83, 0.96] <0.01	0.92 [0.84, 0.96] <0.01
Range of Roll	0.93 [0.85, 0.97] <0.01	0.98 [0.96, 0.99] <0.01	0.50 [0.10, 0.76] 0.01	0.79 [0.58, 0.90] <0.01	0.97 [0.93, 0.99] <0.01	0.85 [0.65, 0.94] <0.01	0.95 [0.84, 0.98] <0.01	0.97 [0.94, 0.99] <0.01	0.94 [0.88, 0.97] <0.01
Peak Velocity to Mouth	0.24 [−0.10, 0.59] <0.01	0.21 [−0.06, 0.57] <0.01	0.07 [−0.25, 0.43] 0.35	0.21 [−0.10, 0.52] 0.04	0.05 [−0.15, 0.32] 0.33	0.37 [−0.08, 0.71] <0.01	0.07 [−0.28, 0.42] 0.36	−0.03 [−0.14, 0.16] 0.66	0.00 [−0.10, 0.16] 0.50
Peak Velocity Down	0.07 [−0.10, 0.30] 0.21	0.09 [−0.05, 0.35] 0.01	0.06 [−0.07, 0.29] 0.16	0.06 [−0.07, 0.26] 0.16	0.07 [−0.07, 0.30] 0.11	0.28 [−0.11, 0.65] <0.01	0.10 [−0.09, 0.36] 0.12	0.07 [−0.08, 0.30] 0.13	0.08 [−0.08, 0.30] 0.13
Fluency - Acceleration Zero Crossing (total)	0.14 [−0.26, 0.50] 0.25	0.41 [0.04, 0.68] <0.01	0.21 [−0.13, 0.54] 0.12	0.00 [−0.32, 0.35] 0.50	0.45 [0.07, 0.72] 0.01	0.26 [−0.21, 0.63] 0.14	−0.11 [−0.48, 0.29] 0.70	0.36 [−0.02, 0.65] <0.01	0.14 [−0.13, 0.43] 0.15

Table 3. Median and IQR values for the different kinematic measures for trakSTAR and DataSpoon, mean difference (bias) and 95% limits of agreement (±1.96 standard deviations) between trakSTAR and DataSpoon measurements.

Measure	Units	trakSTAR	DataSpoon	Mean Bias	95% Limits of Agreement
Duration of Movement to Mouth	Seconds	2.10 (0.71)	2.20 (0.70)	−0.07	[−0.51, 0.38]
Duration of Movement from Mouth	Seconds	1.32 (0.43)	1.31 (0.53)	−0.01	[−0.54, 0.53]
Duration of Movement (total)	Seconds	3.46 (0.91)	3.51 (1.08)	−0.07	[−0.63, 0.48]
Range of Pitch	Degrees	43.74 (21.79)	45.60 (18.58)	−0.27	[−23.47, 22.93]
Range of Roll	Degrees	54.95 (28.50)	56.81 (27.37)	−1.32	[−27.16, 24.51]
Peak Velocity to Mouth	m/s	0.42 (0.19)	0.23 (0.13)	0.18	[−0.20, 0.56]
Peak Velocity Down	m/s	0.49 (0.31)	0.19 (0.09)	0.31	[−0.07, 0.68]
Fluency - Acceleration Zero Crossing (total)	Number	4.67 (6.54)	3.00 (3.33)	1.8	[−7.23, 10.82]

4. Discussion

The current work presents an instrumented spoon which uses a simple, affordable inertial movement sensor and extracts kinematic features of movement that may be clinically important for feeding kinematics in general, and specifically for children and adults with motor disorders. The present results indicate that for most kinematic measures, concurrent validity of movement quality measures, which were extracted automatically from both systems, was fair to excellent when evaluated for young, healthy individuals. These results were similar for different grips, a natural grip and two types of power grip, designed to impose a constraint on movement kinematics which requires a modification of the motor plan. Agreement was low for measures based on tangential velocity and acceleration, as compared with yaw, pitch and roll measures. For some measures, agreement was lower for faster movements. This result merits further investigation, as it may be that for faster movements, differences between devices such as sampling rate or sensitivity may become more significant. However, one of the main differences of self-feeding patterns in people with motor impairments such as stroke [17,32] or cerebral palsy [20,22] is slowness of movement. This suggests that the larger ICCs identified for slower movements may be advantageous when evaluating self-feeding in people with motor impairments.

Recent years have witnessed the increased integration of technology into clinical practice via measures of movement quality, specifically for upper limb movement [33] and functional activities such as handwriting [34]. Affordable systems allow for objective and accurate assessment of movement quality using wearable sensors for mobility as well as upper limb movement [1]. However, a review of objective measures of upper limb functional task performance demonstrated that measurement of upper limb kinematics relies on inertial sensors in only 2.2% of the cases, whereas in 64.5% of papers published between 2002 and 2013, the instrument used was an opto-electric or magnetic motion capture system [24]. These instruments are typically expensive and require special operating conditions (e.g., somewhere to place the cameras). In order for this ratio to change, assessment of movement quality based on relatively cheap inertial sensors should rely on valid and reliable measurement. The current work demonstrated that the validity of the outcome greatly varies, specifically the validity of outcomes based on angular velocity vs. linear acceleration. Most kinematic outcomes involve measures of position (e.g., path length) or velocity (e.g., peak velocity, time to peak velocity) [24]. However, the computation of velocity and position from an inertial sensor is not a trivial problem. Inertial sensor data are characterized by drift, which accumulates when integrating acceleration to velocity and further to position. Potential solutions to this problem may include periodic recalibration of the data at rest [29] which requires manual identification of rest periods, a technique that is labor intensive. The current work takes a different approach, and shows that by using measures based on yaw, pitch and roll, better agreement can be reached between DataSpoon and a gold standard motion capture system. More work is required in order to verify whether these measures accurately capture features of self-feeding in children and adults with motor impairments. To date, we have demonstrated the initial feasibility of DataSpoon with children of different ages with and without CP [23], and future work is required to address its feasibility in other clinical populations. Indeed, in the process of designing DataSpoon, input from clinicians suggested that some of these measures (e.g., duration, smoothness) are clinically meaningful to experts in the field [13]. It should be noted, however, that the placement of the inertial sensor within the spoon itself (and not on the arm/hand complex) limits the ability to capture essential aspects of motor performance during self-feeding, such as the type of grip, or compensations related to movements of the wrist, elbow, shoulder and/or trunk [22]. This limitation is common to wearable sensors which are placed on the end-effector, but may be overcome by adding additional sensors on proximal body segments. In the current study, a single trakSTAR sensor was placed on the spoon itself in order to compare its movement with that computed by DataSpoon. Although the sensor and cable may potentially affect movement kinematics, the cable was exceptionally lightweight and the use of similar setups in many studies involving various arm movements [35,36] suggest that the effect on kinematics is minimal. An additional limitation of the current work is the somewhat heavier (38 g) weight of the DataSpoon compared with an ordinary spoon due to the added board and batteries in the handle. We expect that the effect of this weight change on external torques during self-feeding to be minimal since most of the spoon's weight is located in the handle which is placed close to the anchor point (i.e., the hand). It is thus unlikely that the kinematics of using the DataSpoon differed significantly from that of an ordinary spoon. Furthermore, in preliminary feasibility testing with children [23] the spoon's weight was not subjectively reported to be an issue. We therefore suggest that deviations from typical self-feeding kinematics are minimal, supporting future use of the DataSpoon by people with motor impairment.

5. Conclusions

Eating with a spoon is characterized by several salient kinematic features: a unidirectional change in yaw angle for each movement phase, a short-duration change in roll and in pitch angle during the initial scooping phase, followed by relatively stable roll and pitch angles during the transport to mouth phase [14]. This work shows that automatic identification of these salient events is possible. We suggest that using these measures to describe self-feeding kinematics, makes it possible to tap into functionally-relevant variables associated with efficient performance. In the future, these measures can potentially be used to provide targeted knowledge of performance feedback [32] in order to modify self-feeding performance in people with motor impairments.

In this study, measures of duration and of angular range of motion demonstrated excellent validity. Furthermore, we demonstrate here that kinematic measures based on angular velocity have higher concurrent validity compared with measures based on linear velocity and acceleration (peak velocity, fluency) when extracted from a low-cost inertial sensor, possibly due to the larger computational cost and errors associated with obtaining the latter measures. Future studies will be performed to validate this approach with atypical populations (e.g., cerebral palsy) where self-feeding kinematics are impaired [28]. In addition, additional information will be integrated to complement the DataSpoon, such as a time-coupled trunk movement sensor [22] which will assist identification of postural compensations.

Author Contributions: Conceptualization, T.K., P.L.W., O.Z., T.K.-C.; methodology, T.K., J.F., P.L.W., O.Z.; software, T.K., A.B., J.F.; validation, P.L.W., O.Z., J.F.; formal analysis, T.K.; investigation, T.K.-C., T.K., J.F.; resources, J.F.; data curation, T.K.; writing—original draft preparation, T.K., J.F. writing—review and editing, P.L.W., O.Z., T.K.-C., A.B.; visualization, T.K., J.F.; supervision, P.L.W.; project administration, J.F.; funding acquisition, J.F. All authors have read and agreed to the published version of the manuscript.

Funding: Funding was partially provided by the Israeli Center of Research Excellence "Learning in a Networked Society", grant number 1716/12.

Conflicts of Interest: The authors declare no conflict of interest.

References

1. Patel, S.; Park, H.; Bonato, P.; Chan, L.; Rodgers, M. A review of wearable sensors and systems with application in rehabilitation. *J. Neuro Eng. Rehabil.* **2012**, *9*, 21. [CrossRef]
2. Horak, F.; King, L.; Mancini, M. Role of body-worn movement monitor technology for balance and gait rehabilitation. *Phys. Ther.* **2015**, *95*, 461–470. [CrossRef]
3. Mariani, B.; Hoskovec, C.; Rochat, S.; Büla, C.; Penders, J.; Aminian, K. gait assessment in young and elderly subjects using foot-worn inertial sensors. *J. Biomech.* **2010**, *43*, 2999–3006. [CrossRef]
4. Horak, F.B.; Mancini, M. Objective biomarkers of balance and gait for Parkinson's disease using body-worn sensors. *Mov. Disord.* **2013**, *28*, 1544–1551. [CrossRef]
5. Trojaniello, D.; Ravaschio, A.; Hausdorff, J.M.; Cereatti, A. Comparative assessment of different methods for the estimation of gait temporal parameters using a single inertial sensor: Application to elderly, post-stroke, Parkinson's disease and Huntington's disease subjects. *Gait Posture* **2015**, *42*, 310–316. [CrossRef]
6. Van Den Noort, J.C.; Ferrari, A.; Cutti, A.G.; Becher, J.G.; Harlaar, J. Gait analysis in children with cerebral palsy via inertial and magnetic sensors. *Med. Biol. Eng. Comput.* **2013**, *51*, 377–386. [CrossRef]
7. Ertzgaard, P.; Öhberg, F.; Gerdle, B.; Grip, H. A new way of assessing arm function in activity using kinematic Exposure Variation Analysis and portable inertial sensors–A validity study. *Manual. Ther.* **2016**, *21*, 241–249. [CrossRef]
8. Li, H.T.; Huang, J.J.; Pan, C.W.; Chi, H.I.; Pan, M.C. Inertial sensing based assessment methods to quantify the effectiveness of post-stroke rehabilitation. *Sensors* **2015**, *15*, 16196–16209. [CrossRef]
9. Lemmens, R.J.; Janssen-Potten, Y.J.; Timmermans, A.A.; Smeets, R.J.; Seelen, H.A. Recognizing Complex Upper Extremity Activities Using Body Worn Sensors. *PLoS ONE* **2015**, *10*, e0118642. [CrossRef]
10. Portnoy, S.; Halaby, O.; Dekel-Chen, D.; Dierick, F. Effect of an auditory feedback substitution, tactilo-kinesthetic, or visual feedback on kinematics of pouring water from kettle into cup. *Appl. Ergon.* **2015**, *51*, 44–49. [CrossRef]

11. Kadomura, A.; Li, C.Y.; Chen, Y.C.; Tsukada, K.; Siio, I.; Chu, H.H. Sensing fork: Eating behavior detection utensil and mobile persuasive game. In *CHI'13 Extended Abstracts on Human Factors in Computing Systems*; ACM: New York, NY, USA, 2013; pp. 1551–1556.

12. Chia, F.-Y.; Saakes, D. Interactive training chopsticks to improve fine motor skills. In Proceedings of the 11th Conference on Advances in Computer Entertainment Technology, Funchal, Portugal, 11–14 November 2014; ACM: New York, NY, USA, 2014; p. 57.

13. Zuckerman, O.; Slyper, R.; Keren-Capelovitch, T.; Gal-Oz, A.; Gal, T.; Weiss, P.L. Assisting Caregivers of Children with Cerebral Palsy: Towards a Self-Feeding Assessment Spoon. In Proceedings of the Ninth International Conference on Tangible, Embedded, and Embodied Interaction, Stanford, CA, USA, 15–19 January 2015; ACM: New York, NY, USA, 2015; pp. 557–562.

14. Zuckerman, O.; Gal, T.; Keren-Capelovitch, T.; Karsovsky, T.; Gal-Oz, A.; Weiss, P.L.T. DataSpoon: Overcoming Design Challenges in Tangible and Embedded Assistive Technologies. In Proceedings of the TEI '16: Tenth International Conference on Tangible, Embedded, and Embodied Interaction, Eindhoven, The Netherlands, 14–17 February 2016; ACM: New York, NY, USA, 2016; pp. 30–37.

15. Henderson, A. Self-care and hand skill. In *Hand Function in the Child*; Elsevier: Amsterdam, The Netherlands, 2006; pp. 193–216.

16. Ma, H.I.; Hwang, W.J.; Chen-Sea, M.J.; Sheu, C.F. Handle size as a task constraint in spoon-use movement in patients with Parkinson's disease. *Clin. Rehabil.* **2008**, *22*, 520–528. [CrossRef] [PubMed]

17. Demartino, A.M.; Rodrigues, L.C.; Gomes, R.P.; Michaelsen, S.M. Hand function and type of grasp used by chronic stroke individuals in actual environment. *Top. Stroke Rehabil.* **2019**, *26*, 247–254. [CrossRef] [PubMed]

18. Iyengar, V.; Santos, M.J.; Ko, M.; Aruin, A.S. Grip Force Control in Individuals With Multiple Sclerosis. *Neurorehabil. Neural Repair* **2009**, *23*, 855–861. [CrossRef] [PubMed]

19. Van Roon, D.; Steenbergen, B. The use of ergonomic spoons by people with cerebral palsy: Effects on food spilling and movement kinematics. *Dev. Med. Child Neurol.* **2006**, *48*, 888–891. [CrossRef]

20. Artilheiro, M.C.; Correa, J.C.F.; Cimolin, V.; Lima, M.O.; Galli, M.; de Godoy, W.; Lucareli, P.R.G. Three-dimensional analysis of performance of an upper limb functional task among adults with dyskinetic cerebral palsy. *Gait Posture* **2014**, *39*, 875–881. [CrossRef]

21. Van Roon, D.; van der Kemp, J.; Steenbergen, B. (Eds.) Constraints in children's learning to use spoons. In *Development of Movement Coordination in Children: Applications in the Field of Ergonomics, Health Sciences and Sport*; Routledge: Abingdon, UK, 2013.

22. Roon, D.; van Steenbergen, B.; Meulenbroek, R.G.J. Trunk recruitment during spoon use in tetraparetic cerebral palsy. *Exp. Brain Res.* **2003**, *55*, 186–195.

23. Keren-Capelovitch, T.; Krasovsky, T.; Friedman, J.; Weiss, P.T. Identifying Kinematics of Self-Feeding by Young Children: Foundation for Assessment Using an Instrumented Spoon. *Arch. Phys. Med. Rehabil.* **2019**, *100*, e83. [CrossRef]

24. de los Reyes-Guzmán, A.; Dimbwadyo-Terrer, I.; Trincado-Alonso, F.; Monasterio-Huelin, F.; Torricelli, D.; Gil-Agudo, A. Quantitative assessment based on kinematic measures of functional impairments during upper extremity movements: A review. *Clin. Biomech.* **2014**, *29*, 719–727. [CrossRef]

25. van Roon, D.; van der Kamp, J.; Steenbergen, B. Learning to use spoons. In *Development of Movement Coordination in Children*; Routledge: Abingdon, UK, 2003; p. 75.

26. Steenbergen, B.; Van der Kamp, J.; Smithsman, A.W.; Carson, R.G. Spoon Handling in Two-to-Four-Year-Old Children. *Ecol. Psychol.* **1997**, *9*, 113–129. [CrossRef]

27. Connolly, K.; Dalgleish, M. The emergence of a tool-using skill in infancy. *Dev. Psychol.* **1989**, *25*, 894–912. [CrossRef]

28. Mackey, A.H.; Walt, S.E.; Stott, N.S. Deficits in Upper-Limb Task Performance in Children with Hemiplegic Cerebral Palsy as Defined by 3-Dimensional Kinematics. *Arch. Phys. Med. Rehabil.* **2006**, *87*, 207–215. [CrossRef] [PubMed]

29. Cahill-Rowley, K.; Rose, J. Temporal–spatial reach parameters derived from inertial sensors: Comparison to 3D marker-based motion capture. *J. Biomech.* **2017**, *52*, 11–16. [CrossRef] [PubMed]

30. Flash, T.; Hogan, N. The coordination of arm movements: An experimentally confirmed mathematical model. *J. Neurosci.* **1985**, *5*, 1688–1703. [CrossRef] [PubMed]

31. Shrout, P.E.; Fleiss, J.L. Intraclass correlations: Uses in assessing rater reliability. *Psychol. Bull.* **1979**, *86*, 420. [CrossRef] [PubMed]

32. Cirstea, M.; Levin, M.F. Improvement of arm movement patterns and endpoint control depends on type of feedback during practice in stroke survivors. *Neuro Rehabil. Neural Repair* **2007**, *21*, 398–411. [CrossRef] [PubMed]
33. Brochard, S.; Robertson, J.; Medee, B.; Remy-Neris, O. What's new in new technologies for upper extremity rehabilitation? *Curr. Opin. Neurol.* **2010**, *23*, 683–687. [CrossRef] [PubMed]
34. Rosenblum, S.; Parush, S.; Weiss, P.L. Computerized temporal handwriting characteristics of proficient and non-proficient handwriters. *Am. J. Occup. Ther.* **2003**, *57*, 129–138. [CrossRef]
35. Friedman, J.; Korman, M. Offline optimization of the relative timing of movements in a sequence is blocked by retroactive behavioral interference. *Front. Hum. Neurosci.* **2016**, *10*, 623. [CrossRef]
36. Bullock, I.M.; Feix, T.; Dollar, A.M. Analyzing human fingertip usage in dexterous precision manipulation: Implications for robotic finger design. In Proceedings of the 2014 IEEE/RSJ International Conference on Intelligent Robots and Systems, Chicago, IL, USA, 14–18 September 2014; IEEE: Piscataway, NJ, USA, 2014; pp. 1622–1628.

Article

A Systematic Approach to the Design and Characterization of a Smart Insole for Detecting Vertical Ground Reaction Force (vGRF) in Gait Analysis

Anas M. Tahir [1], Muhammad E. H. Chowdhury [1], Amith Khandakar [1], Sara Al-Hamouz [1], Merna Abdalla [1], Sara Awadallah [1], Mamun Bin Ibne Reaz [2] and Nasser Al-Emadi [1,*]

[1] Department of Electrical Engineering, Qatar University, Doha 2713, Qatar; a.tahir@qu.edu.qa (A.M.T.); mchowdhury@qu.edu.qa (M.E.H.C.); amitk@qu.edu.qa (A.K.); sa1507714@student.qu.edu.qa (S.A.-H.); ma1508307@student.qu.edu.qa (M.A.); sa1509524@student.qu.edu.qa (S.A.)

[2] Department of Electrical, Electronic & Systems Engineering, Universiti Kebangsaan Malaysia, Bangi, Selangor 43600, Malaysia; mamun@ukm.edu.my

* Correspondence: alemadin@qu.edu.qa; Tel.: +974-4403-4213

Received: 7 December 2019; Accepted: 28 January 2020; Published: 11 February 2020

Abstract: Gait analysis is a systematic study of human locomotion, which can be utilized in various applications, such as rehabilitation, clinical diagnostics and sports activities. The various limitations such as cost, non-portability, long setup time, post-processing time etc., of the current gait analysis techniques have made them unfeasible for individual use. This led to an increase in research interest in developing smart insoles where wearable sensors can be employed to detect vertical ground reaction forces (vGRF) and other gait variables. Smart insoles are flexible, portable and comfortable for gait analysis, and can monitor plantar pressure frequently through embedded sensors that convert the applied pressure to an electrical signal that can be displayed and analyzed further. Several research teams are still working to improve the insoles' features such as size, sensitivity of insoles sensors, durability, and the intelligence of insoles to monitor and control subjects' gait by detecting various complications providing recommendation to enhance walking performance. Even though systematic sensor calibration approaches have been followed by different teams to calibrate insoles' sensor, expensive calibration devices were used for calibration such as universal testing machines or infrared motion capture cameras equipped in motion analysis labs. This paper provides a systematic design and characterization procedure for three different pressure sensors: force-sensitive resistors (FSRs), ceramic piezoelectric sensors, and flexible piezoelectric sensors that can be used for detecting vGRF using a smart insole. A simple calibration method based on a load cell is presented as an alternative to the expensive calibration techniques. In addition, to evaluate the performance of the different sensors as a component for the smart insole, the acquired vGRF from different insoles were used to compare them. The results showed that the FSR is the most effective sensor among the three sensors for smart insole applications, whereas the piezoelectric sensors can be utilized in detecting the start and end of the gait cycle. This study will be useful for any research group in replicating the design of a customized smart insole for gait analysis.

Keywords: gait analysis; characterization; smart insole; vertical ground reaction forces; force sensitive resistors; piezoelectric sensors; sensor calibration

---

1. Introduction

Gait analysis offers an opportunity for assessment of the act of walking, one of the most important features of the individual's use pattern that displays posture in action. By identifying gait kinetics, gait kinematics and musculoskeletal activity, gait analysis can be utilized in various applications, such as rehabilitation, clinical diagnostics and sport activities [1]. Gait kinetics studies the forces and moments that results in movement of lower extremities during gait cycle. Vertical ground reaction forces (vGRFs) are the forces between the foot and ground which can be obtained by wearable sensors [2] and are considered as the main measurement in kinetic analysis. Gait kinetics have recently become a convenient tool for biomedical research and clinical practice. Different research teams studied the ability to diagnose or early detection of various diseases using gait analysis [3–5]. Some research teams used gait analysis in fall detection of elderly people, one of the most common domestic accidents among the elderly. With smart insoles, the fall event can be detected and doctors or personal who takes care of the elderly can be notified to take action. In athletic sports where walking, running, jumping and throwing are involved, gait analysis can be utilized to recognize an athlete's faulty movement and, accordingly, enhance it. In addition, gait analysis can play positive role in the rehabilitation process for several diseases and complications.

Recently, with the development in sensor technologies, gait analysis using wearable systems became an effective approach [6–8]. Various types of wearable sensors such as force sensors, strain gauges, magneto-resistive sensors, accelerometers, gyroscopes, inclinometers etc. can analyze different gait characteristics. Accelerometers were used to conduct gait analysis studies, in which they were attached to feet or legs to measure the acceleration or velocity of human lateral movements during gait cycles [4]. Gyroscopes were used in gait analysis to measure the changes in orientation of lower body extremes with respect to the vertical axis. Goniometers measured the relative rotational motion between different body segments [2]. Electromagnetic tracking systems were developed as 3D measurement device that can be applied in the kinematic study of body movements [9].

Gait analysis is typically carried out using a force plate system or multi-camera-based system to capture the ground reaction forces (GRF) during different gait cycles. However, this method requires a costly set up and long post-processing time and can measure only limited number of strides. Therefore, it is not affordable by individuals for personal use [3,8,10]. Instrumented trade mills with few force plates laid on the trade mill are used by different research groups to mitigate the limitations of conventional force plates [2], but with treadmills restrictions are still present as subjects need to walk in a straight line where direction changes and turning cannot be realized. This led to an increase in research interest towards developing smart insoles, where wearable sensors can be employed to detect vGRF, joint movements, acceleration of lower extremities, and other gait variables [3,4,11,12]. vGRF is a useful tool to assess the health conditions of the patient, to enhance the performance of athletes [13–15]. Among different solutions for vGRF measurement, smart insoles have several extra advantages over force plates and multi-camera systems. Although force plates can measure shear forces and pressure changes, smart insoles are portable and capable of tracking motions and measuring pressure without rigid mounting, whereas the camera-based system requires large space for set-up along with long post-processing time. The smart insole offers flexible, portable, and comfortable solution for vGRF measurement. It is designed to monitor, process and display plantar pressure using pressure sensors embedded in the insole [3,4,11,12]. Recently, several off-the-shelf smart insoles have been offered by some companies (e.g., F-scan [16], MoveSole [17], Bonbouton [18], FeetMe [19] etc.), however, the commercial systems are very expensive for individual use, making it difficult for a home setting.

The aim of this study is to design and characterize smart insoles to detect vGRF during gait, with three different types of low-cost commercial force sensor: force-sensitive resistors (FSRs) [20], ceramic piezoelectric sensors [21], and flexible piezoelectric sensors [22]. All three types of sensor were calibrated before checking their suitability for smart insole application. A simple low-cost calibration method based on load cells is presented, mitigating the need to use expensive calibration devices or Motion Analysis Labs as a calibration reference. This work provides a systematic approach for sensor calibration guides, which can be replicated easily by other researchers to perform studies on smart insoles or other body-sensing technologies. To the best of our knowledge, this is the first article to compare three different low-cost commercially available force sensors for smart insole application.

The remainder of the article is organized into five sections. In Section 2, a comprehensive review of the recent works with smart insoles to detect vGRF in gait cycles are summarized. In Section 3, the experimental details for sensors calibration and insole characterization are presented. In Section 4, the mathematical analysis of each insole characterization and sensor calibration are explained. Results and a discussion are presented in Section 5. Finally, we conclude with future recommendations in Section 6.

2. Literature Review

Several research teams focused on fabricating and synthesizing the sensing parts or sensing fabrics of the smart insoles [23–25]. Sensing fabrics are fibers/yarns with sensing technologies or electrical components made of fabric materials, offering a flexible alternative to comfortably measuring human movement. Usually, piezoelectric, piezoresistive and piezo-capacitive materials are used to fabricate the sensing parts of the sensing fabrics, due to their elastic properties [26,27]. Shu et al. [26] implemented a low-cost insole with high pressure sensitivity using a fabric pressure sensing array made by the researchers with a pressure range of 10 Pa to 1000 kPa. It is attached to six locations corresponding to a polyimide film circuit board that takes the shape of the foot. They were able to measure the peak pressure, mean pressure, center of pressure (COP), and illustrate different pressure levels occurring at the six-targeted areas. However, the quality of the gait cycle records was poor, with irregular peak values, where the common gait shape with two peaks of the heel strike and toe off cannot be distinguished. Kessler et al. [27] demonstrated a low-cost flexible insole, made with Velostat and conductive ink electrodes printed on polyethylene terephthalate (PET) substrate. However, repeatability was a major problem and they proposed an averaging method to reduce the repeatability issue. However, the proposed method does not provide a generic solution for the force-sensing problem, it can be utilized only with periodic forces where spatial information is the key. On the other hand, some research teams used low-cost flexible force sensors to design the smart insoles [28–30] using commercially available piezoresistive [20], piezoelectric [21,22], capacitive transducers [2], fiber brag grating [5,31] sensors.

Piezoelectric force sensors are materials that generate electric charges when stressed. However, there are a few factors which limit the usage of piezoelectric sensors in smart insoles. The parasitic effect of piezo materials neutralizes the generated charge within a short time. Therefore, sophisticated electronics are needed to extract resultant charges, and this makes it difficult to use these sensors in measuring static or slow varying forces. In addition, protection circuits are needed, since piezo sensors generate high voltage values, which might reach above 100 V with peak vGRF values. Capacitive force sensors are another alternative force sensor, consisting of parallel capacitor plates that changes the capacitance in correspondence to applied force/weight. However, they need complex conditioning circuits and are highly subject to noise [20].

A commonly used body-sensing technology is the piezo-resistive sensor or FSR, which changes its conductivity based on the applied force. FSR is a polymer thick film (PTF) that is used to measure the applied force in different applications such as human touch and medical applications, industrial and robotics applications, and automotive electronics. The main advantages of FSRs are: thin size, very good shock resistance, low power requirement, fast response to force changes, robustness against noise, simple conditioning circuits, ability to fabricate using flexible materials, and low unit cost compared to other commercial force sensors [20]. However, these sensors have some disadvantages that need to be compensated for, such as non-linear behavior and repeatability error [3].

Bamberg et al. [4] used a combination of different FSRs, piezo electric sensors, accelerometers and gyroscopes to determine the vGRF. The main advantage of this approach is that it enables the detection of heel strike and toe off events in each gait cycle. In addition, it helps in estimating foot orientation and position. Even though gait variability can be analyzed by walking in a straight line, gait analysis concentrating merely on straight walking or running may not be adequate to interpret gait variability, since changing walking directions or turning have effects on extrinsic gait variability [11]. Similar research was done recently in [32], where the research group used the FSR sensor to develop the smart foot sole which transmits wirelessly the vGRF to a computer, and the patients were asked to walk on treadmill during the signal acquisition. Liu et al. [11] developed a wearable measuring insole using five triaxial force sensors in each shoe capable of measuring GRF and center of pressure (COP) on insole. The GRF results showed a great correspondence between the insole and the reference data. Kim et al. [33] conducted a similar study, where they have used similar triaxial force sensors and the sensors performance were tested on seven healthy male subjects. An in-shoe plantar measurement sensor with 64 sensing points made from an optoelectronics transducer covered with silicon in a matrix form covering 80% of contact region between the foot and the insole and handling capability of 1MPa was implemented by De Rossi et al. to measure COP and vGRF [5]. Howel et al. [3] demonstrated the design of a wearable smart insole using low-cost FSRs for gait analysis. This provided subject-specific linear regression models to determine the vGRF accurately using simultaneous collected data from motion analysis laboratory. However, insufficient information was given about the sensors calibration and the hardware design of the insole and the wireless system to transmit the data to host PC, making it difficult for other researchers to replicate the work.

Even though systematic sensor calibration with clear steps was followed by different research teams, expensive calibration devices were used to calibrate the force sensors. Some research teams carried out the experiments on the smart insoles in motion analysis labs, where simultaneous data collection from infrared motion capture cameras/RGB depth camera and force plates were done as reference measurement for the collected insole data [34,35]. In addition, some research teams used a universal testing machine to apply incremental weight values to sensor active area during calibration. Barnea et al. [36] used the CETR Universal Micro-Tribometer (UMT)-2 micro tribometer) device for calibrations, that can apply precise weights in X, Y and Z directions. Marco et al. [5] performed the sensor calibrations using robotic platform that can precisely apply controllable loads to the desired positions. Parmar et al. [37] evaluated the performance of 5 different commercial FSRs during static and dynamic loading with reliable test setups that can mimic realistic conditions when applying pressure on human limbs. The sensors were evaluated quantitatively based on their accuracy, drift, and repeatability behaviors. The tested sensors showed lower accuracy levels with static pressures compared to the dynamic pressure test, with high drift values. This necessitates the need for further study and analysis on the use of FSRs for static pressure applications.

3. Methodology

This section demonstrates the design of a complete system describing the main blocks of the smart insole along with illustrations of sensor calibration and insole characterization process.

3.1. Smart Insole Sub-System

Figure 1 shows the complete block diagram of the system, where the pressure sensor array was placed in a customized shoe above the control circuit. Pressure data were digitized through a microcontroller before they were sent wirelessly to a host computer for post processing and analysis. This subsystem was powered by a battery with the help of a power management unit. Pressure data were analyzed to extract various gait characteristics for different gait applications.

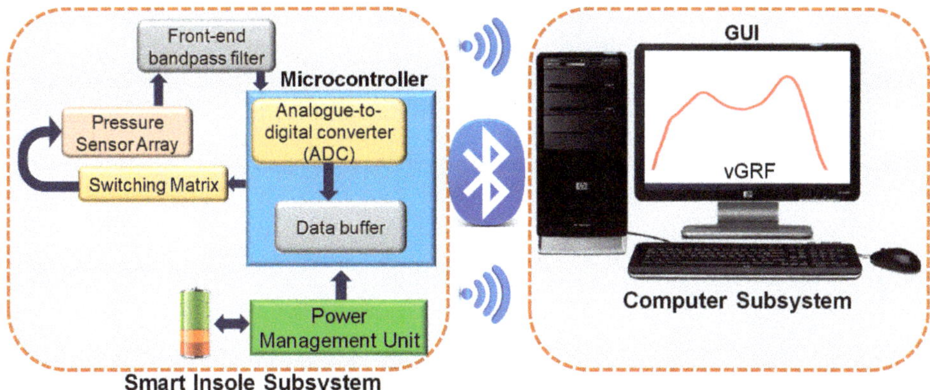

Figure 1. Smart insole block diagram.

3.1.1. Pressure-Sensing Array

The vGRF during gait cycles can be sensed using one of three alternatives:

A.   Force-Sensitive Resistor (FSR)

The FSR exhibits a decrease in resistance as the applied force to the surface of the sensor increases. FSR sensors from Interlink Electronics [20] were used in this study as shown in Figure 2A. The sensors have a flexible round active area of diameter 12.7 mm to detect the applied force, with a two flexible lead wires to connect the sensor to the acquisition circuit. A FSR exhibits a non-linear relation between the applied force and the sensor's resistance. In addition, no direct relationship is provided in the sensor's datasheet. Therefore, proper calibration must be done prior to the sensor usage.

B.   Ceramic Piezoelectric Sensor

A piezoelectric element is a sensor that produces an alternating voltage in response to an applied dynamic pressure or vibration. With applications related to dynamic forces, the piezoelectric sensor is highly recommended. When a force applied to the piezoelectric crystal element, the net movement of both positive and negative ions occurs. When there is a constant or zero pressure, the dipole is not formed [38]. It is important to mention that the force plate is originally made of piezoelectric material mounted between two metal plates to produce three-dimensional forces with a special mechanical arrangement [39]. This comes in different sizes; however, a ceramic piezoelectric element with 12.8 mm electrode diameter would be suitable to obtain a high-resolution pressure map as shown in Figure 2B.

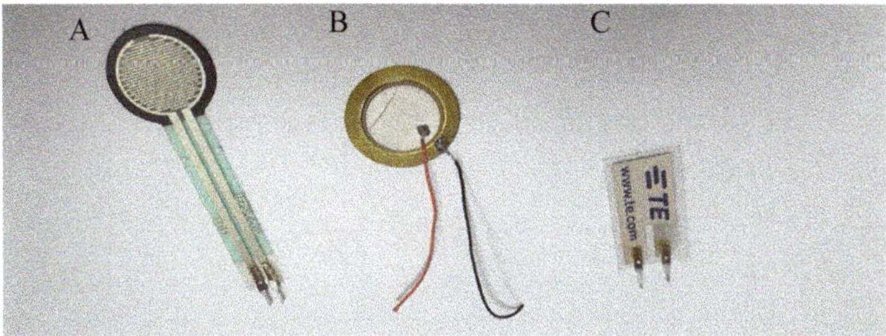

Figure 2. (**A**) Force-sensitive resistor (FSR) sensor from Interlink Electronics [20], (**B**) piezo-electric sensor from Murata Manufacturing Co. [38], (**C**) micro-electromechanical systems (MEMS) sensor LDT0-028K from Measurement Specialties Inc. [40].

C.   Micro-Electromechanical Systems (MEMS) Sensor

The micro-electromechanical systems (MEMS) sensor is a new member of piezoelectric sensors family (Figure 2C). Similar to ceramic piezo electric sensors, it converts mechanical forces into electrical signals. However, the MEMS sensor can detect forces in x, y or z axes generating electrical impulses with positive or negative amplitudes depending on the force direction on a certain axis [40]. MEMS sensors are useful for detecting human motion sensor due to their flexibility, wide frequency range (0.001 Hz to 10 MHz), low acoustic impedance, high mechanical strengths, and high stability resisting moisture, etc. [40].

3.1.2. Data Acquisition System

A.   Microcontroller (MCU):

A microcontroller (MCU) was used to collect the data from the sensor and to send to the computer for classification. Simblee is a very compact and powerful ARM Cortex-M0 MCU with a six channels 10-bit analog-to-digital converter (ADC). It is featured with an inter-integrated circuit (I^2C) and serial peripheral interface (SPI) communication interface, which were required for 9-degree of freedom (DOF) module. Moreover, it has an incorporated Bluetooth low energy (BLE) 4.0 module, which can be utilized to send data to the computer. This MCU operates on a power supply between +2.1 to 3.6 V.

B.   Multiplexer (MUX)

Since MCU has a limited number of ADC channels whereas the number of sensors is needed for better spatial resolution of smart insole, it is suggested to use multiplexers (MUX) to reduce the number of required channels in MCU. A MUX allows several inputs in parallel to be routed into a single output depending on the input combinations of the data selectors. Active area of these sensors are close and sixteen sensors were used to create sensors' array for each leg insole to obtain a high-resolution pressure map. Therefore, the CD74HC4067 multiplexer from Texas Instruments with 16 input channels was used in this study [41].

3.1.3. Transmission Techniques

Three commonly used transmission techniques for connected biomedical sensors are ZigBee, Bluetooth Low Energy (BLE) and Wi-Fi. ZigBee is a two-way wireless communication technique developed for sensors and control networks, which need a wider range, low latency, low energy consumption at lower data rates. BLE is an alternative to the classical Bluetooth with higher data rate and low power consumption within a limited area with low latency at 2.4 GHz. Wi-Fi makes a good candidate for transmitting data with a data rate of up to 450 Mbps for indoor applications. However, it imposes latency on the system of more than 25 ms and higher power consumption. Table 1 shows a comparison between three different communication interfaces.

Table 1. Transmission methods comparison [42–46].

	Latency	Speed	Power Consumption	Range
ZigBee	15 ms	250 Kbps	9.3 mA	291 m
Bluetooth Low Energy	6 ms	1–11 Mbps	4.5 mA	10 m
Wi-Fi	≥25 ms	1.3 Gbps over 5 GHz and 450 Mbps over 2.4 GHz	35 mA	50 m

Since the smart insole was intended for indoor application, BLE and WiFi both were suitable for communication interface; however, the higher power consumption and latency made WiFi non-suitable for smart insole application. Moreover, Simblee MCU has in-built BLE in its small form factor. Therefore, BLE has been chosen as communication interface.

3.1.4. Power Management Unit (PMU)

Power supplies were chosen depending on the operating voltage of the system components. The microcontroller and multiplexer both can operate at 3.3 V. The power management unit (PMU) is LiPo Charger/Booster module MCP73831 [47] and AMS1117 voltage regulator connected to a Lithium Polymer (LiPo) battery of 3.7 V (1000 mAh), which was regulated to 3.3 V. The PMU is not only delivering regulated 3.3 V to the system but also capable to charge LiPo battery.

3.1.5. Host Computer

The acquired data from smart insole can be sent wirelessly to a host computer, where post processing, thereby displaying the vGRF as pressure maps during gait cycle, was carried out. The obtained data can be used in different gait analysis applications such as medical diagnostics, rehabilitation and athlete's performance assessment.

3.2. Sensors' Calibration

The first step in designing the smart insole is to calibrate the force sensors that are going to be used to detect the vGRF during the gait cycle. Three different force sensors were calibrated: FSR [20], piezo-electric sensor [21] and piezo -vibration sensor [22].

3.2.1. Force-Sensitive Resistor (FSR) Calibration

Firstly, a voltage divider circuit must be used with the sensor to convert the resistance change (due to applied force) of the sensor to a voltage value, which can be acquired by microcontrollers. Secondly, a load cell of 5kg from HT sensor technology company [48], with HX711 amplifier modules [15] was used as a weight reference for FSR calibration (Figure 3). The load cells consist of straight metal bar with two strain gauge sensors and two normal resistors arranged in a Whitestone bridge configuration, a constant excitation voltage (3–5 V) can be applied as an input to the circuit and the balanced configuration of the circuit replicates a zero output voltage in normal conditions when no force is applied. Any force applied to the load cell results in an unbalanced condition of the bridge leading to small voltage values in the output that can be detected and converted to force [48]. The load cell has high sensitivity and can detect as small as 1 gm of weight variation. However, the output voltage from the load cell is very small, with a maximum value of 5 mV. Therefore, a HX711 amplifier module was used. The amplifier module has instrumentational amplifier to amplify the signal with a 24-bit ADC that converts the analog signal from the load cell bridge to digital value that is readable by a microcontroller. The HX711 transmits data to the microcontroller using I^2C communication protocol with 10Hz sampling rate [15].

The bar-type load cell was mounted with screws and spacers so that the strain can be measured correctly (refer Figure 3B). The load cell was placed between two plates with only one side screwed into each plate/board. This setup provides a moment of force on the strain gauges rather than just a single compression force, resulting in higher sensitivity to applied forces. The output voltage from the load cell exhibits a linear relationship with the applied force. This can be calibrated easily with any small object of known mass such as a coin that weighs a few grams.

A known weight object (ex. a coin) was placed on load cell plate; the calibration factor was adjusted until the output reading matches the known weight. Once the correct calibration factor is obtained, it was used to convert the load cell voltages to corresponding weights. The calibration factor is the slope of output voltage of load cell vs. real weights' graph. The FSR was attached to adhesive material on the back face of the active area, which was used to fix the FSR on the scale. A cylindrical acrylic of 12.7 mm diameter, matching the active area of FSR, was used to apply force on the sensor only. In addition, a square shaped acrylic plate was glued on top of the cylindrical acrylic to support the weights, as shown in Figure 3. Then, 500 g weights are placed every 4 to 5 s until 5000 g is reached. Readings from load cell and FSR circuit are acquired simultaneously by Arduino, which were saved in a text file in a computer.

Finally, the FSR output voltage was plotted with respect to the load-cell weight and a mathematical relationship was derived. The equation was used to convert smarts insole FSR readings into the corresponding applied pressure by the foot.

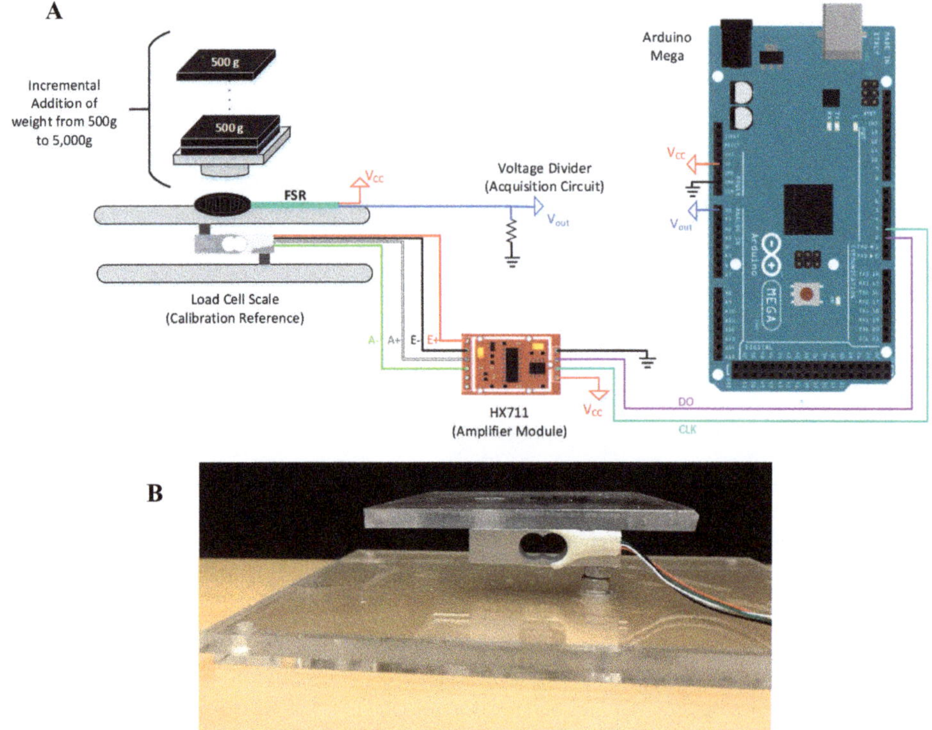

Figure 3. FSR calibration setup (**A**) and load-cell scale (**B**).

3.2.2. Piezo-Electric Sensor Calibration

The same load-cell module was used for piezo calibration with some modifications (as shown in Figure 4). Piezo transducers convert the applied mechanical forces into electrical impulses. Therefore, a high sampling frequency (above 50 Hz) is needed to acquire both the piezo output and the applied weights from the load cells. HX711 amplifier module samples the data with a low sampling frequency of 10 Hz. Therefore, the data was acquired directly by the 10-bit ADC of Arduino MCU with a sampling frequency of 1 kHz. However, AD620AN instrumentational amplifiers [39] were used before the acquisition step to amplify the small load cell outputs (maximum of 5 mV).

Firstly, a voltage divider circuit was used to reduce the high piezo voltage outputs, which can go up to 20 V. Secondly, the load cell was calibrated again due to the modification. Three dead weights of known masses were used: 500 kg, 2500 kg 5000 kg (maximum load for load cell). AD620AN instrumentational amplifier gain was adjusted to give an output of voltage when maximum load is applied. This ensures that the full range of the Arduino ADC was utilized. Three dead weights were added one by one on the scale and the output voltage from load cells were acquired by Arduino. A linear relationship was fitted between the load cell voltages and applied weights. This relationship was used to convert the load cell voltage to a corresponding weight.

Unlike the FSR, weights cannot be used to calibrate the piezo sensor, since piezo sensors are sensitive to dynamic forces only. Therefore, a fast finger press and release is suggested as an alternative. The calibration can be done by pressing the active area/ceramic of the sensor with various strengths and recording the generated electrical signals for each press as shown in Figure 4. Readings from the load cell and piezo voltage divider circuit were acquired simultaneously by the Arduino MCU. Serial terminal software was used to store the data in the computer. Load cell readings were plotted against the piezo output voltage and a linear relationship was derived. The equation was used to convert smarts insole readings into the corresponding applied weight by the foot.

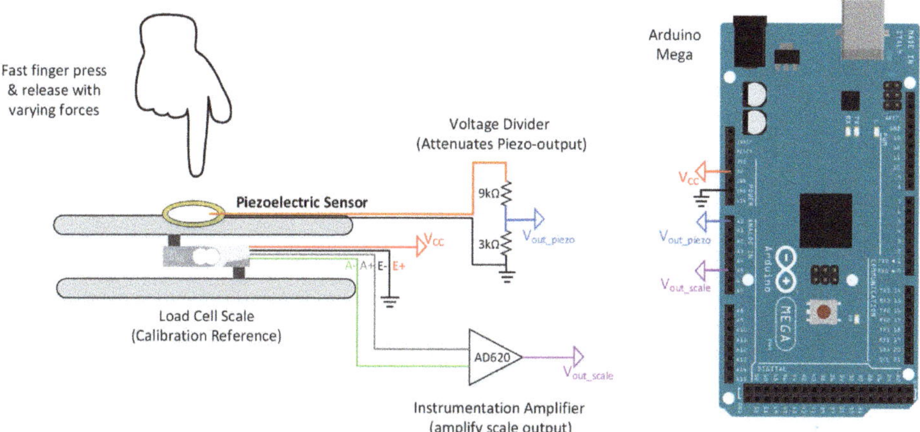

Figure 4. Piezoelectric sensor calibration setup.

3.2.3. Micro-Electromechanical Systems (MEMS) Sensor Calibration

MEMS sensors produce alternating current (AC) impulses with both positive and negative peaks. Therefore, little modification was required for the setup of piezo electric vibration sensor, refer to Figure 5. An offset circuit was added for the piezo-electric acquisition circuit. The piezo-vibration output voltage was reduced by a voltage divider circuit to $\pm 1/2$ V_{cc}, then adder amplifier was used to add an offset of $+1/2$ V_{cc}, so the new AC signal will be centered around $+1/2$ V_{cc} with maximum value of V_{cc} and minimum of 0 V. After modifying the acquisition circuit, the piezo electric sensors' calibration steps were used to calibrate the piezo-vibration sensor.

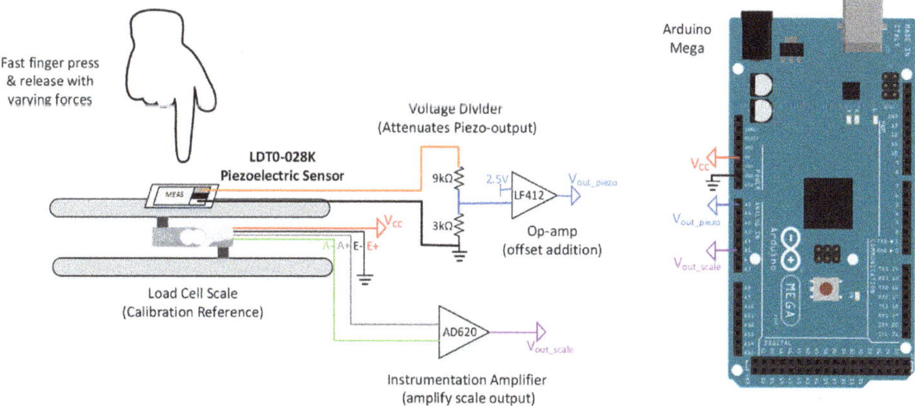

Figure 5. LDT0-028k MEMS sensor calibration setup.

3.3. Insole Fabrication

Once the sensors were calibrated, these sensors were separately used to construct the smart insole for vGRF detection during gait cycles. The FSR sensors and piezo-electric sensors were chosen to construct two different insoles. While the piezo-vibration sensor was found not suitable for vGRF detection, the reason of not selecting the piezo-vibration sensor is discussed in a later section. As shown in Figure 6, the most common place of the foot plane, where most of the pressure is exerted during gait are the heel, metatarsal heads, hallux and toe.

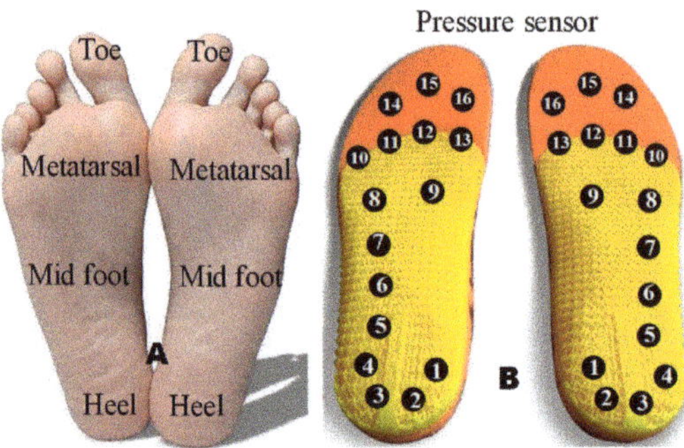

Figure 6. Area of foot selected for sensors (**A**), and array of pressure sensor (**B**) in those areas.

Sixteen sensors were placed on each insole to record pressure values in these areas. While no sensors were placed on the medial arch area of the foot as most people exert very low/no pressure on that area due to it is arch shape [49]. Smart insole data were collected from 16 FSRs/piezo-electric sensors. Sixteen inputs were multiplexed to one output through a 16-to-1 multiplexer and applied to an ADC input of the microcontroller then sent to host computer. All subjects were asked to place the sensor's insole inside their shoes, then placing another layer of insole on top of it to ensure comfort of the subject while walking. The acquisition and transmission circuit were connected through a conductive pathway that can help in minimizing the size of the wire and avoiding any electrical hazard. The insoles were worn by the subject inside his/her own shoe while the acquisition and transmission circuits were placed inside a 7cm × 7cm box attached to the subject's leg by an adhesive strap belt while acquiring the data. Acquired data were sent via Bluetooth to a computer, where they were plotted and analyzed.

3.3.1. FSR Insole Characterization

Twelve healthy subjects (Table 2) were asked to walk a straight 10 m walkway with self-selected cadence six times with an average walking speed of 3–4 mile per hour (MPH) and data acquired at 60 Hz sampling frequency using the smart insole made up of 16 FSRs (Figure 7). On treadmills, participants are restricted to walk in straight line as direction changes and turning cannot be realized; however, in the proposed study, the user walked freely in a 10-m walkway and they were asked to walk in a corridor which has a length of 10m and width of 1.5 m and they did not need to walk completely in a straight path and the user can walk in self-cadence, which is not possible on a treadmill. Subjects were asked to place the smart insole in their shoe while wearing cotton socks to avoid any sweat leakage that might damage the sensors or affect data acquisition from the sensors. Although walking speed is an important factor in some applications, it is not needed in many gait studies where the main focus is to detect the vertical ground reaction forces and asses the gait variables. The statistical gait variables were the symmetry between both feet, percentage of different phases (stance and swing phase) and sub phases (heel strike, mid-stance, toe off etc.) in a full gait cycle. Those statistical variables were used in various studies including sports or medical applications for gait analysis, without the need for walking speed measurement. However, the walking speed was recorded to see the impact of walking speed in the vGRF for a gait cycle. The FSR data were converted into force values by the relationship obtained in the calibration stage. Then 16 sensors' data were added at each time instance to obtain one value that represents the full force exerted by the body while walking (i.e., vGRF).

Table 2. Demographic variables of participants.

Number of Subjects	Age (Year)	Weight (kg)	Height (cm)	Body Mass Index (kg/m^2)	Gender
7	30.1 ± 13.1	77.3 ± 21.2	159.8 ± 4.9	30.3 ± 7.9	Female
5	52.3 ± 4.8	83.5 ± 3.34	172.7 ± 11.7	28.3 ± 2.9	Male

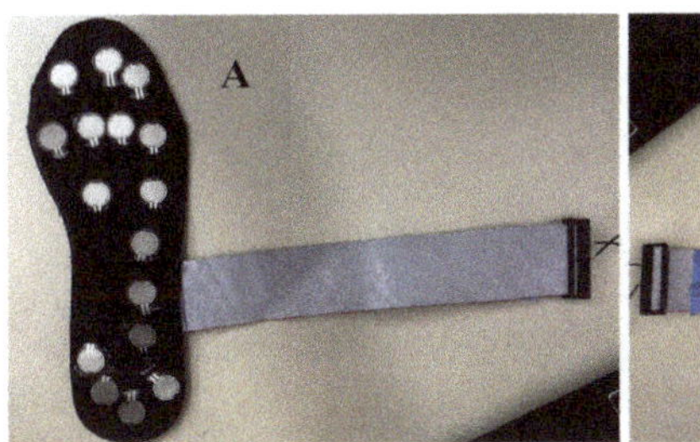
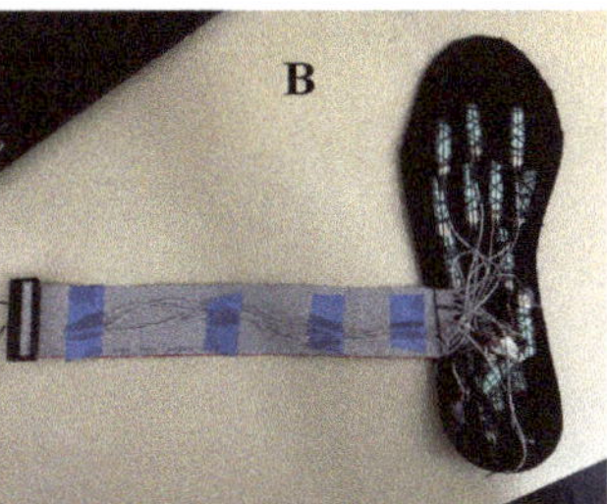

Figure 7. Smart Insole using FSR sensor: top (**A**) and bottom (**B**).

3.3.2. Piezo-Electric Insole Characterization

A similar test was carried out with the piezo-electric sensor based smart insoles (Figure 8). Three subjects were asked to walk in a 10 m walkway in normal cadence, with three trials carried out by each subject. The data were acquired with a sampling frequency of 60 Hz.

Figure 8. (**A**) Piezo insole with 16 piezo sensors, (**B**) additional insole layer placed on top on piezo insole to ensure comfortability.

3.4. Performance Evaluation of the Prototype System

A commercial F-scan smart insole system (Figure 9A) was used to validate the designed insole. The F-scan system is one of the best insoles currently available on the market. The insole comes with ultra-thin (0.18 mm) flexible printed circuit with 960 sensing nodes. Each sensing element was recorded with 8-bit resolution with a scanning speed up to 750 Hz. However, the overall cost of the system is 13,000 $ for the wired system and 17,000 $ for the wireless system. On the other hand, the instrumented insole costs only ~500 $. Usually, the vGRF peak is around ±10% of the subject's weight. Therefore, using F-scan software, data collected from each subject was calibrated based on subject's weight. The user needs to stand on one foot applying his/her full weight on the insole for 4 to 5 s, then the average applied weight was calculated. If the value obtained was less than the subject's weight, the F-scan software adjusted the output by a multiplication factor. Similar approach was used in the prototyped FSR insole as well. Figure 9B,C show the F-scan and prototyped system worn by the same subject to compare the vGRF signal acquired by the individual system.

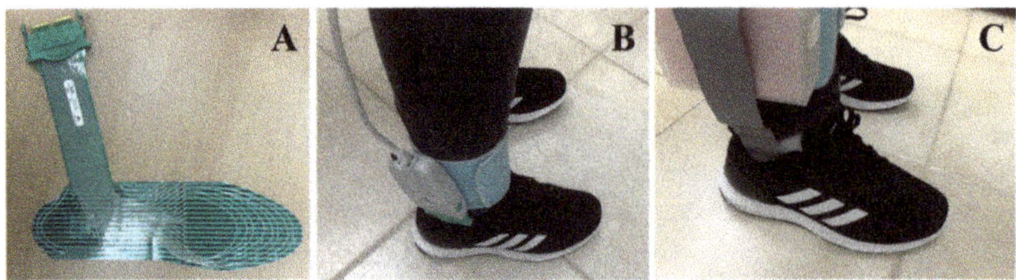

Figure 9. F-scan commercial system (**A**), F-scan system worn by Subject 01 (**B**) and FSR-based prototype system worn by Subject 01 (**C**).

4. Analysis

This section explains the mathematical calculations and analyses used for the sensor calibrations and insole characterization.

4.1. Sensors' Calibration

4.1.1. FSR Sensor Calibration

The FSR sensors exhibits resistance change in correspondence to the applied force. Therefore, a voltage divider circuit was used to convert the resistance changes to voltage values to be acquired by microcontroller.

$$V_{out} = V_{CC} \times \frac{R}{R + FSR} = 5V \times \frac{11\,k\Omega}{11\,k\Omega + FSR} \tag{1}$$

As the applied force increases, the FSR resistance also decreases, showing an increased output voltage according to Equation (1). The acquired voltages were then converted to their equivalent FSR resistance values by substitution of Equation (1).

$$FSR = \frac{5V \times 11k\Omega}{V_{out}} - 11k\Omega \tag{2}$$

4.1.2. Piezo-Electric Sensor Calibration

The piezo-electric sensor generates high voltage values, as high as 20 V with weights less than 5 kg, which requires using a voltage divider circuit before data acquisition by microcontroller.

$$V_{max\ input} = Voltage\ divider\ Gain \times V_{Piezo\ max} \tag{3}$$

$$\Rightarrow Voltage\ divider\ Gain = \frac{V_{max\ input}}{V_{Piezo\ max}} = \frac{V_{CC}}{V_{Piezo\ max}} = \frac{5V}{20V} = 0.25 \tag{4}$$

Therefore, the voltage divider circuit were chosen as follows:

$$V_{out} = \frac{R1}{R1 + R2} \times V_{Piezo} = \frac{3\,M\Omega}{3\,M\Omega + 9\,M\Omega} \times V_{Piezo} = 0.25 V_{Piezo} \tag{5}$$

Substituting the maximum piezo voltage in Equation (5) gives:

$$V_{out\ max} = 0.25 V_{Piezo\ max} = 0.25(20V) = 5V \tag{6}$$

This ensures that maximum microcontroller input voltage (5 V) was not exceeded. The acquired voltages were then converted to their equivalent Piezo sensor voltage outputs by subject substitution of Equation (5).

$$V_{Piezo} = \frac{1}{0.25} \times V_{out} = 4 V_{out} \tag{7}$$

There are different equivalent electrical models for the piezo-electric sensors [28]. A simplified common model is a voltage source/generator with a capacitance, which was used in this study. Usually, the capacitance values are in Nano Farad range. The equivalent capacitance is typically measured using a parallel connection of a capacitance meter to the sensor. Connecting the piezo-electric sensor to the voltage divider circuit forms a first order high-pass filter. Therefore, high resistance values in mega ohms were used to ensure that most of the generated frequencies by the applied forces would pass. Assuming equivalent capacitance of piezo-electric sensor equal to 9 nF, the cut-off frequency can be written as:

$$f_{cutt-off} = \frac{1}{2\pi RC} = \frac{1}{2\pi(3M + 9M)9nF} = 1.47 Hz \tag{8}$$

Apart from DC and very low frequency components, other signal components were expected to be applied to the MCU input. AD620AN instrumentational amplifiers were used to amplify the low amplitude load cell signals, before it was applied to the microcontroller. The load cells give an output of maximum 40 mV, which can be amplified to the full-scale range of the analog channel. Therefore, the gain of the amplifier and the amplifier gain resistor were chosen as follows:

$$G = \frac{V_{cc}}{V_{load\ max}} = \frac{5\ V}{40 mV} = 125 \tag{9}$$

$$R_G = \frac{49.9k}{G - 1} = \frac{49.9k\Omega}{124} = 402\ \Omega \tag{10}$$

4.1.3. MEMS Sensor Calibration

As mentioned previously, the MEMS generates positive or negative amplitude signals based on the applied force in x, y or z directions. This requires an offset circuit along with a voltage divider circuit to reduce the signal amplitude. It is assumed the piezo-vibration output can go up to 10 V with the maximum applied force.

$$V_{max\ input} = Voltage\ divider\ Gain \times V_{Piezo\ max} \tag{11}$$

$$Voltage\ divider\ Gain = \frac{V_{max\ input}}{V_{Piezo\ max}} = \frac{V_{CC}/2}{V_{Piezo\ max}} = \frac{5V/2}{10V} = 0.25 \tag{12}$$

Therefore, the voltage divider circuit were chosen as follows:

$$V_{out} = \frac{R1}{R1 + R2} \times V_{Piezo} = \frac{3\ M\Omega}{3\ M\Omega + 9\ M\Omega} \times V_{Piezo} = 0.25 V_{Piezo}, \tag{13}$$

Substituting the maximum and minimum piezo voltage in Equation (13) gives:

$$V_{out\ max/min} = 0.25 V_{Piezo\ max/min} = 0.25(\pm 10V) = \pm 2.5V \tag{14}$$

The next step is to add an offset of 1/2Vcc to ensure that the signal was within 0 V to V_{cc} range.

4.2. Piezo-Electric Sensor Response

The piezo-electric sensors can detect the applied forces efficiently, by converting the mechanical movements into electrical signals. However, the movements need to be dynamic. The piezo-electric sensor generated electrical pulses that mimicked the applied mechanical movement. If the mechanical movement was a fast press and release of finger on the active area of the piezo-electric sensor, the pulse was shrunk to an impulse-liked shape.

On the other hand, if a gentle force was applied by a slow press and remove by the palm of a hand, the generated signal had irregular pulse shape with longer duration compared to the fast finger press. Even though the piezo-electric sensor's output can mimic dynamically changing force, it fails to detect a static force. Therefore, when the applied force is a mixture of dynamic and static force such as smart insole application, the piezo-electric sensors cannot be used to acquire static pressure. However, the piezo-electric sensors can be used to detect heel strike or toe off with good accuracy. Figure 10 illustrates the individual sensor output for the different applied forces.

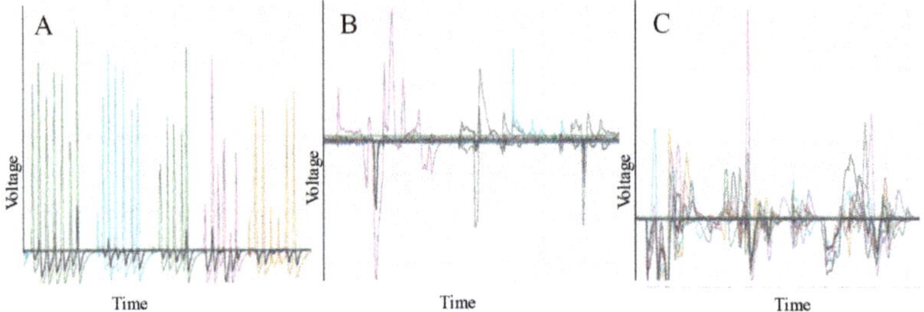

Figure 10. Piezo insole sensors output with Arduino serial plotter: (**A**) fast finger press and release, (**B**) slow palm press and release, (**C**) sensors output for two-step walking.

Figure 11 shows the output from a single piezo-electric sensor of an insole for few gait cycles. When a force was applied vertically on the sensor's active area (ceramic), it compressed, exerting an electrical impulse with a positive peak that mimicked the mechanical force applied. The electrical signal went back to zero. As the applied force was released, the signal continued to some negative values as the piezo ceramic bounced to the opposite direction of the applied force. Finally, the signal returned to zero. The microcontroller clipped the negative part of the signal. However, some part of the negative signal was still there, due to the offset added in the acquisition circuit as illustrated in Figure 11.

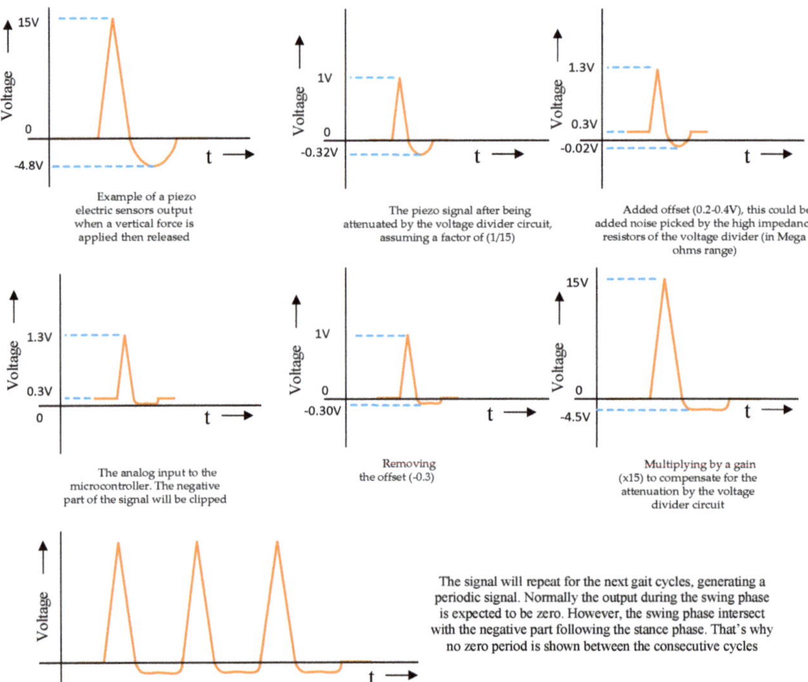

Figure 11. Mimicking piezoelectric sensor output during gait cycle.

5. Results and Discussion

This section illustrates and discuss the results obtained from calibration and characterization tests.

5.1. Sensors' Calibration

5.1.1. FSR Sensor Calibration

Three calibration trials were undertaken for one FSR sensor from Interlink Electronics [22], following the calibration procedure explained previously; 500 g weights where placed one by one every 3–4 s until it reached 5000 g, followed by unloading process from 5000 g down to 0 g. In the loading experiment, output voltage from the voltage divider circuit showed increasing values reflecting the decrease in FSR resistance as shown in Figure 12A. When the applied weight was constant, the output voltage remained constant with small variations.

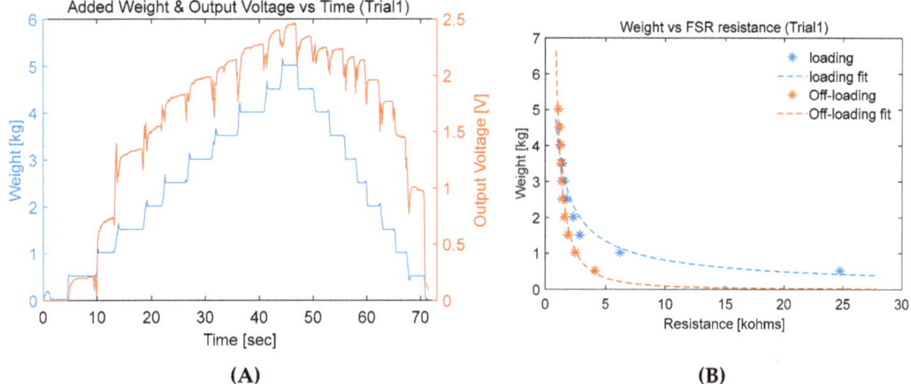

Figure 12. FSR calibration test (**A**) applied weight and FSR circuit output vs. time (**B**) applied weight vs. FSR resistance.

In addition, if the constant weight was kept for a longer time (a few minutes), the sensor voltage stabilized to a steady value. However, the aim of this study was to investigate the dynamic response of the FSR. Therefore, the average output voltage for the sample were calculated and plotted against the corresponding applied weights. Figure 12B shows the plotted data with the fitted waveform. The calibration showed slight difference between the loading and unloading curves, which was expected due to the hysteresis behavior of FSRs. However, the error was caused by the FSR hysteresis, which can be neglected, as the difference was not significantly high. This can be justified if the response from the smart insole using FSR sensors resembles typical vGRF reported in the literature.

Off-Loading tests best fit relations:

$$Weight_{Trial1} = 5035.2 * Resistance^{-1.72}$$

$$Weight_{Trial2} = 3436.5 * Resistance^{-1.895} \quad Weight_{Trial3} = 8111.8 * Resistance^{-2.589}$$

It is evident that the first and third trial relationships were close to each other (Figure 13). Therefore, either of them can be chosen for the FSR insole. The second trial showed a steeper curve due to higher hystersis error.

For each of the three trials, loading and off-loading relationships where obtained. Loading tests best fit relations:

$$Weight_{Trial1} = 4233.3 * Resistance^{-0.714}$$
$$Weight_{Trial2} = 3657.4 * Resistance^{-1.022}$$
$$Weight_{Trial3} = 5117.4 * Resistance^{-0.748}$$

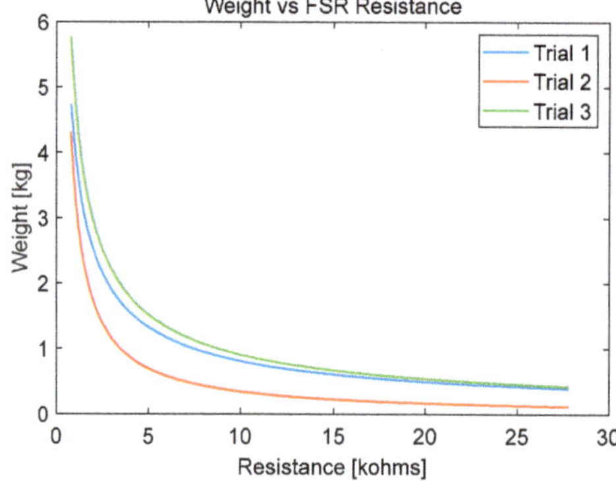

Figure 13. FSR calibration test: best fit curves between applied weight and FSR resistance for three trials.

5.1.2. Piezo-Electric Sensor Calibration

Two piezoelectric sensors were used in the calibration process. Three trials were conducted on the first sensors with four trials for the second sensor. The piezoelectric sensors showed a linear relationship with the applied weights (Figure 14).

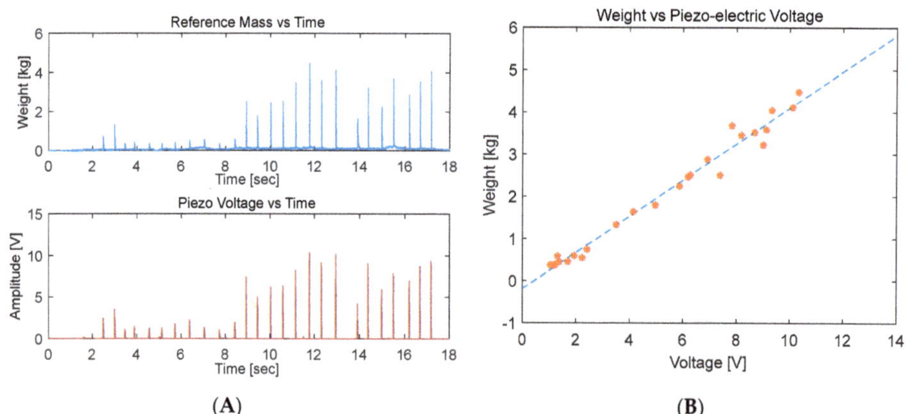

Figure 14. Piezo-electric sensor calibration test: (**A**) applied weight vs. time and piezoelectric output voltage vs. time (**B**) applied weight vs. piezoelectric output voltage.

The second and fourth trials for the 2nd piezo sensor had different slopes compared to the remaining trials. This could be related to the calibration process itself, as the weights were applied by fast presses and releases on the active area of the sensor. Therefore, applying the force on the exact same areas is not guaranteed between successive readings. This issue can be overcome by using a machine to apply the weights. However, this would defeat the purpose of the study in providing a low-cost setup (Figure 15).

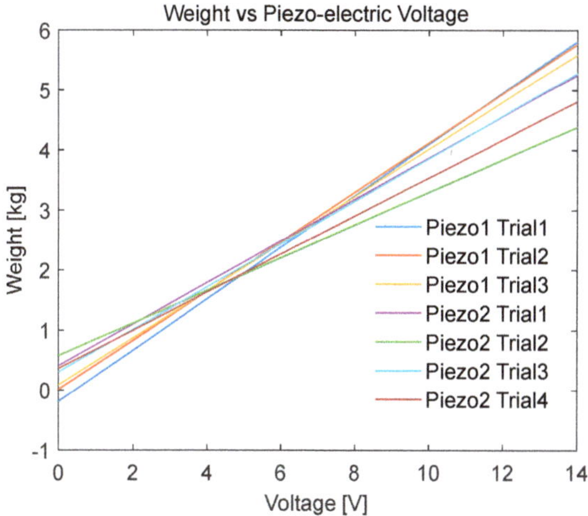

Figure 15. Piezoelectric sensor calibration test: seven trials shows relationship between the applied weight and piezoelectric output voltage.

Obtained lines of best fit:

$$Weight_{Piezo1Trial1} = 0.42867 * PiezoVoltage - 0.19123$$

$$Weight_{Piezo1Trial2} = 0.41110 * PiezoVoltage + 0.0081012$$

$$Weight_{Piezo1Trial3} = 0.39321 * PiezoVoltage + 0.084656$$

$$Weight_{Piezo2Trial1} = 0.34619 * PiezoVoltage + 0.4105$$

$$Weight_{Piezo2Trial2} = 0.27242 * PiezoVoltage + 0.57351$$

$$Weight_{Piezo2Trial3} = 0.35564 * PiezoVoltage + 0.30325$$

$$Weight_{Piezo2Trial4} = 0.31765 * PiezoVoltage + 0.36416$$

5.1.3. MEMS Sensor Calibration

Twenty different calibration trials were conducted on a piezo-vibration sensor. However, high repeatability error persisted, making it difficult to obtain a clear relation between sensor output voltage and the applied weight. The applied weight showed a direct proportional relation with output voltage for some successive readings and an inverse relation with some other successive readings. This is because of the MEMS sensitivity to the applied force in 3-D space (x, y or z directions). It generates 1-D output voltage with positive or negative amplitude depending on the applied force in certain direction. Therefore, if the applied force is a summation of forces in 2 or 3 axes, the output voltage might go to zero or attenuated with the addition of different sign amplitudes.

A linear relation was not clearly obtained by the application of vertical forces, as the applied force might not be applied in one axis only (Figure 16).

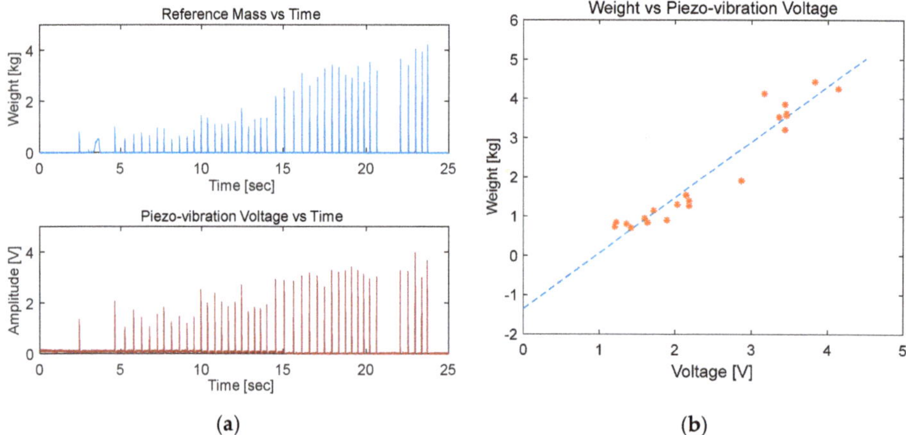

Figure 16. MEMS sensor calibration test: (**A**) applied weight vs. time and MEMS sensor output voltage vs. time (**B**) applied weight vs. MEMS sensor output voltage.

The mathematical relations obtained in the calibration phase cannot be used to design a piezo-vibration sensor-based smart insole, since the applied force in gait can be in any of the x, y or z directions (Figure 17). Therefore, the piezo-vibration sensor was discarded from the sensor list for designing smart insole. However, it can be utilized in other biomedical applications where the force directions are limited to a certain axis or a fixed plane. Moreover, it can be used to detect initial timing of the applied force. The lines of best fit obtained were:

$$Weight_{Trial1} = 1.4145 * PiezoVoltage - 1.3447$$

$$Weight_{Trial2} = 1.5215 * PiezoVoltage - 1.8581$$

$$Weight_{Trial3} = 0.80343 * PiezoVoltage - 0.22721$$

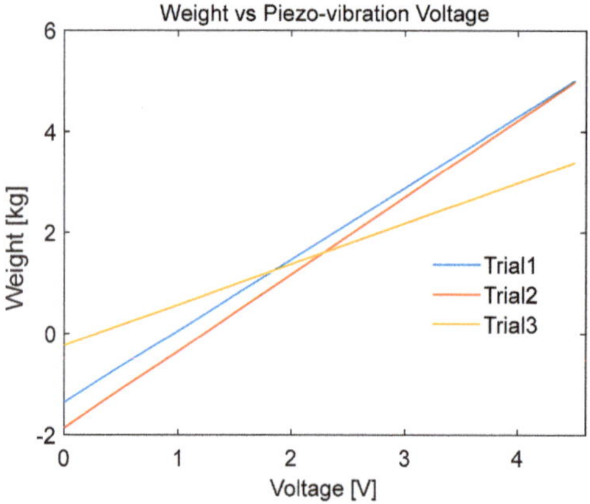

Figure 17. MEMS sensor calibration test: 3 best trials shows relationship between the applied weight and MEMS sensor's output voltage.

5.2. Insole Charecterization

5.2.1. FSR-Based Insole Characterization

The gait cycle of 12 subjects were recorded while walking on a 10 m walkway in self-selected walking manner. Each subject had 6 trials recorded, where the first and last few (1 to 3) cycles were discarded from each trial, and the remaining part of the gait cycles for both feet were considered for analysis. The gait cycle of one of the subjects is analyzed in the following lines, illustrating a simple analysis technique that can be replicated in different application by researchers working with wearable insoles.

In normal gait cycles both heel peak (first peak) and toe off (peak) must show close values, with both feet having symmetrical signals. Even though the right foot signal showed close peak values (Figure 18), the left foot signals showed a big variance between the heel-strike and toe-off peaks. This can be explained, by the sensitivity difference between the FSRs of the insole and their hysteresis effect. As explained earlier, this issue was mitigated by some research teams using a regression models that calibrates the FSR insole readings against a reference signal, recorded simultaneously in motion analysis labs [34,35]. This expensive approach can be neglected in some applications, where the quality of the acquired signal is sufficient to achieve the desired goal. For instance, smart detection application, where the machine-learning algorithm can differentiate between different groups of people even with low- or medium-quality recorded gait cycles (vGRF).

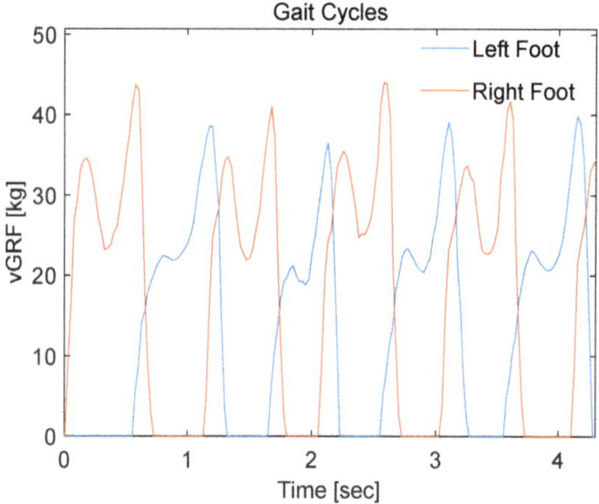

Figure 18. Gait cycles readings for left and right foot with FSR smart insole.

The full gait record was segmented into distinct gait cycles. Then it was resampled into to 512 sample. Segmentation is a common practice to facilitate the comparison between all the gait cycles. The segmented gait cycles are used in smart detection algorithms where segments of equal length are used to train specific machine learning algorithms to classify different groups of people based on their gait. In addition, statistical data of the segmented cycles such as mean, standard deviation, time to peaks, and percentage of stance phase in a full cycle/stride (stance phase plus swing phase) can be utilized as a gait analysis tool in sports and medical applications.

The segmentation was carried out by a customized MATLAB code that detects groups of consecutive non-zero samples. Then it segments those signals into individual stance phases, each starting with a heel-strike and ending with a toe off. Figure 18 shows the first 10-m trial of one of the participants, where four gait cycles were extracted after excluding the first and last two gait cycles. Left foot vGRF was segmented into four stance phases (Figure 19A).

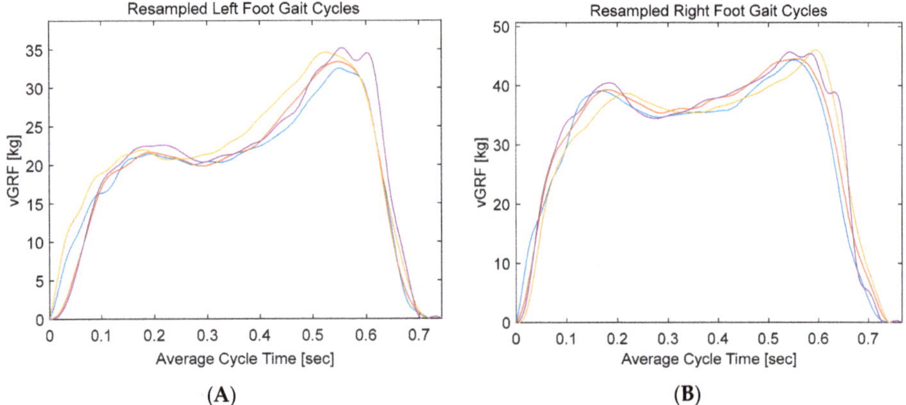

Figure 19. (**A**) Segmented left-foot gait cycles, (**B**) segmented right-foot cycles.

The data were sampled with a sampling rate of 60 samples/second, where each segment (stance phase) takes around 0.7 s. Therefore, each segment consists of around 42 samples, which were then resampled into 512 samples. The mean values and standard deviations of each of the 512 samples with respect to the 4 segmented signals were calculated. Then the mean values along with the deviation from the means (means plus and minus the deviation) for the left foot was calculated and plotted (Figure 20). Similar steps were repeated with left foot vGRF (Figure 20). This provides an illustrative figure that can be used by in different sport and medical applications to asses walking behaviors or complications.

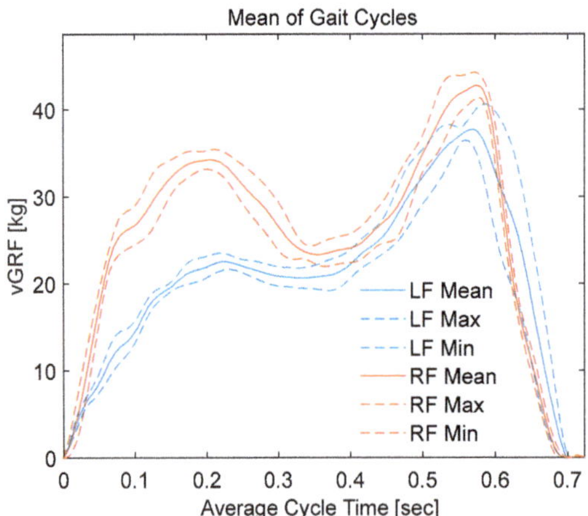

Figure 20. Means and standard deviations of gait cycles; blue curves represents the mean gait value of the left foot with dashed line representing the deviation from the mean, while orange curves represents the mean gait value of the right foot with dashed line representing the deviation from the mean value.

The vGRF of a subject mainly depends on his/her health condition and the footwear used. In this study, all participants were advised to wear comfortable walking shoes avoiding high-heel shoes, especially for female subjects. This ensured that all subjects went through similar condition while conducting the experiment. It was observed that the collected data did not show any significant statistical difference based on gender.

5.2.2. Piezoelectric Sensor Based-Insole Characterization

Three subjects participated in the piezo-electric insole test in the same manner as the testing of FSR based insole. The piezoelectric insoles were expected to detect the gait cycle, with impulse signals in heel-area sensors during the heel strike phase and lower amplitude impulses from all the sensors during the mid-stance phase. Finally, impulse signals from the toe and metatarsal heads sensors were taken in the toe-off phase. However, the readings were not promising, showing single irregular shape pulses per sensor for each individual gait cycle. The addition of different sensors output showed periodic impulses, one impulse per period (Figure 21). This indicates that the full stance period was detected as one event only. Meanwhile, the correct vGRF must show two distinct peaks between the mid-stance phase, summing up to three main phases: heel strike, mid stance and toe off. The rigid nature of the piezo sensor made it difficult to detect different gait phases. Therefore, it can be summarized that it is not suitable for the smart insole application which requires to produce reliable vGRF signal due to gait.

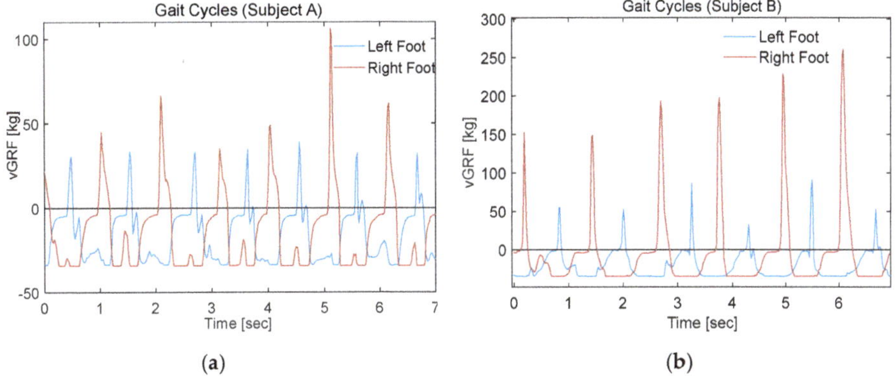

Figure 21. Gait cycles for left and right foot with piezoelectric smart insole (**a**) subject 1, (**b**) subject 2.

In this study, the authors have characterized three samples from each sensor category randomly; however, the smart insole was implemented using 16-sensors. Therefore, it was expected that there would be a small variation of the vGRF recorded from the smart insole in different trials and in different subjects. However, Figure 20 clearly depicts that the vGRF from an individual foot has a unique pattern and this finding matches with the vGRF recorded by the commercial smart insole and force plate. This reflects the fact that the smart insole designed using FSR is capable of acquiring vGRF reliably and the designed system is robust enough to adapt to the age, gender and body mass index (BMI) variation of the participants. However, carbon piezoresistive material (like Velostat), which authors have tested in preliminary experiments (not reported here in order to avoid unnecessary length of the manuscript), showed very high hysteresis and this type of material is not suitable for human dynamicity monitoring. On the other hand, piezoelectric sensors can monitor dynamic pressure variation however, they are very sensitive to small pressure change and incapable to reliably produce mean vGRF. Moreover, the vGRF changed over trials and over subjects significantly and, therefore, temporal feature of vGRF cannot be identified using piezoelectric sensor-based smart insole.

5.3. *Performance Evaluation of the FSR-Based System*

Comparing the mean and standard deviation of vGRF for a gait cycle of the same subject recorded using the two systems, commercial F-scan and the proposed FSR insoles, it can be seen that both showed good quality signals except slight differences in peak values (Figure 22). The FSR insole showed smaller vGRF during left-foot heel strike phase compared to the F-scan insole. This is an expected behavior as each sensor is somewhat unique due to the manufacturing process and we cannot calibrate individual sensors, which can lead to some variation. In addition, due to the presence of an insole, the sensitivity of some FSRs decreases more than the others in the shoe. To mitigate this problem, a highly uniform pressure should be applied across individual sensors. Each sensor should produce uniform output. When this is not the case for a specific sensor, the software should determine a unique scale factor to compensate for the output variation. Currently, there are a few companies such as T-scan, that provide a special piece of equipment (equilibration device) which applies a uniform pressure on the full insole using a thin flexible membrane to perform such calibration. Moreover, compared to the F-scan system, the FSR readings showed smaller differences between vGRF peaks and mid stance values. This was mainly due to the superior number of sensors for the F-scan system (960 sensing areas) compared to the proposed insole (16 FSRs). In addition, the F-scan sensing elements were uniformly distributed on the full foot area, while the FSR sensors were placed on the foot areas where most of the pressure is exerted with no sensors placed on the low-pressure areas (medial arch). Adding a few FSR sensors to the medial arch can improve the quality of the signal obtained, especially for subjects with flat foot, who exerts considerable amount of pressure on medial arch areas.

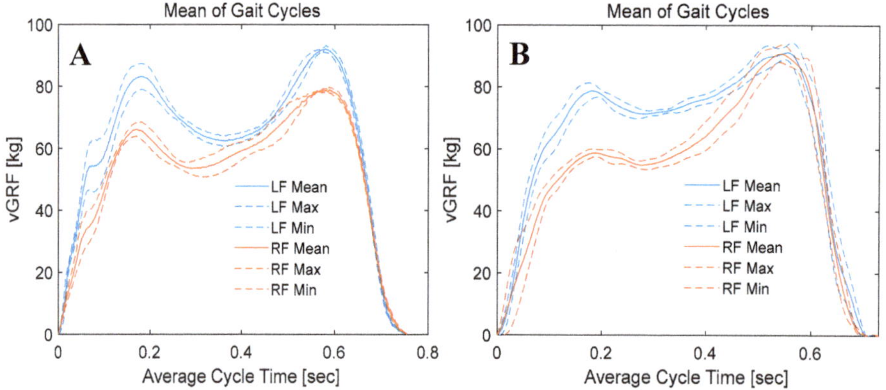

Figure 22. Comparison between the mean and standard deviation of vertical ground reaction forces (vGRF) from left (blue) and right (orange) foot using F-scan system (**A**) and FSR-system (**B**).

6. Conclusions

In this study, the authors have proposed and designed low-cost calibration setups for calibrating three different force sensors: FSR, ceramic piezoelectric and flexible piezoelectric sensors. The experiments conducted showed the effectiveness of the proposed setup in calibrating FSR and piezoelectric sensors, which are mainly affected by 1D force. It was found that the flexible piezoelectric sensors were performing poor in terms of calibration due to their sensitivity to 3D forces. Special force calibration machines are required to control the applied force in x, y or z directions. In addition, a systematic procedure for designing and characterizing two different smart insoles were illustrated. The vGRF signal acquired and segmented to obtain mean vGRF and its standard deviation for a gait cycle were calculated, which can be used to measure different statistical metrics (such as mean standard deviation, time to peak, etc.) that can help in assessing the walking behavior of athletes, patients or normal people. The FSR-based smart insole was able to acquire high quality vGRF for different gait cycles. On the other hand, the piezoelectric sensor-based insole failed to detect distinct gait phases. It cannot be utilized as an alternative to FSR in smart insole application. However, the calibrated piezo sensors can be utilized in other bio-sensing technologies such as detecting the start and end of each gait cycle.

Author Contributions: Experiments were designed by M.E.H.C., N.E. and A.K. Experiments were performed and experimental data were acquired by A.M.T., S.A.-H., M.A., and S.A. Data were analyzed by A.M.T., M.E.H.C., A.K. and N.A.-E. All authors were involved in interpretation of data and writing the paper. All authors have read and agreed to the published version of the manuscript.

Funding: This research was partially funded by Qatar National Research Foundation (QNRF), grant number NPRP12S-0227-190164 and Research University Grant DIP-2018-017. The publication of this article was funded by the Qatar National Library.

Acknowledgments: The authors would like to thank Engr. Ayman Ammar, Electrical Engineering, Qatar University for helping in printing the printed circuit boards (PCBs).

Conflicts of Interest: The authors declare no conflict of interest.

References

1. Kowalski, E.; Catelli, D.S.; Lamontagne, M. Side does not matter in healthy young and older individuals–Examining the importance of how we match limbs during gait studies. *Gait Posture* **2019**, *67*, 133–136.

2. Tao, W.; Liu, T.; Zheng, R.; Feng, H. Gait analysis using wearable sensors. *Sensors* **2012**, *12*, 2255–2283. [CrossRef] [PubMed]

3. Howell, A.M.; Kobayashi, T.; Hayes, H.A.; Foreman, K.B.; Bamberg, S.J.M. Kinetic gait analysis using a low-cost insole. *IEEE Trans. Biomed. Eng.* **2013**, *60*, 3284–3290. [CrossRef] [PubMed]

4. Bamberg, S.J.M.; Benbasat, A.Y.; Scarborough, D.M.; Krebs, D.E.; Paradiso, J.A. Gait analysis using a shoe-integrated wireless sensor system. *IEEE Trans. Inf. Technol. Biomed.* **2008**, *12*, 413–423. [CrossRef] [PubMed]

5. De Rossi, S.; Lenzi, T.; Vitiello, N.; Donati, M.; Persichetti, A.; Giovacchini, F.; Vecchi, F.; Carrozza, M.C. Development of an in-shoe pressure-sensitive device for gait analysis. In Proceedings of the 2011 Annual International Conference of the IEEE Engineering in Medicine and Biology Society, Boston, MA, USA, 30 August–3 September 2011; IEEE: Piscataway, NJ, USA, 2011.

6. Smart Tracking and Wearables: Techniques in Gait Analysis and Movement in Pathological Aging. Available online: https://www.intechopen.com/books/smart-healthcare/smart-tracking-and-wearables-techniques-in-gait-analysis-and-movement-in-pathological-aging (accessed on 9 February 2020).

7. Gurchiek, R.D.; Choquette, R.H.; Beynnon, B.D.; Slauterbeck, J.R.; Tourville, T.W.; Toth, M.J.; McGinnis, R.S. Remote gait analysis using wearable sensors detects asymmetric gait patterns in patients recovering from ACL reconstruction. In Proceedings of the 2019 IEEE 16th International Conference on Wearable and Implantable Body Sensor Networks (BSN), Chicago, IL, USA, 19–22 May 2019; IEEE: Piscataway, NJ, USA, 2019.

8. Brognara, L.; Palumbo, P.; Grimm, B.; Palmerini, L. Assessing gait in parkinson's disease using wearable motion sensors: A systematic review. *Diseases* **2019**, *7*, 18. [CrossRef] [PubMed]

9. Mills, P.M.; Morrison, S.; Lloyd, D.G.; Barrett, R.S. Repeatability of 3D gait kinematics obtained from an electromagnetic tracking system during treadmill locomotion. *J. Biomech.* **2007**, *40*, 1504–1511. [CrossRef]

10. Huang, E.; Sharp, M.T.; Osborn, E.; MacLellan, A.; Mlynash, M.; Kemp, S.; Buckwalter, M.S.; Lansberg, M.G. Abstract TP173: Feasibility and Utility of Home-Based Gait Analysis Using Body-Worn Sensors. *Stroke* **2019**, *50*, ATP173. [CrossRef]

11. Liu, T.; Inoue, Y.; Shibata, K. A wearable ground reaction force sensor system and its application to the measurement of extrinsic gait variability. *Sensors* **2010**, *10*, 10240–10255. [CrossRef] [PubMed]

12. Razak, A.; Hadi, A.; Zayegh, A.; Begg, R.K.; Wahab, Y. Foot plantar pressure measurement system: A review. *Sensors* **2012**, *12*, 9884–9912. [CrossRef] [PubMed]

13. Alam, U.; Riley, D.R.; Jugdey, R.S.; Azmi, S.; Rajbhandari, S.; D'Août, K.; Malik, R.A. Diabetic neuropathy and gait: A review. *Diabetes Ther.* **2017**, *8*, 1253–1264. [CrossRef] [PubMed]

14. Farahpour, N.; Jafarnezhad, A.; Damavandi, M.; Bakhtiari, A.; Allard, P. Gait ground reaction force characteristics of low back pain patients with pronated foot and able-bodied individuals with and without foot pronation. *J. Biomech.* **2016**, *49*, 1705–1710. [CrossRef] [PubMed]

15. Sacco, I.C.; Picon, A.P.; Macedo, D.O.; Butugan, M.K.; Watari, R.; Sartor, C.D. Alterations in the lower limb joint moments precede the peripheral neuropathy diagnosis in diabetes patients. *Diabetes Technol. Ther.* **2015**, *17*, 405–412.

16. Dabbah, M.A.; Graham, J.; Petropoulos, I.N.; Tavakoli, M.; Malik, R.A. Automatic analysis of diabetic peripheral neuropathy using multi-scale quantitative morphology of nerve fibres in corneal confocal microscopy imaging. *Med. Image Anal.* **2011**, *15*, 738–747. [CrossRef] [PubMed]

17. International Diabetes Federation: IDF Diabetes Atlas. Available online: https://www.idf.org/e-library/epidemiology-research/diabetes-atlas/19-atlas-6th-edition.html (accessed on 9 February 2020).

18. Dworkin, R.H.; Malone, D.C.; Panarites, C.J.; Armstrong, E.P.; Pham, S.V. Impact of postherpetic neuralgia and painful diabetic peripheral neuropathy on health care costs. *J. Pain* **2010**, *11*, 360–368. [CrossRef]

19. Dyck, P.J.; Overland, C.J.; Low, P.A.; Litchy, W.J.; Davies, J.L.; Dyck, P.J.B.; O'brien, P.C. Signs and symptoms versus nerve conduction studies to diagnose diabetic sensorimotor polyneuropathy: Cl vs. NPhys trial. *Muscle Nerve* **2010**, *42*, 157–164.

20. Taksande, B.; Ansari, S.; Jaikishan, A.; Karwasara, V. The diagnostic sensitivity, specificity and reproducibility of the clinical physical examination signs in patients of diabetes mellitus for making diagnosis of peripheral neuropathy. *J. Endocrinol. Metab.* **2011**, *1*, 21–26. [CrossRef]

21. Watari, R.; Sartor, C.D.; Picon, A.P.; Butugan, M.K.; Amorim, C.F.; Ortega, N.R.; Sacco, I.C. Effect of diabetic neuropathy severity classified by a fuzzy model in muscle dynamics during gait. *J. Neuroeng. Rehabil.* **2014**, *11*, 11. [CrossRef]

22. England, J.; Gronseth, G.; Franklin, G.; Carter, G.; Kinsella, L.; Cohen, J.; Asbury, A.; Szigeti, K.; Lupski, J.; Latov, N. Practice Parameter: Evaluation of distal symmetric polyneuropathy: Role of laboratory and genetic testing (an evidence-based review) Report of the American Academy of Neurology, American Association of Neuromuscular and Electrodiagnostic Medicine, and American Academy of Physical Medicine and Rehabilitation. *Neurology* **2009**, *72*, 185–192.

23. Zhang, Q.; Wang, Y.L.; Xia, Y.; Wu, X.; Kirk, T.V.; Chen, X.D. A low-cost and highly integrated sensing insole for plantar pressure measurement. *Sens. Bio-Sens. Res.* **2019**, *26*, 100298. [CrossRef]

24. Zhang, W.; Yin, B.; Wang, J.; Mohamed, A.; Jia, H. Ultrasensitive and wearable strain sensors based on natural rubber/graphene foam. *J. Alloys Compd.* **2019**, *785*, 1001–1008. [CrossRef]

25. Sengupta, D.; Pei, Y.; Kottapalli, A.G.P. Ultralightweight and 3D Squeezable Graphene-Polydimethylsiloxane Composite Foams as Piezoresistive Sensors. *ACS Appl. Mater. Interfaces* **2019**, *11*, 35201–35211. [CrossRef] [PubMed]

26. Shu, L.; Hua, T.; Wang, Y.; Li, Q.; Feng, D.D.; Tao, X. In-shoe plantar pressure measurement and analysis system based on fabric pressure sensing array. *IEEE Trans. Inf. Technol. Biomed.* **2010**, *14*, 767–775. [PubMed]

27. Giovanelli, D.; Farella, E. Force sensing resistor and evaluation of technology for wearable body pressure sensing. *J. Sens.* **2016**, *2016*. [CrossRef]

28. Feasibility of Force Detection in 3D Printed Flexible Material Using Embedded Sensors. Available online: https://www.spiedigitallibrary.org/conference-proceedings-of-spie/10970/109702F/Feasibility-of-force-detection-in-3D-printed-flexible-material-using/10.1117/12.2514051.short?SSO=1 (accessed on 9 February 2020).

29. Heng, W.; Pang, G.; Xu, F.; Huang, X.; Pang, Z.; Yang, G. Flexible Insole Sensors with Stably Connected Electrodes for Gait Phase Detection. *Sensors* **2019**, *19*, 5197. [CrossRef]

30. Lee, W.; Hong, S.-H.; Oh, H.-W. Characterization of Elastic Polymer-Based Smart Insole and a Simple Foot Plantar Pressure Visualization Method Using 16 Electrodes. *Sensors* **2019**, *19*, 44. [CrossRef]

31. Theodosiou, A.; Kalli, K. Recent trends and advances of fibre Bragg grating sensors in CYTOP polymer optical fibres. *Opt. Fiber Technol.* **2020**, *54*, 102079. [CrossRef]

32. Abdelhady, M.; van den Bogert, A.; Simon, D. A High-Fidelity Wearable System for Measuring Lower-Limb Kinetics and Kinematics. *IEEE Sens. J.* **2019**. [CrossRef]

33. Kim, J.; Bae, M.N.; Lee, K.B.; Hong, S.G. Gait event detection algorithm based on smart insoles. *ETRI J.* **2019**. [CrossRef]

34. Vilas-Boas, M.d.C.; Rocha, A.P.; Choupina, H.M.P.; Cardoso, M.; Fernandes, J.M.; Coelho, T.; Cunha, J.P.S. TTR-FAP Progression Evaluation Based on Gait Analysis Using a Single RGB-D Camera. In Proceedings of the 2019 41st Annual International Conference of the IEEE Engineering in Medicine and Biology Society (EMBC), Berlin, Germany, 23–27 July 2019; IEEE: Piscataway, NJ, USA, 2019.

35. Cui, X.; Zhao, Z.; Ma, C.; Chen, F.; Liao, H. A Gait Character Analyzing System for Osteoarthritis Pre-diagnosis Using RGB-D Camera and Supervised Classifier. In Proceedings of the World Congress on Medical Physics and Biomedical Engineering, Prague, Czech Republic, 3–8 June 2018; Springer: Berlin, Germany, 2018.

36. Barnea, A.; Oprisan, C.; Olaru, D. Force Sensitive Resistors Calibration for the Usage in Gripping Devices. *Diagn. Predict. Mech. Eng. Szstems DIPRE* **2012**, *3*, 18–27.

37. Parmar, S.; Khodasevych, I.; Troynikov, O. Evaluation of flexible force sensors for pressure monitoring in treatment of chronic venous disorders. *Sensors* **2017**, *17*, 1923. [CrossRef]

38. Insole Energy Harvesting from Human Movement Using Piezoelectric Generators. Available online: https://trepo.tuni.fi/handle/123456789/25768 (accessed on 9 February 2020).

39. Development of a Smart Insole System for Real-Time Detection of Temporal Gait Parameters and Related Deviations in Unilateral Lower-Limb Amputees. Available online: https://scholarlyrepository.miami.edu/cgi/viewcontent.cgi?article=1393&context=oa_theses (accessed on 9 February 2020).

40. Sacco, I.; Amadio, A. A study of biomechanical parameters in gait analysis and sensitive cronaxie of diabetic neuropathic patients. *Clin. Biomech.* **2000**, *15*, 196–202. [CrossRef]

41. Menz, H.B.; Lord, S.R.; St George, R.; Fitzpatrick, R.C. Walking stability and sensorimotor function in older people with diabetic peripheral neuropathy1. *Arch. Phys. Med. Rehabil.* **2004**, *85*, 245–252. [CrossRef] [PubMed]

42. Dementyev, A.; Hodges, S.; Taylor, S.; Smith, J. Power consumption analysis of Bluetooth Low Energy, ZigBee and ANT sensor nodes in a cyclic sleep scenario. In Proceedings of the 2013 IEEE International Wireless Symposium (IWS), Beijing, China, 14–18 April 2013; IEEE: Piscataway, NJ, USA, 2013.

43. Sacco, I.; Amadio, A. Influence of the diabetic neuropathy on the behavior of electromyographic and sensorial responses in treadmill gait. *Clin. Biomech.* **2003**, *18*, 426–434. [CrossRef]

44. Kwon, O.-Y.; Minor, S.D.; Maluf, K.S.; Mueller, M.J. Comparison of muscle activity during walking in subjects with and without diabetic neuropathy. *Gait Posture* **2003**, *18*, 105–113.

45. Akashi, P.M.; Sacco, I.C.; Watari, R.; Hennig, E. The effect of diabetic neuropathy and previous foot ulceration in EMG and ground reaction forces during gait. *Clin. Biomech.* **2008**, *23*, 584–592. [CrossRef] [PubMed]

46. Sacco, I.C.; Akashi, P.M.; Hennig, E.M. A comparison of lower limb EMG and ground reaction forces between barefoot and shod gait in participants with diabetic neuropathic and healthy controls. *BMC Musculoskelet. Disord.* **2010**, *11*, 24. [CrossRef]

47. Gomes, A.A.; Onodera, A.N.; Otuzi, M.E.; Pripas, D.; Mezzarane, R.A.; Sacco, N.; Isabel, C. Electromyography and kinematic changes of gait cycle at different cadences in diabetic neuropathic individuals. *Muscle Nerve* **2011**, *44*, 258–268.

48. Sawacha, Z.; Spolaor, F.; Guarneri, G.; Contessa, P.; Carraro, E.; Venturin, A.; Avogaro, A.; Cobelli, C. Abnormal muscle activation during gait in diabetes patients with and without neuropathy. *Gait Posture* **2012**, *35*, 101–105.

49. Zhu, H.; Wertsch, J.J.; Harris, G.F.; Loftsgaarden, J.D.; Price, M.B. Foot pressure distribution during walking and shuffling. *Arch. Phys. Med. Rehabil.* **1991**, *72*, 390–397.

Article

Photoplethysmographic Time-Domain Heart Rate Measurement Algorithm for Resource-Constrained Wearable Devices and its Implementationt

Marek Wójcikowski * and Bogdan Pankiewicz

Faculty of Electronics, Telecommunications and Informatics, Gdańsk University of Technology, 80-233 Gdańsk, Poland; bpa@eti.pg.edu.pl
* Correspondence: marwojci1@pg.edu.pl; Tel.: +48-58-347-19-74

Received: 17 February 2020; Accepted: 21 March 2020; Published: 23 March 2020

Abstract: This paper presents an algorithm for the measurement of the human heart rate, using photoplethysmography (PPG), i.e., the detection of the light at the skin surface. The signal from the PPG sensor is processed in time-domain; the peaks in the preprocessed and conditioned PPG waveform are detected by using a peak detection algorithm to find the heart rate in real time. Apart from the PPG sensor, the accelerometer is also used to detect body movement and to indicate the moments in time, for which the PPG waveform can be unreliable. This paper describes in detail the signal conditioning path and the modified algorithm, and it also gives an example of implementation in a resource-constrained wrist-wearable device. The algorithm was evaluated by using the publicly available PPG-DaLia dataset containing samples collected during real-life activities with a PPG sensor and accelerometer and with an ECG signal as ground truth. The quality of the results is comparable to the other algorithms from the literature, while the required hardware resources are lower, which can be significant for wearable applications.

Keywords: heart rate; photoplethysmography; PPG; time-domain; wearable device

---

1. Introduction

Advancements in modern technologies enabled field monitoring of some parameters of human health [1]; for example, heart monitoring is used for off-hospital monitoring and in fitness and professional sport activities. The most commonly measured value is the heart rate (HR), although advanced applications also use other values, e.g., pulse irregularity, as well as biometric identification or analysis of accurate electrical signals that cause heart contraction, i.e., electrocardiography (ECG) [2]. Accurate ECG requires connecting electrodes to the patient's body in several different places, which is inconvenient for the patient, and it can be used only in certain situations. A much more convenient method is measuring the pulse on the wrist by using photoelectric methods. The skin of the wrist is irradiated with single or multicolor light, and then the reflected light is measured. The intensity of the reflected light depends on the absorption of the skin, which depends on the blood volume supplied to the tissues. In this way, the received signal contains information about the current blood supply to the vessels near the measuring device. This method, introduced by Hertzman [3], is known as photoplethysmography (PPG). Unfortunately, PPG signals obtained from a moving person's wrist are weak, distorted, and contain noise. The noise level is often higher than a usable PPG signal. Correct analysis of a low-quality PPG signal is a very challenging task and can consume significant processing time, energy, and resources.

Heart rate estimation from wrist PPG is now a popular area of investigation, and many algorithms for such a task have been proposed [4]. Some of them require significant computing power and memory usage, blocking their application in small portable low-power devices. There are two main approaches used: time-domain sophisticated filtering and frequency-domain processing. Both are often accompanied by a movement sensor (accelerometer) for movement-based artefacts/spectrum removal.

A straightforward approach toward heart rate estimation is peak detection in a periodic signal. One of the simplest possibilities is to use threshold or auto-threshold values in the signal time window [5,6]. Another way is to use transforms such as continuous wavelet [7] or Hilbert [8]. In [9], a nonlinear filter bank was used, with variable cutoff frequencies. In [10], the detection algorithm was based on a time-varying autoregressive model, with a Kalman filter used for autoregressive parameter estimations. Hidden Markov models were used in [11] for combining structural and statistical knowledge of the signal in a single parametric model. A neural network adaptive whitening filter to model the lower frequencies of the signal is presented in [12]. Much recent work is concentrated on methods using time-frequency spectra [13,14] and deep-learning approaches [15–17]. A review of the current state-of-the-art signal-processing techniques for HR estimation from a wrist PPG signal can be found, for example, in [4,18].

The proposed solution of time-domain heart rate measurement algorithm (TDHR) consists of three main blocks: signal conditioning, peak detection, and heart-rate-measuring blocks. In this work, the modified Automatic Multiscale-based Peak Detection (AMPD) algorithm from [19], together with a bandpass filter/limiter, was used for finding the HR from a wrist-based PPG signal. All of the signal processing was done while taking into account the need for low power, low resources, and computing power utilization, necessary for a self-sufficient mobile sensor. The main contributions of this paper are as follows:

- A two-stage input-signal-conditioning digital nonlinear filtering block with limiter;
- Application of the AMPD algorithm for HR peak detection;
- Modification of the AMPD algorithm toward efficient implementation in low-power resource-constrained hardware;
- The proposition of a time-domain heart-rate-measuring algorithm with an accelerometer-based false measurement removal.

This paper is organized as follows. In Section 2, the signal conditioning block is described, which is followed by the peak detection algorithm described in Section 3 and the heart rate calculation block presented in Section 4. The proposed algorithm was evaluated by using a multi-hour dataset; the results of the evaluation and the comparison to the other solutions from the literature are presented in Section 5. Section 6 contains the details of the implementation in a low-power wearable device. Finally, conclusions are provided in Section 7.

2. Signal Conditioning

PPG-signal measuring is usually done with an infrared (IR), red, or green light-emitting diode (LED) as the light source and a photodetector (PD) receiving the reflected or transmitted light. Often, instead of one, a few LEDs and/or PDs are used, providing the possibility to choose the best observed signal or to use additional preprocessing, such as, for example, averaging. Two measure modes can be used: reflectance and transmission. In the first case, the LED and PD are placed on the same skin surface, close to each other; the typical spacing between the PD and LED is in the range of 5–15 mm. The PD measures the light reflected from the tissue. In transmission mode, the LED and PD are placed on the opposite sides of a human body part, and the light reaching the PD goes through the whole body part. The commonly used places on a human body for such PPG devices are the fingertip, wrist, earlobe, forehead, torso, ankle, and nose [18,20]. The wrist is the most convenient place to mount the monitoring device for everyday life, but this is not optimal regarding the strength of the signal. Due to the relatively large thickness and the presence of bones, only the reflectance method of measurement is practical on the wrist. In Figure 1, the waveforms of the signal received from the sensor in reflectance mode on the index finger and on the outer wrist, where a watch is usually worn, are presented. A useful signal carrying information about the pulse is present in the form of the peaks with the period of about 30–40 samples superimposed on the curve. As shown in Figure 1, the amplitude of these peaks for the signal measured at the finger is about 200, while the amplitude measured at the wrist is much lower, about 30–50, while the average signal level (baseline) is about 11,100.

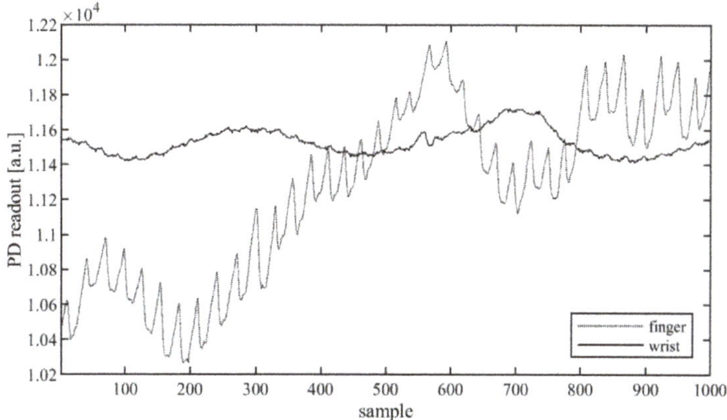

Figure 1. Photodetector(PD) waveforms: raw signals obtained from photoplethysmography (PPG) sensor in reflective configuration placed on the index tip (dotted line) and on the wrist (continuous line).

As can be seen from the waveforms from Figure 1, the signal from the wrist is much weaker; moreover, each signal has a variable offset, and it contains noise and interference. To extract the heart rate in a time-domain, the signal must be preprocessed, so that the peak detection algorithm can find the peaks. The simplest approach is to use a band-pass filter that can help to reduce the noise and eliminate the constant component of the signal. Such a solution is in common use [4]. However, the signal at the input of the filter can have significant changes of the constant component, as well as signal fluctuations caused by wrist movement, resulting in large peaks and oscillations at the output of the band-pass filter, as can be seen in Figure 2a,b.

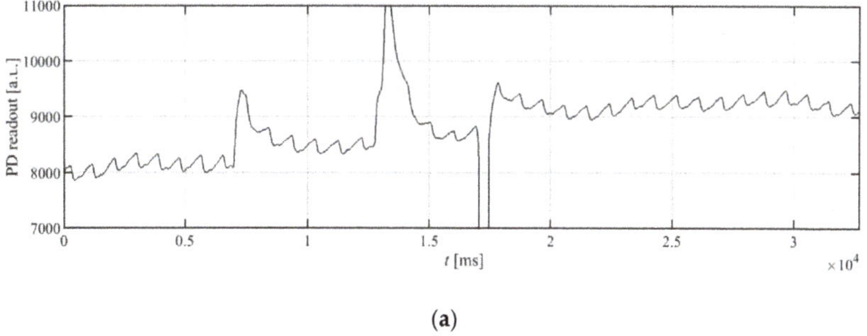

(a)

Figure 2. *Cont.*

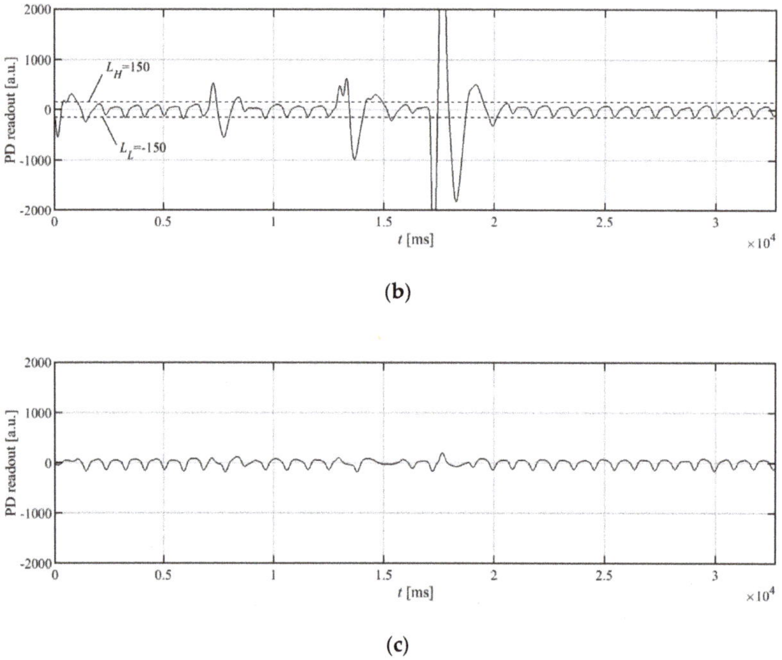

(b)

(c)

Figure 2. Waveforms of the signals from the PPG detector: (**a**) the raw signal; (**b**) the raw signal from (**a**) filtered only by a band-pass filter; and (**c**) the raw signal from (**a**) filtered first with the limiting section described in the paper and then by the same biquadratic filter as used in (**b**).

The proposed approach for PPG signal acquisition and processing is presented in Figure 3. The raw signal obtained from the PD is fed, in the first step, to the band-pass biquad section, with an internal limiter. The output value of the standard biquad section using direct version I is given by Equation (1):

$$y_i = \frac{1}{a_0}(b_0 x_i + b_1 x_{i-1} + b_2 x_{i-2} - a_1 y_{i-1} - a_2 y_{i-2}), \tag{1}$$

where y_i is i-th output sample; x_i is i-th input sample; a_0, a_1, and a_2 and b_0, b_1, and b_2 are, respectively, the denominator and nominator coefficients of the biquad transfer function. The biquad section with the internal limiter works in two steps: First, it calculates candidate $y_{C,i}$ for the output value, according to Equation (1), and then it uses Equation (2) to update the actual output value:

$$y_i = \begin{cases} y_{C,i} \text{ if } L_L \leq y_{C,i} \leq L_H \\ L_L \text{ if } y_{C,i} < L_L \\ L_H \text{ if } y_{C,i} > L_H \end{cases}, \tag{2}$$

where L_L and L_H are the limiter's parameters. These parameters should be selected so that the PPG signal from the heart contractions will remain intact, while distortions resulting from hand movements should be cut off. Figure 2a shows the PPG signal containing distortions taken from a wrist. For this case, the limiter parameters should therefore cut off these high distortions, while the small sawtooth waveform should not be affected. The signal after the bandpass filter does not contain a constant component, as is presented in Figure 2b. In the case of the signal from Figure 2b, good choices for the L_L and L_H parameters' values could be −150 and 150, respectively. Figure 2c shows the results of preprocessing for such a limiter, together with the band-pass biquad section. In general, these values for a given PPG system should be selected as about 100–150% of the negative and positive amplitude of the useful signal, respectively.

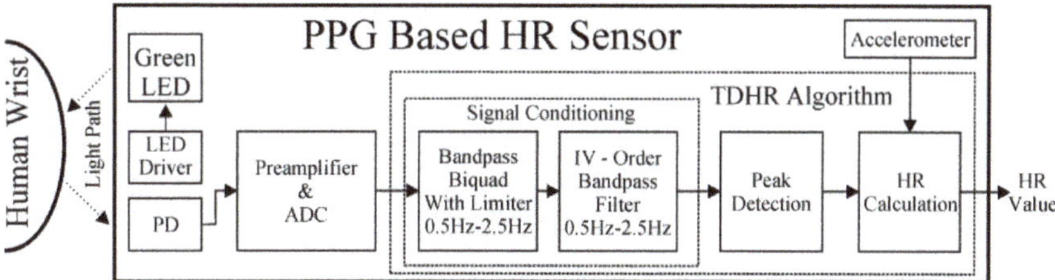

Figure 3. Block diagram of PPG signal acquisition, preprocessing, and final HR estimation by TDHR algorithm.

In the second stage of the proposed preprocessing block, a typical fourth-order band-pass filter built from two biquads was employed. The band-pass of both stages was set to 0.5–2.5 Hz.

The introduction of the limiting section in the form of a single biquad stage significantly reduces the rapid changes of the signal at the input of the band-pass filter and improves its recovery after a rapid change of the constant component in the input signal. An example of distorted signal, together with the accelerometer readouts, is presented later in this paper.

3. Peak Detection

The conditioned signal from the detector was used as an input to the block responsible for finding the peak values of the signal; the heart rate can then be easily calculated from the detected peaks. The detection of peaks is based on the AMPD algorithm [19]. The AMPD algorithm has the capability to work with noisy periodic and quasi-periodic signals. It needs the input signal to be linearly detrended, but the use of the input filter of band-pass characteristic with the limiter described in the previous section satisfies this requirement. The AMPD algorithm performs well for the filtered PPG signal, but it is computationally expensive, which can be unacceptable for wearable devices. The need to calculate a large matrix with real-valued elements, where moving windows are used, can be avoided due to the modifications of the algorithm proposed in further parts of this paper. This section starts with the detailed description of the AMPD algorithm, and then the proposed modifications are introduced. In this way, the authors wanted to make it easier for the reader to track changes that were applied to the original algorithm, without having to refer to the reference.

The main part of the AMPD algorithm is the Local Maxima Scalogram (LMS) matrix M of elements, $m_{k,i}$, which is calculated for the discrete uniformly sampled signal $x = [x_1, x_2, \ldots, x_N]$ in the analyzed window, where N is the constant number of samples, and scale k defines the moving window of varying length, w_k, according to Equation (3):

$$m_{k,i} = \begin{cases} 0 & x_{i-1} > x_{i-k-1} \wedge x_{i-1} > x_{i+k-1} \\ r+1 & \text{otherwise} \end{cases} \tag{3}$$

where $w_k = 2k \mid k = 1,2, \ldots, L$, k is k-th scale of the signal, $L = \text{ceil}(N/2) - 1$, and r is a uniformly distributed random number of values [0,1]. The values of $m_{k,i}$ are calculated for every scale k and $i = k + 2, k + 3, \ldots, N - k + 1$.

Having calculated the LMS matrix, the next step in the AMPD algorithm is to calculate the scale-dependent distribution of zeros in the LMS, by calculating vector $\gamma = [\gamma_1, \gamma_2, \ldots, \gamma_k]$:

$$\gamma_k = \sum_{i=1}^{N} m_{k,i} \text{ for } k \in \{1, 2, \ldots, L\} \tag{4}$$

and the global minimum of γ, $\lambda = \arg\min(\gamma_k)$. The value of γ is used to obtain the matrix M_r, which is the matrix M with deleted bottom rows for $k > \lambda$. The peaks are then found for indexes, i, for which the column-wise standard deviation, σ_i, is equal to zero:

$$\sigma_i = \frac{1}{\lambda-1}\sum_{k=1}^{\lambda}\left[\left(m_{k,i} - \frac{1}{\lambda}\sum_{k=1}^{\lambda}m_{k,i}\right)^2\right]^{\frac{1}{2}}. \tag{5}$$

Parameter λ enables the determination of how many rows of matrix M should be used for calculating the matrix M_r. The noisier the signal from the PPG sensor, the larger the value of λ.

In this paper, the authors propose modifying the AMPD algorithm toward a more efficient implementation. From practical observation, it has been inferred that the signal is too noisy, and it is of no use for peak detection and heart rate calculation, when we have the following:

$$\lambda > \lambda_{\max} \tag{6}$$

The value of $\lambda_{\max} = 17$ was found empirically to be a good choice. Therefore, it is practical to assume a priori, that the signals for which the condition (6) is true are of no use and the full matrix M is not used. To save the memory, only the matrix M_r, instead of M, can be calculated and stored together with the vector γ.

Processing and storing floating point values, $m_{k,i}$, requires storage space and processing resources. To simplify the processing, the authors propose replacing real-valued elements of matrix M_r with the matrix M_r' containing 1-bit binary values, $m'_{k,i}$, as follows:

$$m'_{k,i} = \begin{cases} 0 & x_{i-1} > x_{i-k-1} \wedge x_{i-1} > x_{i+k-1} \\ 1 & \text{otherwise} \end{cases}. \tag{7}$$

This saves a lot of the device's memory, requires only integer operations, and results in another simplification: Instead of calculating the column-wise standard deviation σ_i from Equation (5), calculating the column-wise summation presented in Equation (8) can be used:

$$s_i = \sum_{k=1}^{\lambda} m'_{k,i}. \tag{8}$$

As the $m'_{k,i}$ are 1-bit binary values, s_i can be calculated with fast integer summation. The indices of peaks p_i can be located by finding all indices, i, for which $s_i = 0$. The values of σ_i and s_i, together with the values of γ_k were shown in Figure 4b,c, for an exemplar input signal (Figure 4a).

The PPG signal from a wrist is weak, as can be seen in Figure 2c, and when sampled, the situation of a "flat" peak with two equal values for samples t_{i-1} and t_i, as shown in an example in Figure 5, can sometimes occur. Such a peak would neither be detected by the AMPD algorithm nor its modified version proposed in this paper. To improve the peak detection in such a case, an additional rule was introduced: The peak is also detected at time t'_i.

$$t'_i = \frac{t_{i-1} + t_i}{2} \tag{9}$$

for which the following simple condition (10) holds for s_i from Equation (8):

$$s_{i-2} > 1 \wedge s_{i-1} = 1 \wedge s_i = 1 \wedge s_{i+1} > 1. \tag{10}$$

The proposed simplified peak detection algorithm was compared to the original AMPD algorithm presented in [19]. For this purpose, the PPG signals from the PPG-DaLia database [15] were used, and the detected peaks were compared. As the original AMPD peak detection algorithm uses a random variable to fill the LMS matrix, its results are slightly different from run to run, depending on the seed value of the random number generator. The results are very similar, while the modified version can be implemented more efficiently. An example is presented in Figure 6: the arrows indicate the missing peaks, which were exclusively detected by the other algorithm. It can be seen that only a few peaks are differently detected.

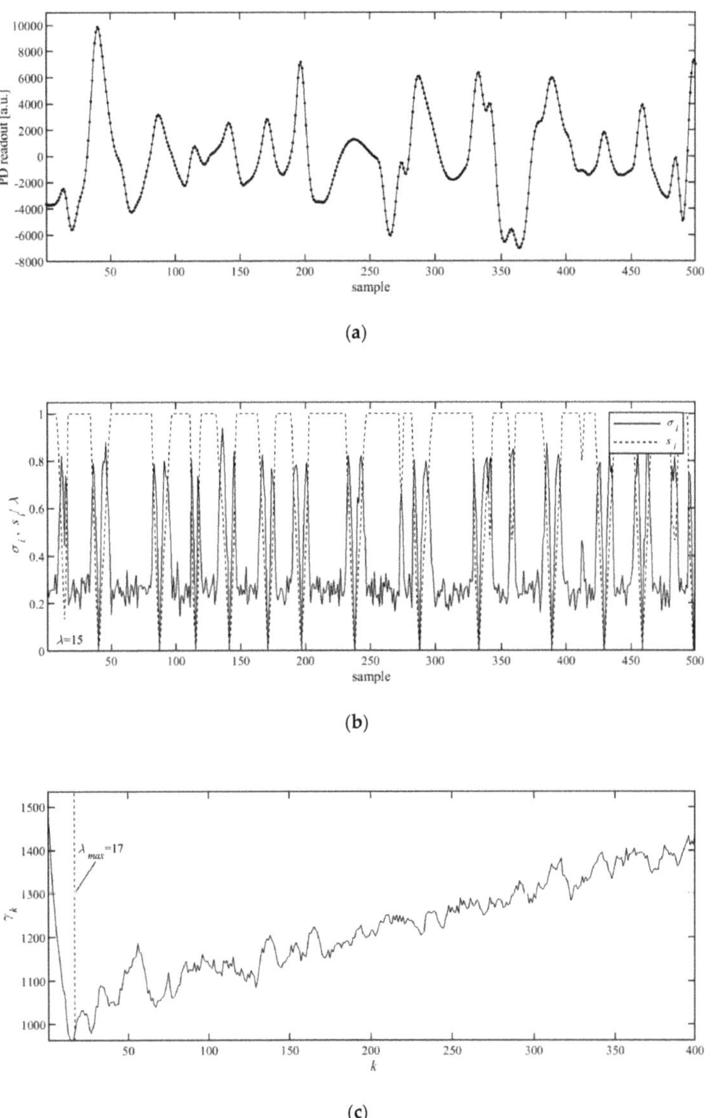

Figure 4. Sample of (**a**) filtered input PPG signal, (**b**) calculated and normalized values of σ_i from Equation (5) and s_i from Equation (8), and (**c**) calculated values of γ_k from Equation (4).

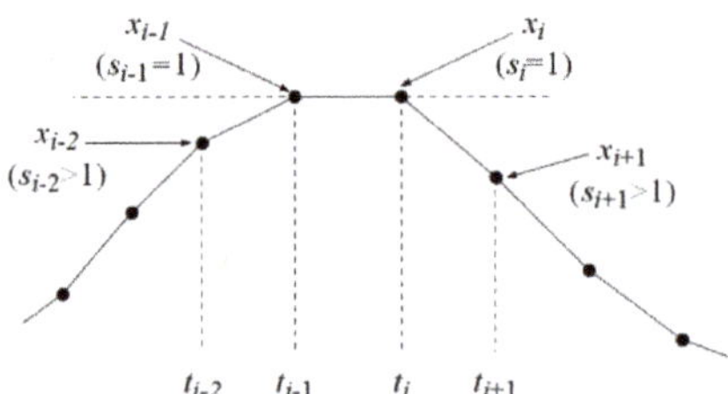

Figure 5. An example of special case of a "flat" peak consisting of two equal values at t_{i-1} and t_i.

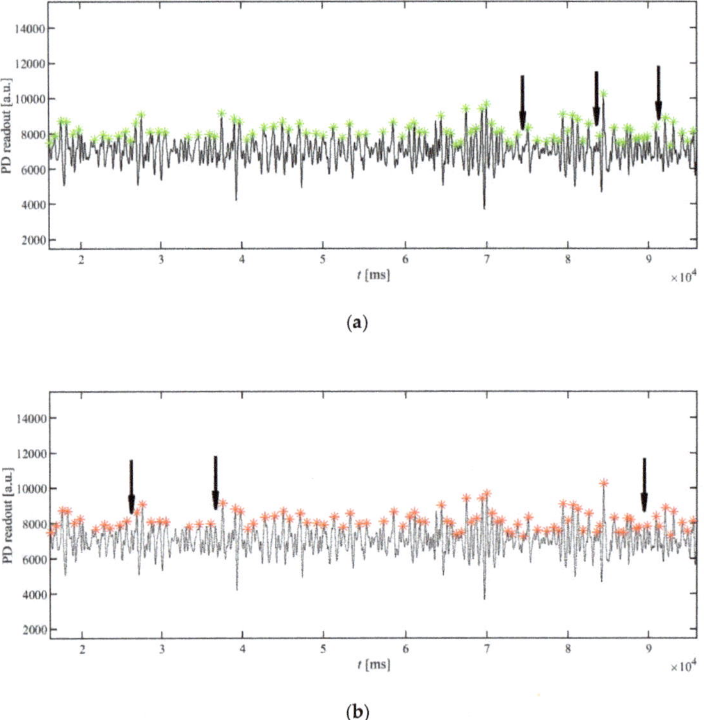

(a)

(b)

Figure 6. Comparison of the results of the peak detection algorithm proposed in this paper (**a**) with the AMPD peak detection algorithm (**b**) from [19] on the same waveform. The arrows indicate the peaks not detected by one of the algorithms but detected by the other one. All other peaks were detected at the same positions, by both methods.

The presented example uses an unfiltered PPG signal to present the robustness of the algorithm. The signal after filtration would be more regular; thus, the differences in the results between the two versions of the algorithm would be even smaller.

4. Heart Rate Calculation

In an ideal situation, where the patient is not moving, the detected peaks from the peak detection algorithm can be directly used to calculate the heart rate. However, the PPG signal can be seriously distorted by the movement of the patient. Each movement of the patient can cause a change in the saturation of the tissues with blood, and a sensor displacement on the wrist, which results in artefacts in the signal received by the optical detector. To mitigate this problem, the accelerometer is used for detecting the movement of the patient's hand. The movement data are aligned with the detected peaks, and the periods between the peaks which are affected by movements are discarded from the calculation of the pulse period. The example of the elimination of the patient's movement is presented in Figure 7.

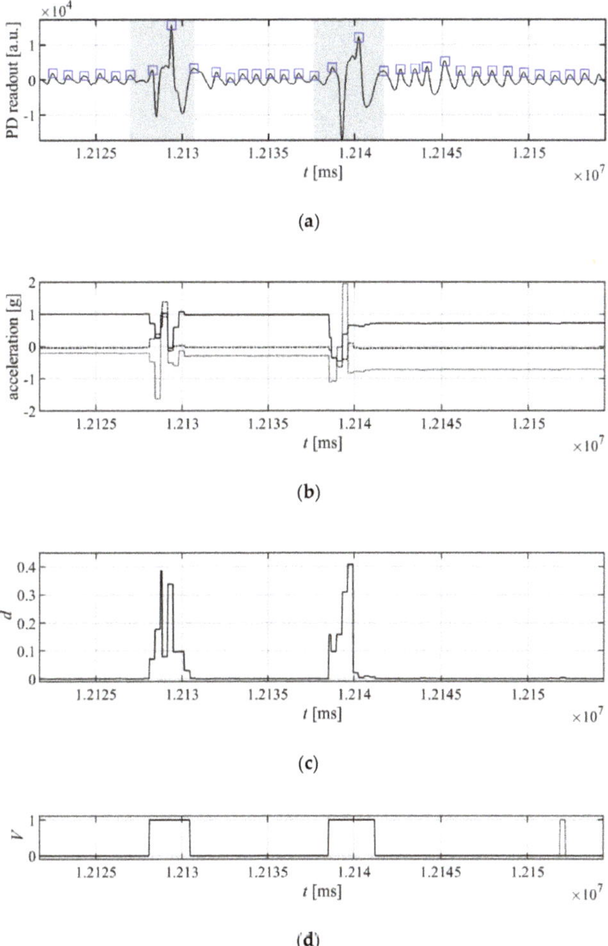

Figure 7. Elimination of the patient's movement from pulse-signal detection: (**a**) the PPG signal distorted with the patient's movement; the areas in gray are excluded from pulse period calculation due to the detected movement; (**b**) acceleration values read from the accelerometer placed together with the heart rate detector on the wrist, (**c**) and the values of movement d_i from Equation (11); (**d**) signal d_i after threshold, as in Equation (12), and with the eliminated short-term peaks (the eliminated peak is shown with a dotted line).

To find the time periods that will be excluded from period calculations, first the acceleration values from the accelerometer are differenced to obtain the movement indicator, d_i, according to the following equation:

$$d_i = \frac{1}{G}\sqrt{\left(g_{x,i} - g_{x,i-1}\right)^2 + \left(g_{y,i} - g_{y,i-1}\right)^2 + \left(g_{z,i} - g_{z,i-1}\right)^2},\tag{11}$$

where i is the index of the sample; $g_{x,i}$, $g_{y,i}$, and $g_{z,i}$ are the acceleration values read from the accelerometer for axes X, Y, and Z, respectively; and G is a constant value used for normalization to obtain $d_i \in [0,1]$ for all i and for the acceleration values in $g_{x,i}$, $g_{y,i}$, and $g_{z,i}$ in the accelerometer's full measuring range. The movement values, d_i, are then compared to the constant, H, to obtain a digital binary signal V_i:

$$V_i = \begin{cases} 1 & d_i > H \\ 0 & \text{otherwise} \end{cases} \tag{12}$$

Signal V_i is then filtered in time, to eliminate spikes longer than T_s. The values of H and T_s were experimentally set to $H = 0.0025$ and $T_s = 500$ ms for the accelerometer range ± 2 g.

For the final result of the heart rate, the inverse of the median of the periods between the peaks not affected by the movement is calculated. For a valid result, a minimal number of peaks, P, is required, and it is calculated as shown in Equation (13):

$$P = floor\left(\frac{N \cdot BPM_{\min}}{60 f_s}\right), \tag{13}$$

where f_s is the sampling frequency, and $BPM_{\min}$ is the minimal heart rate required to be measured by the system.

5. Evaluation on Dataset

For the purpose of evaluating the algorithm, the PPG-DaLia database [15], containing more than 35 h of data recorded from 15 persons, was used. The database contains signals collected from a PPG sensor, accelerometer, and ECG, where the ECG is used as ground truth. As described in [15], ground truth heart rate values were obtained from an ECG signal processed by R-peak detection algorithm [21]. Then, the detection results were manually inspected and corrected, mainly for a few cases where significant motion was observed. The ECG signal was segmented with a shifted window; ground truth heart rate was finally calculated as the mean heart rate within each window. The signals were collected during eight different types of typical daily life activities, under controlled but close-to-real-life conditions. This dataset is the longest one available to the authors. The accuracy of the algorithm was evaluated by using the mean absolute error (*MAE*) metric of beats per minute (bpm), calculated by using the sliding window approach of length 8 s, with a 2 s shift, according to the following equation:

$$MAE = \frac{1}{W} \sum_{j=1}^{W} \left(BPM_{est}(j) - BPM_{ref}(j)\right), \tag{14}$$

where W is the total number of windows, $BPM_{est}(w)$ is the heart rate in bpm for window j, and $BPM_{ref}(j)$ is the reference heart rate obtained from the ECG ground truth signal for the same window j. This evaluation method is commonly used in related work [14,15,22,23].

The quality of the signal and the possibility of the measurement are checked in the TDHR algorithm. When the measurement is not possible due to not satisfying Equation (6), i.e., $\lambda > \lambda_{max}$ or due to movement detected by the accelerometer affecting all of the periods between the detected peaks, the measurement result is indicated as invalid. The evaluation of the algorithm is performed in two ways: (i) When the result is not available, the last valid result is used; (ii) only valid results are used to calculate the performance metric; and then the percentage of the valid samples is also given. The results of the evaluation, together with the performance of the SpaMa [14], SpaMaPlus [15], Schaeck2017 [23], CNN average, and CNN ensemble [15] algorithms are presented in Table 1. The TDHR algorithm was evaluated for several lengths of the sliding window: $N = 1024$ (32 s), $N = 512$ (16 s), $N = 256$ (8 s), and $N = 128$ (4 s).

Table 1. Comparison of the accuracy of the presented algorithm to the algorithms from the literature on the large PPG-DaLia dataset as MAE (bpm). The proposed TDHR algorithm was tested in several versions, with various numbers of analyzed samples N (for all other algorithms $N = 256$), sampled with a frequency of 32 Hz, with a $BPM_{min} = 40$ bpm. The measurements were done every 2 s. For the TDHR algorithm, the accuracy was also calculated for valid measurements, and then the percentage of valid measurements was also given.

	S1	S2	S3	S4	S5	S6	S7	S8	S9	S10	S11	S12	S13	S14	S15	All
SpaMa [14]	11.86	14.75	9.53	17.2	39.28	16.78	15.88	15.2	17.19	9.08	21.63	12.63	9.5	10.73	12.23	15.56 ± 7.5
SpaMaPlus [15]	8.86	9.67	6.4	14.1	24.06	11.34	6.31	11.25	16.04	6.17	15.15	12.03	8.5	7.76	8.29	11.06 ± 4.8
Schaeck 2017 [23]	33.05	27.81	18.49	28.82	12.64	8.72	20.65	21.75	22.25	12.6	21.05	22.74	27.71	12.05	16.4	20.45 ± 7.1
CNN average [15]	8.45	7.92	5.96	7.86	18.97	13.55	5.16	11.49	10.65	6.07	9.87	9.95	5.25	5.85	5.25	8.82 ± 3.8
CNN ensemble [15]	7.73	6.74	4.03	5.9	18.51	12.88	3.91	10.87	8.79	4.03	9.22	9.35	4.29	4.37	4.17	7.65 ± 4.2
TDHR $N = 1024$	8.10	7.98	10.51	11.82	20.60	12.11	7.62	11.71	14.79	4.92	20.05	8.76	7.88	9.15	8.45	10.96 ± 4.49
TDHR $N = 1024$ only valid measurements (%)	5.32 74%	4.99 86%	6.47 71%	6.57 72%	16.26 43%	8.29 55%	5.40 80%	10.28 80%	6.93 64%	3.01 81%	9.17 66%	5.22 61%	4.70 83%	4.84 79%	4.12 75%	6.77 ± 3.26 72%
TDHR $N = 512$	8.61	7.49	11.55	11.93	20.79	14.24	7.96	11.91	14.95	5.83	19.60	8.86	8.22	9.10	8.66	11.31 ± 4.41
TDHR $N = 512$ only valid measurements (%)	6.00 74%	5.60 86%	7.72 73%	7.55 71%	16.05 58%	9.62 65%	6.01 81%	10.50 79%	8.79 65%	3.71 80%	12.15 71%	5.65 62%	5.27 84%	5.34 78%	4.82 75%	7.65 ± 3.30 74%
TDHR $N = 256$	11.08	10.84	11.63	14.06	21.67	15.63	8.86	13.30	15.12	7.27	21.08	10.49	9.54	10.24	9.04	12.66 ± 4.25
TDHR $N = 256$ only valid measurements (%)	8.40 74%	8.21 86%	9.40 75%	10.10 73%	17.02 69%	12.85 72%	7.57 83%	12.04 78%	11.75 68%	5.14 79%	13.86 74%	7.73 62%	7.33 84%	7.45 81%	6.63 74%	9.70 ± 3.21 76%
TDHR $N = 128$	13.13	13.57	13.11	15.59	25.55	18.05	10.25	16.19	19.23	10.32	19.39	12.64	11.63	12.33	12.80	14.92 ± 4.16
TDHR $N = 128$ only valid measurements (%)	12.71 80%	12.25 89%	11.93 82%	14.05 80%	20.84 82%	16.44 79%	9.53 86%	15.65 83%	17.53 78%	8.75 85%	17.13 82%	12.02 72%	10.29 86%	10.66 85%	10.73 80%	13.37 ± 3.46 82%

Table 2 presents the comparison of the computational cost of several algorithms. For the TDHR algorithm, the number of operations per second was estimated from a manual analysis of the C code as the number of arithmetic operations needed to obtain a single heart-rate result. The TDHR algorithm requires only a few parameters, as opposed to the CNN algorithms, but it needs storage memory for calculating the LMS array; the size of this memory depends on the window length, N.

Table 2. Comparison of the performance versus computational cost of the TDHR algorithm and the algorithms from the literature. The computational cost of the TDHR algorithm was calculated as the number of arithmetical operations and the number of memory bytes needed for algorithm realization in a microcontroller.

Algorithm	Performance Mean MAE ± STD (% = Only Valid Measurements)	Computational Cost	
		Number of Parameters/Memory Bytes	Operations Per Second
CNN average	8.82 ± 3.8	8.5 M	34.5 M
CNN ensemble	7.65 ± 4.2	60 M	240 M
CNN constrained	9.99 ± 5.9	26 K	190 K
TDHR $N = 1024$	10.96 ± 4.49	66 k	2.4 M
TDHR $N = 1024$	6.77 ± 3.26 (72%)		
TDHR $N = 512$	11.31 ± 4.41	16 k	598 k
TDHR $N = 512$	7.65 ± 3.30 (74%)		
TDHR $N = 256$	12.66 ± 4.25	4 k	152 k
TDHR $N = 256$	9.70 ± 3.21 (76%)		
TDHR $N = 128$	14.92 ± 4.16	1 k	40 k
TDHR $N = 128$	13.37 ± 3.46(82%)		

As can be seen from Tables 1 and 2, the performance of the proposed TDHR algorithm is similar to *SpaMa*, *SpaMaPlus*, and *Schaeck2017*, while it is worse than *CNN average*, *CNN ensemble*, and *CNN constrained*. The computational costs of *SpaMa*, *SpaMaPlus*, and *Schaeck2017* are not known, but they can be high, as those algorithms require the calculation of the power spectral density and the analysis of the PPG spectrum. The CNN-based algorithms, even *CNN constrained*, require larger computational costs than most versions of the TDHR algorithm. The proposed algorithm uses the mechanism of removing the time periods with body motion registered by the accelerometer, so there can be gaps between valid measurements in the case of long-lasting motions. The presented results in Tables 1 and 2 can help to select N to achieve a compromise between the accuracy and computational cost.

6. Implementation

The proposed algorithm was implemented in low-power wearable hardware and tested. The hardware consists of the main board, power supply PCB, with external coil for wireless inductive charging, and a LiPo battery of capacity 110 mAh. A block diagram of the hardware is presented in Figure 8. As the processing unit, a PSoC6 microcontroller from Cypress was used. This is an ultra-low-power microcontroller with dual processor architecture: Arm Cortex M4 and M0+ cores. A BH1790GLC optical sensor from Rohm detects the PPG signal and also drives the four green 527 nm LEDs, equally placed in a circle of diameter of 6 mm, with the optical sensor placed in the center. The detector detects the light emitted by the LEDS and reflected from the patient's skin. An LSM6DSL accelerometer from STMicroelectronics is mounted on the same board as the optical sensor. The optical sensor and the accelerometer are connected to the microcontroller via an I^{2}C bus, used for configuration and data transfer.

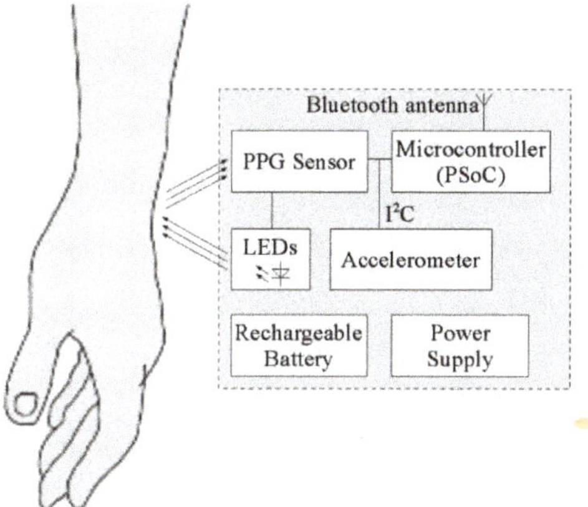

Figure 8. Block diagram of the device implementing the algorithm for measuring the heart rate.

The software for heart rate measurements runs on the Cortex M4 processing core of the PSoC6 microcontroller. The algorithm presented in this paper was written in C as two tasks running on the FreeRTOS operating system: The first task constantly reads data from the sensors, realizes filtering, and buffers data. The second task starts periodically and executes the peak detection algorithm, using data from the buffer. The calculated heart rate is sent wirelessly, using the Bluetooth Low Energy protocol. For this purpose, a built-in BLE transceiver in PSoC6 device was used. According to the BLE nomenclature, in the proposed solution, the BLE transceiver block was configured to perform a peripheral (a device constrained in resources such as energy and computing power) and server role (a device working as a data source and sending that data to the remote master device). As a data format, standard BLE Heart Rate Profile was used. The values of the measured heart rate are transmitted periodically; the user can receive the transmitted values by connecting any Bluetooth receiver compatible with Heart Rate Profile.

The PPG signal is sampled with 14-bit resolution and with frequency f_s = 32 Hz, using a buffer of $N = 128 \ldots 1024$ samples, which enables data from 4 to 32 s to be analyzed, depending on the selected value of N, compromising accuracy versus computational cost and memory usage.

The prototype device was built and installed inside a custom-made 3D-printed case with a rubber strap, as shown in Figure 9. The size of the printed case (without the rubber strap) is $11.7 \times 26 \times 46$ mm. The cost of the components, including the battery, was about \$60, at retail prices, for 15 pieces. The proposed implementation was capable of continuous measuring of the heart rate for more than 24 h.

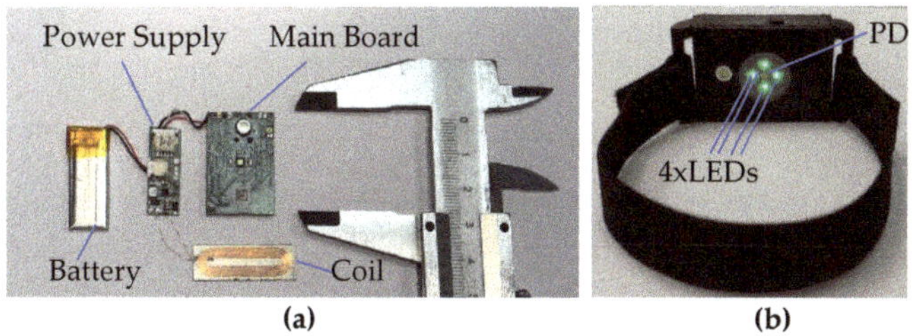

Figure 9. The prototype implementation of a wrist sensor for heart rate measurements with the implemented TDHR algorithm: (**a**) the picture of the internal modules and parts; (**b**) the picture of the final device with strap. The main board has the dimensions of 20×30 mm.

7. Discussion and Conclusions

Nowadays, there are many smart watches on the market that are capable of measuring the heart rate. The details regarding the optical part of the measurement, such as the number of sensors or the wavelengths of the LEDs used during measurement, are often revealed. However, the details regarding the algorithms used are not available. Most of the publications focus on measuring the accuracy of the popular devices. In [24], the authors measured the performance of Apple's iWatch Apple Watch Sport 42 mm (first generation), during cardiopulmonary exercise test (CPET). They observed MAE from 6.34 to 7.55. It is difficult to exactly compare this to our results, as we use different and longer test conditions. In fact, the details of the PPG algorithms can only be found in the scientific publications, where the authors try to increase the accuracy of the measurement by using novel ideas and powerful techniques.

In this paper, a time-domain algorithm for the real-time detection of the heart rate was presented. The algorithm is aimed at wearable, resource-constrained devices, where battery capacity, processor speed and memory size are constrained. The algorithm processes raw data in a time-domain and requires only a few parameters. The approach is simple, the proposed algorithm has a reasonable accuracy, and it can be implemented in a typical (not DSP) microcontroller.

The algorithm consists of a two-stage input-signal-conditioning block with a limiter, the peak detection block, and period calculation block. The two-stage input conditioning block is built out of two digital bandpass filters, where the first filter has been modified to process data nonlinearly, to provide fast recovery after large signal transients. The use of band-pass filters is very simple to implement; it appeared sufficient and very effective for conditioning the signal; therefore, other methods such as wavelet-based baseline removal would not need to be considered.

The peak detection block enables operation at significantly lower processing and implementation costs, compared to the original AMPD algorithm. The period calculation block uses the median to calculate the heart rate based on the time differences between the peaks, with the use of an accelerometer to exclude the time periods affected by body movement.

In this study, most of the parameters were set experimentally, as we targeted on a simple and economical hardware implementation. It would be interesting to provide a method of automatic and dynamic adjustment of the parameters to further reduce the computational cost, basing on the input signal quality and the movement readouts. This will be a topic of our further research.

The proposed algorithm was evaluated in several variants, for different sliding window lengths, N, providing the possibility to compromise the accuracy versus lower operational costs. The authors mainly used $N = 1024$, which seems to be a good compromise between the calculation cost and the accuracy. The proposed algorithm was compared to the other algorithms from the literature. The achieved accuracy is comparable to the other algorithms at smaller computational costs. The proposed solution was also implemented in a low-power wrist-wearable device.

Author Contributions: Conceptualization, B.P. and M.W.; methodology, M.W.; software, M.W.; validation, M.W.; formal analysis, M.W.; investigation, B.P. and M.W.; resources, M.W.; data curation, M.W.; writing—original draft preparation, M.W. and B.P.; writing—review and editing, M.W. and B.P.; visualization, M.W.; supervision, M.W.; project administration, M.W.; funding acquisition, M.W. All authors have read and agreed to the published version of the manuscript.

Funding: This research was partially supported by the DS Programs of Faculty of Electronics, Telecommunications and Informatics, as well as National Centre for Research and Development, Poland, project "E-Pionier—using the potential of universities to improve the innovation of ICT solutions in the public sector", No. 17/02/2018/UD.

Acknowledgments: The authors would like to thank Małgorzata Szczerska for her initiative in organizing research and raising funds for research.

Conflicts of Interest: The authors declare no conflict of interest.

References

1. Rault, T.; Bouabdallah, A.; Challal, Y.; Marin, F. A survey of energy-efficient context recognition systems using wearable sensors for healthcare applications. *Pervasive Mob. Comput.* **2017**, *37*, 23–44. [CrossRef]
2. Szczepański, A.; Saeed, K. A Mobile Device System for Early Warning of ECG Anomalies. *Sensors* **2014**, *14*, 11031–11044. [CrossRef]
3. Hertzman, A.B. Observations on the finger volume pulse recorded photoelectrically. *Am. J. Physiol.* **1937**, *119*, 334–335.
4. Biswas, D.; Simões-Capela, N.; Van Hoof, C.; Van Helleputte, N.; Simues-Capela, N. Heart Rate Estimation from Wrist-Worn Photoplethysmography: A Review. *IEEE Sensors J.* **2019**, *19*, 6560–6570. [CrossRef]
5. Jacobson, M. Auto-threshold peak detection in physiological signals. In Proceedings of the 2001 Conference Proceedings of the 23rd Annual International Conference of the IEEE Engineering in Medicine and Biology Society, Istanbul, Turkey, 25–28 October 2001. [CrossRef]
6. Pan, J.; Tompkins, W.J. A Real-Time QRS Detection Algorithm. *IEEE Trans. Biomed. Eng.* **1985**, *32*, 230–236. [CrossRef] [PubMed]
7. Du, P.; Kibbe, W.A.; Lin, S.M. Improved peak detection in mass spectrum by incorporating continuous wavelet transform-based pattern matching. *Bioinform* **2006**, *22*, 2059–2065. [CrossRef] [PubMed]
8. Benitez, D.; Gaydecki, P.; Zaidi, A.; Fitzpatrick, A.P. The use of the Hilbert transform in ECG signal analysis. *Comput. Boil. Med.* **2001**, *31*, 399–406. [CrossRef]
9. Aboy, M.; McNames, J.; Thong, T.; Tsunami, D.; Ellenby, M.; Goldstein, B. An Automatic Beat Detection Algorithm for Pressure Signals. *IEEE Trans. Biomed. Eng.* **2005**, *52*, 1662–1670. [CrossRef] [PubMed]
10. Tzallas, A.T.; Oikonomou, V.P.; I Fotiadis, D. Epileptic Spike Detection Using a Kalman Filter Based Approach. In Proceedings of the 2006 International Conference of the IEEE Engineering in Medicine and Biology Society, New York, NY, USA, 30 August–3 September 2006; Volume 1, pp. 501–504. [CrossRef]
11. Coast, D.A.; Stern, R.; Cano, G.G.; A Briller, S. An approach to cardiac arrhythmia analysis using hidden Markov models. *IEEE Trans. Biomed. Eng.* **1990**, *37*, 826–836. [CrossRef] [PubMed]
12. Xue, Q.; Hu, Y.H.; Tompkins, W.J. Neural-network-based adaptive matched filtering for QRS detection. *IEEE Trans. Biomed. Eng.* **1992**, *39*, 317–329. [CrossRef] [PubMed]
13. Mashhadi, M.B.; Asadi, E.; Eskandari, M.; Kiani, S.; Marvasti, F.; Boloursaz, M. Heart Rate Tracking using Wrist-Type Photoplethysmographic (PPG) Signals during Physical Exercise with Simultaneous Accelerometry. *IEEE Signal Process. Lett.* **2015**, *23*, 227–231. [CrossRef]
14. Salehizadeh, S.M.A.; Dao, D.; Bolkhovsky, J.; Cho, C.; Mendelson, Y.; Chon, K.H. A Novel Time-Varying Spectral Filtering Algorithm for Reconstruction of Motion Artifact Corrupted Heart Rate Signals During Intense Physical Activities Using a Wearable Photoplethysmogram Sensor. *Sensors* **2016**, *16*, 10. [CrossRef] [PubMed]
15. Reiss, A.; Indlekofer, I.; Schmidt, P.; Van Laerhoven, K. Deep PPG: Large-Scale Heart Rate Estimation with Convolutional Neural Networks. *Sensors* **2019**, *19*, 3079. [CrossRef] [PubMed]
16. Sumukha, B.N.; Kumar, R.C.; Bharadwaj, S.S.; George, K. Online peak detection in photoplethysmogram signals using sequential learning algorithm. In Proceedings of the 2017 International Joint Conference on Neural Networks (IJCNN), Anchorage, AK, USA, 14–19 May 2017; pp. 1313–1320. [CrossRef]
17. Vijaya, G.; Kumar, V.; Verma, H.K. ANN-based QRS-complex analysis of ECG. *J. Med Eng. Technol.* **1998**, *22*, 160–167. [CrossRef] [PubMed]
18. Ghamari, M.; Castaneda, D.; Esparza, A.; Soltanpur, C.; Nazeran, H. A review on wearable photoplethysmography sensors and their potential future applications in health care. *Int. J. Biosens. Bioelectron.* **2018**, *4*, 195–202. [CrossRef] [PubMed]
19. Scholkmann, F.; Boss, J.; Wolf, M. An Efficient Algorithm for Automatic Peak Detection in Noisy Periodic and Quasi-Periodic Signals. *Algorithms* **2012**, *5*, 588–603. [CrossRef]
20. Nasal ALAR SpO2 Pulse Oximetry Sensor. Available online: https://www.pentlandmedical.co.uk/critical-care/nasal-alar-spo2-pulse-oximetry-sensor/ (accessed on 4 February 2020).
21. Hamilton, P. Open source ECG analysis. *Computers in Cardiology* **2003**, 101–104. [CrossRef]
22. Zhang, Z. Photoplethysmography-Based Heart Rate Monitoring in Physical Activities via Joint Sparse Spectrum Reconstruction. *IEEE Trans. Biomed. Eng.* **2015**, *62*, 1902–1910. [CrossRef] [PubMed]

Sensors **2020**, *20*, 1783

23. Schack, T.; Muma, M.; Zoubir, A.M. Computationally efficient heart rate estimation during physical exercise using photoplethysmographic signals. In Proceedings of the 2017 25th European Signal Processing Conference (EUSIPCO), Kos, Greece, 28 August–2 September 2017; pp. 2478–2481.
24. Falter, M.; Budts, W.; Goetschalckx, K.; Cornelissen, V.; Buys, R.; Shcherbina, A.; Goessler, K.; Boudreaux, B.; Goris, J. Accuracy of Apple Watch Measurements for Heart Rate and Energy Expenditure in Patients With Cardiovascular Disease: Cross-Sectional Study. *JMIR mHealth uHealth* **2019**, *7*, e11889. [CrossRef] [PubMed]

 *sensors*

Article

Variability of Coordination in Typically Developing Children Versus Children with Autism Spectrum Disorder with and without Rhythmic Signal

Lidia V. Gabis [1,2], **Shahar Shefer** [2] and **Sigal Portnoy** [1,*]

[1]   Department of Occupational Therapy, Sackler Faculty of Medicine, Tel Aviv University,
      Tel Aviv 6997801, Israel; Lidia.Gabis@sheba.health.gov.il
[2]   The Weinberg Child Development Center, Safra Children's Hospital, Sheba Medical Center,
      Ramat Gan 52621, Israel; drshahar.shefer@sheba.health.gov.il
*   Correspondence: portnoys@tauex.tau.ac.il; Tel.: +972-3-640-5441

Received: 14 April 2020; Accepted: 11 May 2020; Published: 13 May 2020

Abstract: Motor coordination deficit is a cardinal feature of autism spectrum disorder (ASD). The evaluation of coordination of children with ASD is either lengthy, subjective (via observational analysis), or requires cumbersome post analysis. We therefore aimed to use tri-axial accelerometers to compare inter-limb coordination measures between typically developed (TD) children and children ASD, while jumping with and without a rhythmic signal. Children aged 5–6 years were recruited to the ASD group (n = 9) and the TD group (n = 19). Four sensors were strapped to their ankles and wrist and they performed at least eight consecutive jumping jacks twice: at a self-selected rhythm and with a metronome. The primary outcome measures were the timing lag (TL), the timing difference of the maximal acceleration of the left and right limbs, and the lag variability (LV), the variation of TL across the 5 jumps. The LV of the legs of children with ASD was higher compared to the LV of the legs of TD children during self-selected rhythm jumping ($p < 0.01$). Additionally, the LV of the arms of children with ASD, jumping with the rhythmic signal, was higher compared to that of the TD children ($p < 0.05$). There were no between-group differences in the TL parameter. Our preliminary findings suggest that the simple protocol presented in this study might allow an objective and accurate quantification of the intra-subject variability of children with ASD via actigraphy.

Keywords: motor variability; actigraphy; triaxial accelerometers; jumping

---

1. Introduction

Autism spectrum disorder (ASD) is defined by impairments in social communication and/or the presence of repetitive or restricted behaviors. The prevalence of ASD in the United States is reported as one in 68 children [1]. Aside from social difficulties, motor impairments are prevalent in individuals with ASD and worsen with age [2]. Motor impairments in individuals with ASD might affect both gross and fine motor functions, e.g., manual dexterity and balance [3]. Specifically, motor coordination deficit was characterized as a cardinal feature of ASD [3].

To date, the accuracy level of the diagnostic tests of ASD are relatively limited [4]. Specifically, standardized motor assessments for children with ASD take between 15 min to over one hour to complete [5]. It has been recently suggested that simple quantitative measures of motor coordination may assist in the identification of subtle motor impairments in individuals with ASD [5,6]. Early detection of abnormalities in the coordination abilities of the child may assist clinicians in devising an optimal treatment plan, e.g., engaging children with ASD in ball games [7]. Therefore, devising a quick and simple protocol for an examination that produces accurate quantitative measures of the child's coordination capabilities is a challenge for future studies.

Previous studies characterizing coordination abnormalities in children have attempted to use rhythmic signals, e.g., via a metronome, to analyze movement synchronization. For example, coordination abnormalities in children with ASD were measured while performing various multi-limb actions with a metronome, such as marching and clapping [8]. The coefficient of variation (CV) of the inter-event duration was obtained as a variability measure by analyzing recorded video data. The authors reported that children with ASD exhibit higher CV compared to typically developed (TD) children, but there was no difference in the CV between two groups of children with ASD, with lower and higher intelligence quotient (IQ) [8]. However, using video recordings of the child in the clinical setting might not be appreciated by their parents. Moreover, the post analysis is cumbersome and does not produce quick results.

Wearable tri-axial accelerometry is a simple and effective mean to record activity in individuals with ASD. Actigraphy was previously used to report physical activity levels in individuals with ASD [9] and the effects of various factors (for example age [10], social engagement with adults [11], and household structure [12]) on the physical activity levels of these individuals. Additionally, actigraphy has been used to monitor sleep patterns in individuals with ASD [13,14], demonstrating, for example, that sleep latency, as measured by actigraphy, was longer in individuals with ASD compared to controls [15]. Another study showed that an accelerometer worn by youth with ASD can predict aggression to others, one minute before it occurs [16]. Overall, the literature supports the usage of wearable tri-axial accelerometers with individuals with ASD. However, to the best of our knowledge, no studies utilized these sensors to assess lower and upper limb coordination in children with ASD while performing a quick and simple jumping activity. Furthermore, rhythmic auditory cueing was suggested as a technique to stabilize the variability in the movement pattern and facilitate a motor plan for individuals with ASD [17]. Since the effect of a rhythmic signal on the coordination measures while performing jumping activity has not been reported, our aim was to compare inter-limb coordination measures between TD children and children ASD, while jumping with and without a rhythmic signal.

2. Methods

2.1. Population

We recruited children, aged 5–6 years, diagnosed with ASD according to the criteria listed in the Diagnostic and Statistical Manual of Mental Disorders, Fifth Edition (DSM-V). Exclusion criteria were inability to understand and/or comply with simple instructions, other pathologies, e.g., epilepsy, orthopedic impairments, uncorrected auditory or visual impairments, cognitive disability. Age-matched TD children were recruited as controls. The study received the approval of the Helsinki committee of the hospital (approval #0119-13-SMC).

2.2. Tools

Four Actigraph™ sensors (GT3X; ActiGraph, Pensacola, FL, USA) were used in this study. These are small (3.8 × 3.7 × 1.8 cm) and lightweight (27 g) sensors, easily donned on the limbs of the subjects using elastic belts. The sensors provide accelerations in three axes and can be activated at a frequency of 30 Hz. These sensors have been extensively used in various populations of different ages, as recently reviewed in [18]. Specifically, the sensors have been used in children with ASD [12]. For the rhythmic signal, a digital metronome was set to 1 Hz. This frequency was chosen following a pilot with TD children, by setting the metronome to various frequencies and asking the children which was their preferred comfortable choice of jumping frequency.

2.3. Protocol

The parents signed an informed consent form and the child gave verbal consent to participate in the trial. The parents filled out a demographic questionnaire. Then, two sensors were attached to the wrists of the child and two to the ankles using elastic belts. The child was asked to ambulate with the sensors and report any inconvenience. Then, the child was asked to perform at least 8 consecutive jumping jacks (also called star jumps). The jumps were demonstrated before data recording began. This was performed twice: once at a self-selected rhythm and once guided by the metronome, located 1 m behind the subject.

2.4. Post Analysis

The data from each accelerometer were downloaded to a personal computer and then exported as CSV files using Actilife™ software version 5.10.0 (ActiGraph, Pensacola, FL, USA). A custom code was created in LabView (v2015, National Instruments, Austin, TX, USA). The acceleration magnitude was calculated using the accelerations of the 3 axes, A_x, A_y, and A_z, as:

$$Limb\ acceleration\left[\frac{m}{s^2}\right] = \sqrt{A_x^2 + A_y^2 + A_z^2} \tag{1}$$

Five consecutive jumps were taken (the first and last jumps were excluded from the analysis as initiation and termination of movement). We designed this protocol to include a small number of jumps because a longer jumping sequence might involve fatigue, which will produce bias in the results of limb accelerations. Also, the cooperation levels and concentration span of 5-year-olds in the clinical settings might be low, so that a longer examination might not be possible. The maximal acceleration of each limb of each jump was calculated. Also, the timing of the peak acceleration (in seconds) of each jump was calculated and the following inter-limb timing measures were computed using the following formulas:

$$TL\ [sec] = \frac{\sum(t_L - t_R)}{5} \tag{2}$$

$$LV\ [sec] = \sqrt{\frac{\sum TL^2 - \frac{(\sum TL)^2}{5}}{5}} \tag{3}$$

The timing lag (*TL*) is the difference in the timing of the maximal acceleration of the left lower or upper limb (t_L) and right lower or upper limb (t_R), averaged for five consecutive jumps. It is defined similarly as constant error of two measures (left and right limb herein) that are expected to be identical during symmetric movement [19,20]. A positive TL value denotes that the left limb reached maximal acceleration sooner compared to the right limb. As the TL value decreases towards zero, the coordination of the two limbs in reaching maximal acceleration is higher, meaning that they are more in-phase. The lag variability (*LV*) is the variation of TL across the 5 jumps. It is defined in a similar manner as variable error of two measures and produces the average of the standard deviation [19]. High LV is indicative of low consistency between jumps. These measures are calculated separately for each condition, with and without the metronome rhythm, and presented in seconds.

All of the statistical analyses were performed in IBM SPSS Statistics 25. Mann–Whitney U test was used to test for between-group differences in age and body mass index (BMI), and the Chi-square test was used to test for between-group differences in sex. ANOVA analysis of group (ASD and TD) × condition (with and without metronome) was performed. Post hoc tests were administered according to the findings. Effect size estimates, *r*, for Mann–Whitney non-parametric tests were calculated according to [21]:

$$r = \frac{Z}{\sqrt{N}} \tag{4}$$

Statistical significance was set to $p < 0.05$.

3. Results

The personal characteristics of the two groups are detailed in Table 1. There were no significant differences between the two groups in age, sex, and BMI.

Table 1. Median and interquartile ranges of demographic characteristics of the subjects.

	TD (n = 19)	ASD (n = 8)	*p* Value
Age (years)	5.0 (4.4–6.0)	5.2 (5.0–6.6)	0.322
Sex	8 girls, 11 boys	2 girls, 6 boys	0.395
BMI (kg/m^2)	14.4 (13.6–16.7)	15.4 (13.4–16.7)	0.710

TD: Typically Developed; ASD: Autism Spectrum Disorder; BMI: Body Mass Index.

The maximal limb accelerations of each group in each condition, as well as the coordination parameters of the upper and lower limbs are presented in Table 2. There was a significant main effect of group, but there was no main effect of condition. Specifically, the TD children reached higher accelerations of their left limbs during self-selected rhythm jumping compared to children with ASD (Figure 1). In that condition, the LV of their legs were lower compared with the LV of the legs of children with ASD (Figure 2). Additionally, the LV of the arms in TD children, jumping with the rhythmic signal, was lower compared to that of the ASD group (Figure 2).

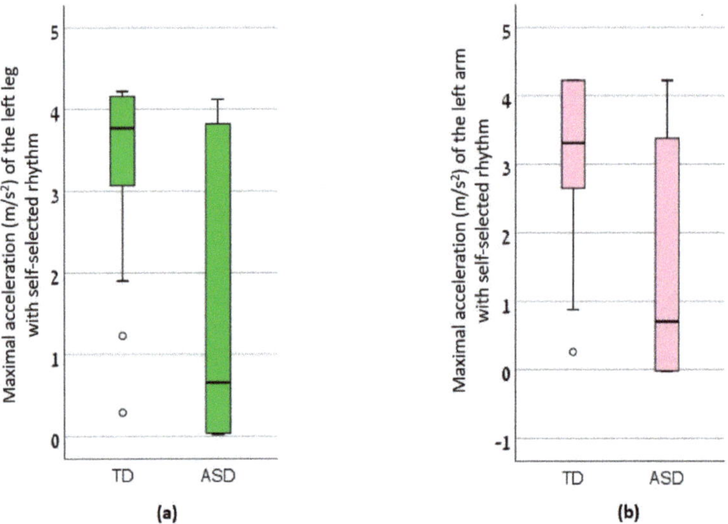

Figure 1. The maximal acceleration (m/s^2) of the (**a**) left leg with self-selected rhythm and (**b**) left arm with self-selected rhythm in Typically Developed (TD) children and children with Autism Spectrum Disorder (ASD).

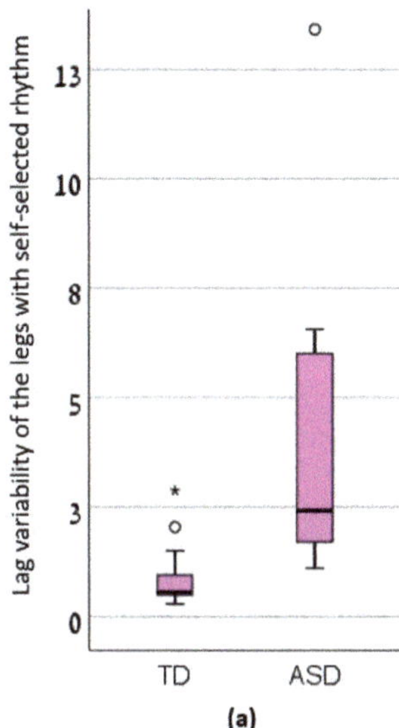
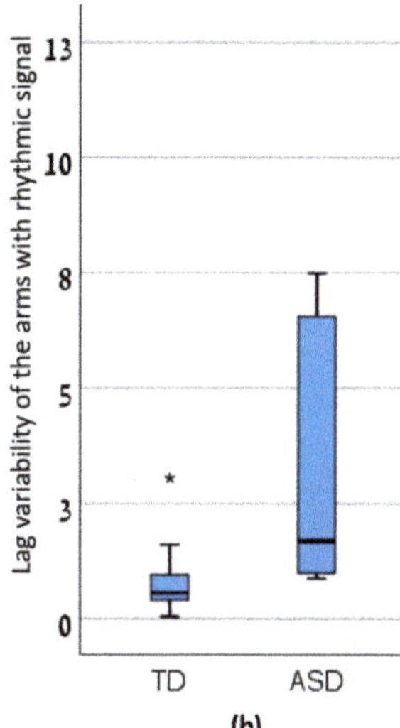

Figure 2. The lag variability of the (**a**) legs with self-selected rhythm and (**b**) arms with rhythmic signal in Typically Developed (TD) children and children with Autism Spectrum Disorder (ASD).

Table 2. Medial and interquartile percentage of the maximal acceleration values (m/s^2) of each limb and the coordination measures for the time of the peak acceleration (sec) of the arms and legs of each group, with and without the rhythmic signal.

	TD (n = 19)		ASD (n = 8)		Between-Groups F,p	Effect Size r
	Self-Selected Rhythm	Rhythmic Signal	Self-Selected Rhythm	Rhythmic Signal		
Left arm Acc.	**3.31 (2.49–4.22)** *	2.38 (1.35–4.13)	**0.71 (0.02–3.55)** *	0.68 (0.02–3.26)	6.444, 0.018	−0.412
Left leg Acc.	**3.77 (3.06–4.17)** *	3.78 (1.60–4.22)	**0.66 (0.03–3.94)** *	0.69 (0.10–3.51)	6.324, 0.019	−0.445
Right arm Acc.	3.26 (1.45–3.91)	1.93 (0.80–3.69)	1.89 (0.07–3.50)	0.91 (0.09–3.14)	NS	NS
Right leg Acc.	3.51 (3.28–3.96)	3.44 (2.10–4.14)	0.73 (0.02–4.06)	0.54 (0.01–3.78)	NS	NS
TL arms	0.13 (−0.78–0.64)	0.05 (−0.49–1.11)	0.87 (−0.35–2.86)	1.22 (−1.68–2.26)	NS	NS
TL legs	0.22 (−0.87–0.57)	0.61 (−0.08–0.91)	1.30 (−3.00–2.73)	0.64 (−4.28–3.17)	NS	NS
LV arms	0.61 (0.39–1.31)	**0.56 (0.41–0.98)** *	1.53 (0.88–3.15)	**1.68 (0.94–6.55)** *	9.812, 0.004	−0.583
LV legs	**0.55 (0.46–0.97)** **	0.58 (0.34–0.95)	**2.41 (1.68–6.28)** **	1.80 (0.55–3.20)	18.707, <0.001	−0.685

TD: Typically Developed; ASD: Autism Spectrum Disorder; Acc: Acceleration; TL: Timing Lag (see formula 2); LV: Lag Variability (see formula 3); NS: No Significance. Between group post hoc Mann-Whitney results in bold (* $p < 0.05$ and ** $p < 0.01$).

4. Discussion

We compared inter-limb coordination measures between TD children and children with ASD, while performing jumping jacks with and without rhythmic signal. Our main findings show no effect of the rhythmic signal on the coordination measures. However, children with ASD exhibited high variability in limb coordination, as shown by the LV measure, compared to the TD children. This is the first study to report a difference in an inter-limb coordination measure between children with ASD and TD children, performing a simple, quick four-limb jumping activity.

While the maximal value of the limb accelerations was not a primary outcome measure of this study, we report lower accelerations in the ASD group compared to the TD group (statistically significant for the left leg and arm during self-selected rhythm jumps and not significant but showing the same trend for the other limbs and the rhythmic signal condition). This finding is expected, as several publications report slower repetitive hand and foot movement assessed with standardized test batteries in individuals with ASD, as reviewed by Gowen and Hamilton [22]. Also, drumming movements of children with ASD were reported as slower compare to TD children [23].

There was no statistically significant main effect of the rhythmic signal, provided during the jumping activity. Since there are reports of impaired early auditory pathways in ASD [24], the timing of motor neuron transmission in our ASD group may have been influenced by delayed auditory processing, rendering the cues unhelpful or even disturbing to the task execution. This explanation was suggested in a study that compared the cadence during an auditory-cued two-legged hopping task between TD and ASD groups [25]. While the TD group showed a high performance of synchronizing their jumps with the cues, the ASD group showed a varied deviant response to the cueing [25]. In·our study, however, we did not test for synchronization between the rhythmic signal and the movements of the children. Therefore, we cannot attest to the success or failure of the metronome in regulating the jumping sequence. Our results suggest that the rhythmic signal has no effect on the inter-limb coordination or its variability between jumps in children (with or without ASD) while performing jumping jacks.

Surprisingly, we found no statistically significant differences in the TL outcome measure between the two groups. This finding suggests similar inter-limb coordination between TD and ASD children performing jumping jacks. We assume that this finding can be explained by the simplicity of the chosen activity. Contrarily to marching or drumming activities, which involve out-of-phase inter-limb movement, jumping jacks comprise of in-phase symmetrical limb movements. It was suggested that this type of activity produces a simultaneous activation of homologous muscle groups [26]. Therefore, the complexity of motor planning required for the activity chosen for this study might be smaller compared to activities such as gait. We assume that the deficits in motor planning in children with ASD contributed to the inter-limb coordination deficits in the studies reported in the literature due to the more complex task chosen. Conversely, for the jumping jacks activity, we surmise that the main factor influencing coordination is not motor planning, but the sense of proprioception. The ability to perform inter-limb coordinated movements relies, among other factors, on an intact sense of proprioception [27]. It has been reported that the sensory input of individuals with ASD is intact. Specifically, studies demonstrated no deficit in proprioception in individuals with ASD. For example, the accuracy and precision of the proprioceptive estimates of identifying the angle of the elbow and the position of the fingertip in adolescents with ASD was similar to adolescents without ASD [28]. We therefore conclude that the similarity in the TL between the ASD and TD groups relates to the characteristics of the jumping activity, selected for this study, which relies more on the sense of proprioception then on motor planning.

As expected, children with ASD exhibited high variability in limb coordination, i.e., high LV measures, compared to the TD children. This means that while the inter-limb asynchronization in TD children was consistent across consecutive jumps, the inter-limb asynchronization in children with ASD varied between consecutive jumps, and this difference was statistically significant. High intra-individual variability is considered a marker for ASD [29,30]. The high intra-individual variability in ASD was demonstrated for measures such as reaction time [31,32], hand grip strength [29], finger tapping [29], drumming [23], and walking tasks [29,33,34]. Our research makes an important contribution to the literature as it demonstrates the ability to perform objective quantitative discrimination between TD children and children with ASD using a simple and quick protocol of a four-limb activity. This protocol could be used in future studies to investigate differences in intra-subject variability between groups of children with ASD of different sex, age, and level of IQ.

The main limitation of this study is the small sample size of the ASD group. Future studies should be encouraged by our preliminary results and continue the investigation on a larger population of children with ASD. Another limitation concerns the placement of the sensors, attached to the wrist and ankles of the subject. Although there were no between-group differences in BMI, slight variability of limb length between the children is expected. Accordingly, the maximal acceleration values may have been influenced by this, so that higher acceleration values would be measured when the sensor is located further from the shoulder or hip joint. This limitation, however, has no effect on the values of the TL and the LV since the calculations of these measures consider the difference between both limbs. Also, the metronome frequency, set to 1 Hz, might not have been suited to all participants and they might have ignored it. Finally, although we report coordination deficits in children with ASD, these could be attributed to differences in IQ, motivation, or imitation ability.

In conclusion, our preliminary findings suggest that the simple protocol presented in this study might allow an objective and accurate quantification of the intra-subject variability of children with ASD via actigraphy. This method should be further explored to discern between groups of children with ASD and other populations with motor dysfunction.

Author Contributions: Conceptualization, S.P.; methodology, S.P., L.V.G. and S.S.; software, S.P.; formal analysis, S.P.; investigation, S.S. and L.V.G.; resources, S.S. and L.V.G.; data curation, S.S. and L.V.G.; writing—original draft preparation, S.P.; writing—review and editing, S.P., S.S. and L.V.G.; visualization, S.P.; supervision, S.P.; project administration, S.P. and L.V.G. All authors have read and agreed to the published version of the manuscript.

Funding: This research received no external funding.

Acknowledgments: The authors greatly appreciate the work of the following undergraduate occupational therapy students in data collection: T.S., N.I., B.L., and L.S.

Conflicts of Interest: The authors declare no conflict of interest.

References

1. Baio, J.; Wiggins, L.; Christensen, D.L.; Maenner, M.J.; Daniels, J.; Warren, Z.; Kurzius-Spencer, M.; Zahorodny, W.; Rosenberg, C.R.; White, T.; et al. Prevalence of autism spectrum disorder among children aged 8 Years—Autism and developmental disabilities monitoring network, 11 Sites, United States, 2014. *MMWR Surveill. Summ.* **2018**, *67*, 1–23. [CrossRef]

2. Torres, E.B.; Caballero, C.; Mistry, S. Aging with autism departs greatly from typical aging. *Sensors* **2020**, *20*, 572. [CrossRef]

3. Fournier, K.A.; Hass, C.J.; Naik, S.K.; Lodha, N.; Cauraugh, J.H. Motor coordination in autism spectrum disorders: A synthesis and meta-analysis. *J. Autism Dev. Disord.* **2010**, *40*, 1227–1240. [CrossRef]

4. Randall, M.; Egberts, K.J.; Samtani, A.; Scholten, R.J.P.M.; Hooft, L.; Livingstone, N.; Sterling-Levis, K.; Woolfenden, S.; Williams, K. Diagnostic tests for autism spectrum disorder (ASD) in preschool children. *Cochrane Database Syst. Rev.* **2018**. [CrossRef]

5. Wilson, R.B.; McCracken, J.T.; Rinehart, N.J.; Jeste, S.S. What's missing in autism spectrum disorder motor assessments? *J. Neurodev. Disord.* **2018**, *10*, 33. [CrossRef]

6. Wilson, R.B.; Enticott, P.G.; Rinehart, N.J. Motor development and delay: Advances in assessment of motor skills in autism spectrum disorders. *Curr. Opin. Neurol.* **2018**, *31*, 134–139. [CrossRef]

7. Pan, C.-Y.; Chu, C.-H.; Tsai, C.-L.; Sung, M.-C.; Huang, C.-Y.; Ma, W.-Y. The impacts of physical activity intervention on physical and cognitive outcomes in children with autism spectrum disorder. *Autism* **2017**, *21*, 190–202. [CrossRef]

8.   Kaur, M.; Srinivasan, S.M.; Bhat, A.N. Comparing motor performance, praxis, coordination, and interpersonal synchrony between children with and without Autism Spectrum Disorder (ASD). *Res. Dev. Disabil.* **2018**, *72*, 79–95. [CrossRef]

9.   Stanish, H.I.; Curtin, C.; Must, A.; Phillips, S.; Maslin, M.; Bandini, L.G. Physical Activity Levels, Frequency, and Type Among Adolescents with and Without Autism Spectrum Disorder. *J. Autism Dev. Disord.* **2017**, *47*, 785–794. [CrossRef]

10.   Pan, C.Y.; Frey, G.C. Identifying physical activity determinants in youth with autistic spectrum disorders. *J. Phys. Act. Health* **2005**, *2*, 412–422. [CrossRef]

11.   Pan, C.Y. Age, social engagement, and physical activity in children with autism spectrum disorders. *Res. Autism Spectr. Disord.* **2009**, *3*, 22–31. [CrossRef]

12.   Memari, A.H.; Ghaheri, B.; Ziaee, V.; Kordi, R.; Hafizi, S.; Moshayedi, P. Physical activity in children and adolescents with autism assessed by triaxial accelerometry. *Pediatr. Obes.* **2013**, *8*, 150–158. [CrossRef] [PubMed]

13.   Yavuz-Kodat, E.; Reynaud, E.; Geoffray, M.-M.; Limousin, N.; Franco, P.; Bourgin, P.; Schroder, C.M. Validity of Actigraphy Compared to Polysomnography for Sleep Assessment in Children with Autism Spectrum Disorder. *Front. Psychiatry* **2019**, *10*, 551. [CrossRef] [PubMed]

14.   Moore, M.; Evans, V.; Hanvey, G.; Johnson, C. Assessment of Sleep in Children with Autism Spectrum Disorder. *Children* **2017**, *4*, 72. [CrossRef] [PubMed]

15.   Goldman, S.E.; Alder, M.L.; Burgess, H.J.; Corbett, B.A.; Hundley, R.; Wofford, D.; Fawkes, D.B.; Wang, L.; Laudenslager, M.L.; Malow, B.A. Characterizing sleep in adolescents and adults with autism spectrum disorders. *J. Autism Dev. Disord.* **2017**, *47*, 1682–1695. [CrossRef] [PubMed]

16.   Goodwin, M.S.; Mazefsky, C.A.; Ioannidis, S.; Erdogmus, D.; Siegel, M. Predicting aggression to others in youth with autism using a wearable biosensor. *Autism Res.* **2019**, *12*, 1286–1296. [CrossRef]

17.   Hardy, M.W.; LaGasse, A.B. Rhythm, movement, and autism: Using rhythmic rehabilitation research as a model for autism. *Front. Integr. Neurosci.* **2013**, *7*, 19. [CrossRef]

18.   Migueles, J.H.; Cadenas-Sanchez, C.; Ekelund, U.; Delisle Nyström, C.; Mora-Gonzalez, J.; Löf, M.; Labayen, I.; Ruiz, J.R.; Ortega, F.B. Accelerometer Data Collection and Processing Criteria to Assess Physical Activity and Other Outcomes: A Systematic Review and Practical Considerations. *Sports Med.* **2017**, *47*, 1821–1845. [CrossRef]

19.   Krishnan, V.; Jaric, S. Effects of task complexity on coordination of inter-limb and within-limb forces in static bimanual manipulation. *Motor Control* **2010**, *14*, 528–544. [CrossRef]

20.   Ridderikhoff, A.; Peper, C.; Lieke, E.; Beek, P.J. Unraveling Interlimb Interactions Underlying Bimanual Coordination. *J. Neurophysiol.* **2005**, *94*, 3112–3125. [CrossRef]

21.   Fritz, C.O.; Morris, P.E.; Richler, J.J. Effect size estimates: Current use, calculations, and interpretation. *J. Exp. Psychol. Gen.* **2012**, *141*, 2–18. [CrossRef] [PubMed]

22.   Gowen, E.; Hamilton, A. Motor abilities in autism: A review using a computational context. *J. Autism Dev. Disord.* **2013**, *43*, 323–344. [CrossRef] [PubMed]

23.   Fitzpatrick, P.; Romero, V.; Amaral, J.L.; Duncan, A.; Barnard, H.; Richardson, M.J.; Schmidt, R.C. Evaluating the importance of social motor synchronization and motor skill for understanding autism. *Autism Res.* **2017**, *10*, 1687–1699. [CrossRef] [PubMed]

24.   Marco, E.J.; Hinkley, L.B.N.; Hill, S.S.; Nagarajan, S.S. Sensory processing in autism: A review of neurophysiologic findings. *Pediatr. Res.* **2011**, *69*, 48–54. [CrossRef] [PubMed]

25.   Moran, M.F.; Foley, J.T.; Parker, M.E.; Weiss, M.J. Two-legged hopping in autism spectrum disorders. *Front. Integr. Neurosci.* **2013**, *7*, 14. [CrossRef] [PubMed]

26.   Fuchs, A.; Kelso, J.A. A theoretical note on models of interlimb coordination. *J. Exp. Psychol. Hum. Percept. Perform.* **1994**, *20*, 1088–1097. [CrossRef]

27.   Serrien, D.J.; Li, Y.; Steyvers, M.; Debaere, F.; Swinnen, S.P. Proprioceptive regulation of interlimb behavior: Interference between passive movement and active coordination dynamics. *Exp. Brain Res.* **2001**, *140*, 411–419. [CrossRef]

28.   Fuentes, C.T.; Mostofsky, S.H.; Bastian, A.J. No proprioceptive deficits in autism despite movement-related sensory and execution impairments. *J. Autism Dev. Disord.* **2011**, *41*, 1352–1361. [CrossRef]

29. Morrison, S.; Armitano, C.N.; Raffaele, C.T.; Deutsch, S.I.; Neumann, S.A.; Caracci, H.; Urbano, M.R. Neuromotor and cognitive responses of adults with autism spectrum disorder compared to neurotypical adults. *Exp. Brain Res.* **2018**, *236*, 2321–2332. [CrossRef]
30. Mottron, L.; Belleville, S.; Rouleau, G.A.; Collignon, O. Linking neocortical, cognitive, and genetic variability in autism with alterations of brain plasticity: The Trigger-Threshold-Target model. *Neurosci. Biobehav. Rev.* **2014**, *47*, 735–752. [CrossRef]
31. Geurts, H.M.; Grasman, R.P.P.P.; Verté, S.; Oosterlaan, J.; Roeyers, H.; van Kammen, S.M.; Sergeant, J.A. Intra-individual variability in ADHD, autism spectrum disorders and Tourette's syndrome. *Neuropsychologia* **2008**, *46*, 3030–3041. [CrossRef] [PubMed]
32. Adamo, N.; Huo, L.; Adelsberg, S.; Petkova, E.; Castellanos, F.X.; Di Martino, A. Response time intra-subject variability: Commonalities between children with autism spectrum disorders and children with ADHD. *Eur. Child Adolesc. Psychiatry* **2014**, *23*, 69–79. [CrossRef] [PubMed]
33. Rinehart, N.J.; Tonge, B.J.; Bradshaw, J.L.; Iansek, R.; Enticott, P.G.; McGinley, J. Gait function in high-functioning autism and Asperger's disorder: Evidence for basal-ganglia and cerebellar involvement? *Eur. Child Adolesc. Psychiatry* **2006**, *15*, 256–264. [CrossRef] [PubMed]
34. Manicolo, O.; Brotzmann, M.; Hagmann-von Arx, P.; Grob, A.; Weber, P. Gait in children with infantile/atypical autism: Age-dependent decrease in gait variability and associations with motor skills. *Eur. J. Paediatr. Neurol.* **2019**, *23*, 117–125. [CrossRef] [PubMed]

Article

Differences in Motion Accuracy of Baduanjin between Novice and Senior Students on Inertial Sensor Measurement Systems

Hai Li [1,2], Selina Khoo [1] and Hwa Jen Yap [3,*]

[1] Centre for Sport and Exercise Sciences, University of Malaya, Kuala Lumpur 50603, Malaysia; 10000861@njtc.edu.cn (H.L.); selina@um.edu.my (S.K.)

[2] College of Sport, Neijiang Normal University, Neijiang 641112, China

[3] Department of Mechanical Engineering, Faculty of Engineering University of Malaya, Kuala Lumpur 50603, Malaysia

* Correspondence: hjyap737@um.edu.my; Tel.: +603-79675240

Received: 12 September 2020; Accepted: 30 October 2020; Published: 2 November 2020

Abstract: This study aimed to evaluate the motion accuracy of novice and senior students in Baduanjin (a traditional Chinese sport) using an inertial sensor measurement system (IMU). Study participants were nine novice students, 11 senior students, and a teacher. The motion data of all participants were measured three times with the IMU. Using the motions of the teacher as the standard motions, we used dynamic time warping to calculate the distances between the motion data of the students and the teacher to evaluate the motion accuracy of the students. The distances between the motion data of the novice students and the teacher were higher than that between senior students and the teacher ($p < 0.05$ or $p < 0.01$). These initial results showed that the IMU and the corresponding mathematical methods could effectively distinguish the differences in motion accuracy between novice and senior students of Baduanjin.

Keywords: motion capture; inertial sensor measurement systems; motion accuracy; Baduanjin; physical education

---

1. Introduction

Traditional Chinese sport has been a compulsory component of Physical Education (PE) in universities in China since 2002 [1]. Although there are various traditional Chinese sports to choose from, 76.7% of universities taught martial arts in their PE curriculum [2]. In 2016, the Communist Party of China and the Chinese government adopted the 'Healthy China 2030' national health plan [3]. In this plan, Baduanjin was identified as a traditional Chinese sport that was promoted and supported by the government. This resulted in increased Baduanjin teaching and research in universities throughout the country [4].

Although universities in China must incorporate traditional Chinese sports into their PE curriculum, there have been problems with its implementation. These include a high student-teacher ratio, uninteresting forms of teaching-learning resources, and an incomplete assessment system. These three problems adversely affected the requirements for teaching quality set by the People's Republic of China Ministry of Education [5,6]. Although the high student–teacher ratio has been a problem since 2005, it has yet to be resolved [7,8]. Teachers are not able to provide individual guidance to each student because of the large number of students in the class. As a result, teachers cannot correct all the students' mistakes, and students are not aware of their incorrect movements [9].

In recent years, motion capture (Mocap) has been widely applied in fields such as clinical and sports biomechanics to distinguish between different types of motions or analyze differences between motions [10,11]. Studies have also applied Mocap in PE for adaptive motion analysis to evaluate the motion quality of learners and feedback the information to assist them in detecting and correcting their inaccurate motions. In the study by Koji Yamada et al. [12], a system based on Mocap was developed for Frisbee learners. Researchers used the Kinect device to obtain 3D motion data of learners during exercise, detect their pre-motion/motion/post-motion, and display the feedback information to improve their motions. The results showed that the system developed by the researchers can effectively improve the motions of learners [12]. Chen et al. [13] applied Kinect in Taichi courses in universities. The 3D data of motions from novice students captured by Kinect were compared with an expert in order to evaluate the quality of motions and students were informed of their results. The research showed that the motion evaluation system on Kinect developed by the researchers accelerated learning by novice Taichi students. More recently, Amani Elaoud et al. [14] used Kinect V2 to obtain red, green, blue and depth (RGB-D) data of motion. They used these data to compare the differences between novice students and experts on the central angles of points that affect throwing performance in handball. These experiments show how researchers have used various categories of Mocap.

Based on different technical characteristics, the application of Mocap in PE can be divided into four categories: optoelectronic system (OMS) [15], electromagnetic system (EMS), image processing systems (IMS), and inertial sensor measurement system (IMU) [16]. In these four Mocap categories, OMS is the most accurate and is considered to be the gold standard in motion capture [16,17]. However, OMS requires a large number of high-precision and high-speed cameras that will inevitably result in issues related to cost, coordination, and manual use [18]. Moreover, OMS cannot capture the movement of objects when the marker is obscured [19]. These deficiencies have limited the practical application of OMS in PE. The advantage of EMS over OMS is that it can measure motion data of a specific point of the body regardless of visual shielding [20]. However, EMS is susceptible to interference from the electromagnetic environment which distorts measurement data [21]. Also, EMS has to be kept within a certain distance from the base station, which limits the use range [22]. IMS has better accuracy compared to EMS and an improved range compared to OMS [16]. Most studies have used low-cost IMS (such as the Kinect device) to capture motion for analyzing motion in PE. However, there are some disadvantages in low-cost IMS, namely low-accuracy, insufficient environment adaptability, and limited range of motion because the Kinect sensor has a small field of vision [16]. High-performance IMS does not have these shortcomings. Generally, high-performance IMS has favorable accuracy and a good measurement range. However, high-performance IMS requires expensive high quality and/or high-speed cameras which has limited its application [16].

Based on the disadvantages of low-cost IMS in its application in PE, applying IMU (a motion capture consisting of an accelerometer, gyroscope, and a magnetometer) in PE may mitigate these application problems [23]. In recent years, the development of technology has reduced the cost of IMU, making it possible to be used in PE. The validity of assessing motion accuracy of IMUs has been confirmed. Poitras et al. [24] confirmed the criterion validity of a commercial IMU system (MVN Awinda system, Xsens) by comparing it to a gold standard optoelectronic system (Vicon). Compared to low-cost IMS, IMU has certain advantages in environmental adaptability and a sufficient range of motion. The IMU does not require any base station to work, which means it is the most mobile of the available motion capture systems [16]. Moreover, IMU can measure high-speed movements and is non-invasive for the user, making it an attractive application for PE [16,25].

However, there are a few issues capturing motions using IMU. First, the IMU sensors are sensitive to metal objects nearby which distort the measurement data [16]. Therefore, participants should wear fewer metal objects when capturing motion using IMU. Fortunately, in traditional Chinese martial arts such as Baduanjin, exercisers should, in principle, wear traditional Chinese costumes without any metal, which minimizes the impact of metals on IMU sensors. Second, a common IMU system, Perception Neuron 2.0, was used in our research which uses data cables to connect all sensors with a transmitter. Although users cannot wear this wired IMU system on their own, it does not affect the accuracy of the data. Also, the latest IMU overcomes this problem that users can't wear it on their own by having each sensor transmit the data to the external receiving terminal directly [26].

Therefore, we propose applying IMU in Baduanjin by developing a system that assesses and records the quality of motions to assist teachers and students in determining inaccurate motions. Using IMU, students can learn Baduanjin independently after class and teachers can evaluate students' progress, which is useful for formative assessment. That may alleviate current problems faced in PE classes in Chinese universities. For this purpose, we explored the feasibility of using an IMU to distinguish the difference in motion accuracy of Baduanjin between novice and senior students.

2. Materials and Methods

2.1. Overview

This study consists of three sections, namely recruiting and selecting participants, capturing motion data of Baduanjin participants on IMU, and processing and analyzing the motion data. We invited teachers and students from a university in Southwest China to participate in the study. We divided them into three groups—teachers, novice students, and senior students. We captured motion data of all participants on IMU when they practised Baduanjin. The motion data were converted to quaternions and analysed in two different ways. The first way was based on the quaternions of motion, where dynamic time warping (DTW) was used to calculate the distances between the quaternions of the teacher and the two groups of students (novice and senior). The motion accuracy of the students was expressed by distances. DTW is a classic similarity method to solve the time-warping issue in similarity computation of time series [27]. Compared with the other methods, namely the hidden Markov model (HMM) and symbolic aggregate approximation (SAX), the taken time of DTW is shorter [28,29]. Considering that, in the actual teaching, students need to get feedback information and a large number of student data in realtime, we adopted DTW in the study. The second way used the extracted key-frames to calculate distances. Based on the quaternions of key-frames, DTW was used to calculate the distances. Finally, based on data of the distances, an independent sample T-test or Mann–Whitney U test was used to define whether the motions of the two groups of students (novice and senior) were different in motion accuracy (see Figure 1).

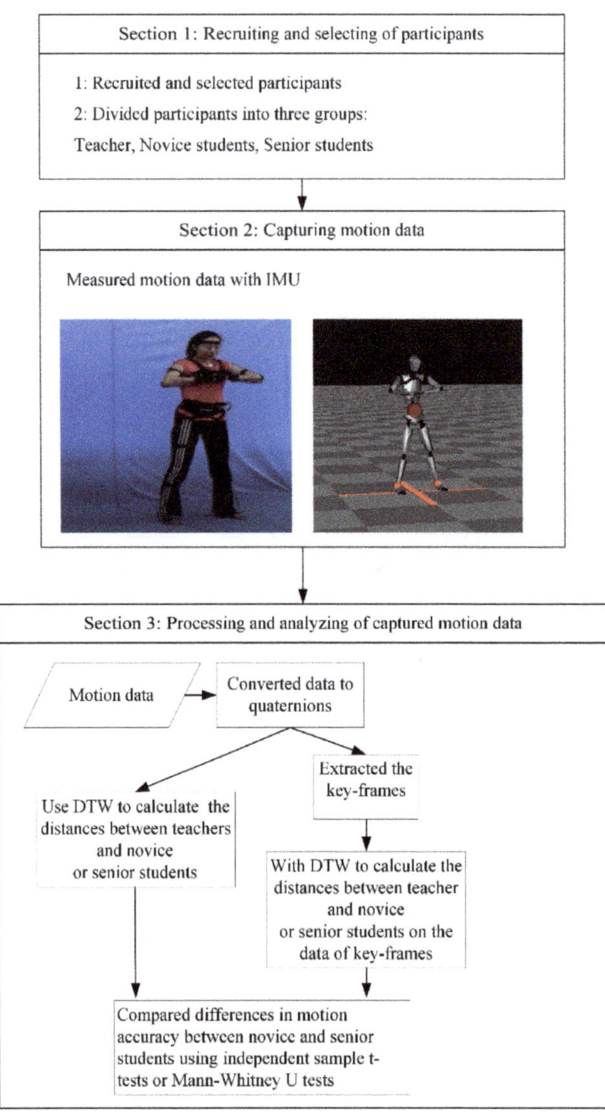

Figure 1. Flow diagram of the study.

2.2. Recruiting and Selecting Participants

In this study, we invited a martial arts PE teacher and undergraduate students to participate in the study. The inclusion criteria for participants are as follows:

Teacher: martial arts PE teacher, former national martial arts athlete, with an undergraduate and master's degree in traditional Chinese sports (martial arts specialization), and more than ten years' experience teaching Baduanjin.

Novice students: undergraduate students in the university with no experience of Baduanjin, without a disability and no clinical or mental illness.

Senior students: undergraduate students in the university who have passed Baduanjin in their PE course, without a disability, and no clinical or mental illness.

Participants read the information sheet that outlined the purpose and procedure of the study. Those who agreed to participate were given the consent form to sign.

2.3. Capturing Motion Data of Participants on IMU

Baduanjin is a traditional Chinese martial art for fitness. The speed of motions is relatively slow [30]. We used IMU to capture the motion data of the teacher, novice, and senior students for eight standard motions of Baduanjin as shown in Figure 2.

Motion 1: Holding the Hands High with Palms Up to Regulate the Internal Organ

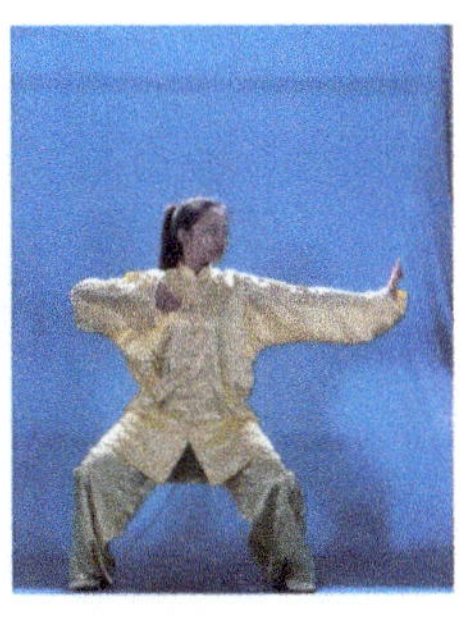

Motion 2: Posing as an Archer Shooting Both Left- and Right-Handed

Motion 4: Looking Backwards to Prevent Sickness and Strain

Motion 5: Swinging the Head and Lowering the Body to Relieve Stress

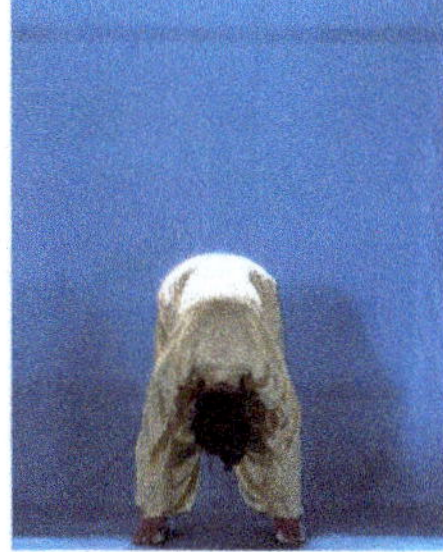

Motion 3: Holding One Arm Aloft to Regulate the Functions of the Spleen and Stomach

Motion 6: Moving the Hands down the Back and Legs, and Touching the Feet to Strengthen the Kidneys

Motion 7: Thrusting the Fists and Making the Eyes Glare to Enhance Strength

Motion 8: Raising and Lowering the Heels to Cure Diseases

Figure 2. Eight standard motions of Baduanjin.

2.3.1. IMU

We used Perception Neuron 2.0, a low-cost IMU developed by Noitom, to capture Baduanjin motion data of participants [31]. This IMU includes 17 inertial sense units and each unit comprised a 3-axis gyroscope, 3-axis accelerate, and 3-axis magnetometer, which measures and records the rotation angle data of 17 position points of human movement. Sers et al. [32] compared the IMU used in our study with a gold standard optoelectronic system (Vicon), and confirmed the IMU's effectiveness in measuring motion accuracy. The supporting software of the IMU, Axis neuron software developed by Noitom, transforms the recorded data into Biovision Hierarchy (BVH) motion files.

2.3.2. Capturing Motion Data

Before measuring the motion data, the teacher and senior students practised Baduanjin for 30 min. As the novice students had not learned Baduanjin, they followed the demonstration of the teacher practising Baduanjin for 30 min. After the practice, the motion data of participants were measured by IMU. No feedback was given to students during practice.

2.4. Data-Analysis

2.4.1. Extracting and Converting Raw Data

The raw data was converted into BVH file by the Axis neuron software. The BVH file is a file format developed by the BVH Company to store skeleton hierarchy information and three-dimensional motion data [33]. The BVH file comprises two parts: one is used to store skeleton hierarchy information and the other to store motion information. The skeleton hierarchy information includes the connection relationship between joint points and the offsets of the child joint points from their parent skeleton points. In the skeleton hierarchy, the first skeleton point is defined as Root. Root is the parent of all other skeleton points in the skeleton hierarchy. Motion information stores the global translation amount and the rotation amount of Root in each frame of the movement. The global translation amount is the position coordinate: X position, Y position, and Z position in the world coordinate system and the rotation amount is the rotation component: X rotation, Y rotation, and Z rotation in the Euler angle [33]. The motion information of other skeleton points is recorded on the rotation amount related to the parent points. The IMU used 17 sensors to measure motion data on 17 points of the body and the recorded order of the rotation amount of each point is Z rotation, Y rotation, and X rotation. The skeleton hierarchy information of BVH on the IMU and the skeleton model are shown in Figure 3.

In the BVH file, the rotation data is recorded on the Euler angle of 17 skeleton points. Some issues with rotation data expressed on the Euler angle (gimbal lock and singularity problems) were overcome using quaternion [34]. Quaternion is a 4-dimensional hyper-complex number, expressing a three-dimensional vector space on real numbers [35]. We used four-tuple notation to represent quaternion as follows:

$$q = [w, x, y, z] \tag{1}$$

In this quaternion, w is the scalar component, and x, y, z are the vectors.

Therefore, the format of the rotation data from BVH files was converted from Euler angle to quaternion. If the order of rotation in Euler angle is z, y, x, we used α, β, γ to represent the rotation angles of the object around $x, y,$ and z axes. The corresponding quaternion can be converted as follows:

$$q = \begin{bmatrix} w \\ x \\ y \\ z \end{bmatrix} = \begin{bmatrix} \cos(\gamma/2) \\ 0 \\ 0 \\ \sin(\gamma/2) \end{bmatrix} \begin{bmatrix} \cos(\beta/2) \\ 0 \\ \sin(\beta/2) \\ 0 \end{bmatrix} \begin{bmatrix} \cos(\alpha/2) \\ \sin(\alpha/2) \\ 0 \\ 0 \end{bmatrix}$$
$$= \begin{bmatrix} \cos(\gamma/2)\cos(\beta/2)\cos(\alpha/2) + \sin(\gamma/2)\sin(\beta/2)\sin(\alpha/2) \\ \cos(\gamma/2)\cos(\beta/2)\sin(\alpha/2) - \sin(\gamma/2)\sin(\beta/2)\cos(\alpha/2) \\ \cos(\gamma/2)\sin(\beta/2)\cos(\alpha/2) + \sin(\gamma/2)\cos(\beta/2)\sin(\alpha/2) \\ \sin(\gamma/2)\cos(\beta/2)\sin(\alpha/2) - \cos(\gamma/2)\sin(\beta/2)\sin(\alpha/2) \end{bmatrix} \tag{2}$$

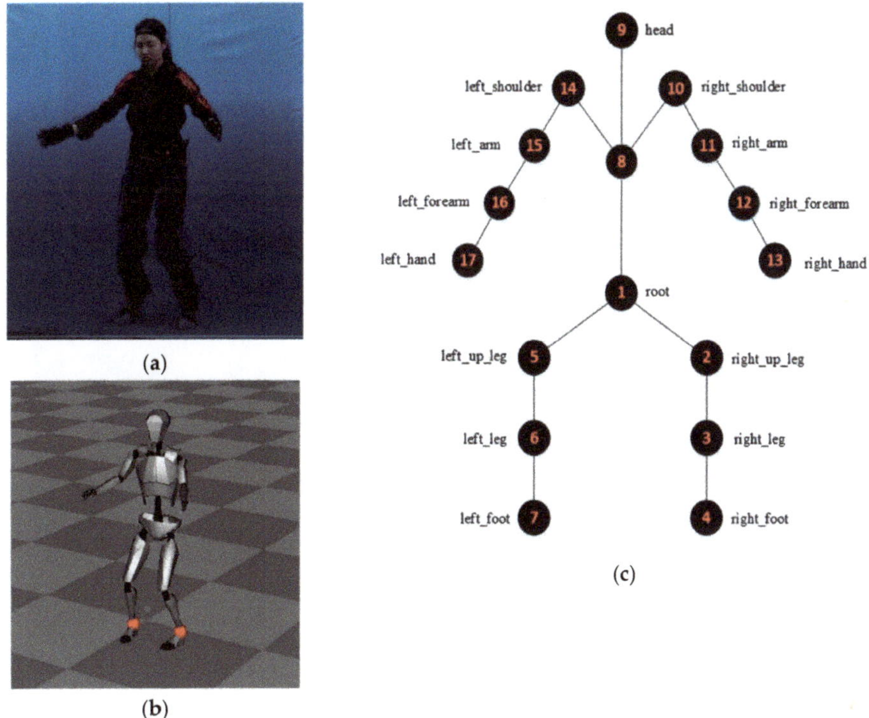

Figure 3. The skeleton hierarchy information of BVH on the IMU (Perception Neuron 2.0): (**a**) A participant wearing the IMU (Perception Neuron 2.0) to measure the motion data; (**b**) The interface of Perception Neuron 2.0; and (**c**) The skeleton model of BVH file for Perception Neuron 2.0.

2.4.2. Extracting Key-Frames

After extracting the motion data, we used key-frames extraction to reduce the motion data. Due to the limited storage and bandwidth capacity available to users, the large amount of motion data collected on Mocap may restrict its application [36]. Key-frames extraction, which extracts a small number of representative key-frames from a long motion sequence, is widely used in motion analysis. This technology can reduce the data amount, which facilitates data storage and subsequent data analysis [36,37].

Extraction of Key-Frames on Inter-Frame Pitch

We used the distance between quaternions to evaluate the inter-frame pitch between frames and set a threshold of inter-frame pitch to extract key-frames [38]. The method is based on the rotation data of each skeleton point which is represented as a quaternion and uses a simple form to evaluate the distance between two quaternions. The inter-frame pitch between the two frames is assessed by the sum of the distances between the quaternions of every point. The process is constructed with three sections: calculating the distance between quaternions, calculating the inter-frame pitch between frames, and extracting key-frames on the set threshold of inter-frame pitch.

1. The distance between quaternions

To evaluate the distance between two quaternions, the conjugate quaternion $q*$ of a quaternion is defined as follows:

$$q* = [w, -x, -y, -z] \tag{3}$$

and the quaternion norm $\|q\|$ is defined as follows:

$$\|q\| = \sqrt{w^2 + x^2 + y^2 + z^2} \tag{4}$$

then:

$$\|q\|^2 = qq* = w^2 + x^2 + y^2 + z^2 \tag{5}$$

when a quaternion norm $\|q\|$ is 1, which means:

$$w^2 + x^2 + y^2 + z^2 = 1 \tag{6}$$

the quaternion is a unit quaternion. A quaternion is converted to a unit quaternion by dividing it by its norm.

From the definitions of conjugate quaternion, quaternion norm, and unit quaternion, we can define the inverse of a quaternion (q^{-1}) as follows [39]:

$$q^{-1} = \frac{1}{\|q\|} = \frac{1}{\|q\|^2}q*, \ \|q\| \neq 0 \tag{7}$$

According to Shunyi et al. [38], if there are two quaternions: q_1, q_2 are unit quaternions and:

$$q_1 q_2^{-1} = [w, \ x, \ y, \ z] \tag{8}$$

the distance between the quaternions q_1 and q_2 is:

$$d(q_1, q_2) = \arccos w \tag{9}$$

Therefore, we converted the rotation of a skeleton point based on Euler angles into quaternion, then normalized and converted the quaternion into unit quaternion, and finally calculated the difference between any two quaternions of the point according to Equation (9).

2. Calculation of Inter-Frame Pitch between Two Frames

We used the sum of the differences between the quaternions at 17 skeleton points to evaluate the inter-frame pitch between two frames. The human motion represented by the BVH file are discrete-time vectors, which is the same after conversion to quaternions [38]. The weightage for different points needs to be taken into account when calculating the inter-frame pitch due to the tree-structure (parent-child) of the BVH format. Referring to the methods used in previous research [38,40], and the relationship structure between the skeleton points on the IMU in this study (see Figure 3), we assigned the weightage values of the 17 skeleton points as shown in Table 1.

If t_1 and t_2 are the two frames in a sequence of frames, we defined the inter-frame pitch between two frames: t_1 and t_2 as the following equation:

$$D(t_1, t_2) = \sum_{i=1}^{n} W_i d(q_i(t_1), q_i(t_2)) \tag{10}$$

In Equation (10), n represents the total number of skeleton points ($n = 17$), W_i represents the weightage of each skeleton point (shown in Table 1), and q_i represents the quaternions of each skeleton point.

Table 1. The weightage of the 17 skeleton points.

Point	Weightage
Hip	16
Right up leg	8
Right leg	4
Right foot	2
Left up leg	8
Left leg	4
Left foot	2
Spine	8
Head	4
Right shoulder	4
Right arm	2
Right forearm	1
Right hand	0.5
Left shoulder	4
Left arm	2
Left forearm	1
Left hand	0.5

3. Key-frames extraction on the set threshold of inter-frame pitch

Based on the inter-frame pitch between two frames, we set: key_frame as an array to store the quaternion corresponding to the key-frames of motion; key_num as a set of vector to store the serial number corresponding to a key-frame; key_num1 presents the time series number corresponding to the first key-frame; current_key as the last frame in the set of key_num. λ is a preset threshold value of inter-frame pitch which is mainly determined based on the demand for a compression rate of frames. The algorithm steps are shown in Figure 4.

Start

Input the motion data: $[m_1, m_2, \ldots, m_i, \ldots, m_n]$ and a threshold: λ. The total length of the time series is n. m_i presents the motion data corresponding to time series i.

Step 1: the first frame of motion is set as the first key_frame. current_key = 1;

Step 2: Enter the loop. The range of loop is: $i = 2 : n\text{-}1$;

    If $D(m_i$, mcurrent_key) => λ ; then

        Add m_i into key_frame, add i into key_num, current_key = i, $i = i +1$;

        Else $i = i + 1$;

    End the loop when $i > n -1$;

Step 3: the last frame of motion is set as the last key-frame, add m_n into key_frame, add n into key_num,

Step 4: Output key_frame, key_num;

End

Figure 4. The algorithm steps of key-frames extraction.

4. Motion reconstruction error

The purpose of motion reconstruction is to rebuild the same number of frames as the original frames based on interpolation reconstruction of non-key-frames between adjacent key-frames [38,41]. First, individually, the position coordinates (in the world coordinate system) of points were calculated on the point hierarchy and relative rotation angle between the points in the BVH file. Second, given that p_{t1} and p_{t2} are the positions of a point of adjacent key-frames in time t_1 and t_2, then p_t (representing the position of a point of non-key-frame in time t) is calculated by linear interpolation between p_{t1} and p_{t2} as follows [41]:

$$\begin{aligned} p_t &= u(t)p_{t1} + (1 - u(t))p_{t2}, \\ u(t) &= \tfrac{t_2 - t}{t_2 - t_1}, \\ t_1 &< t < t_2 \end{aligned} \tag{11}$$

The algorithm steps of motion reconstruction are shown in Figure 5.

Start

Input data of the key-frames: $[P_{t1}, P_{t2}, ..., P_m]$. The total length of the time series is n. P_{ti} presents the positions data of skeleton points corresponding to time series i.

Step 1: Enter the loop. The range of loop is: $i = 1 : n - 1$;

Step 2: P_i of non-key-frames is calculated by interpolating $P_{t[i]}$ and $Pt_{[i+1]}$ of adjacent key-frames linearly

Step 3: Output reconstruction frames

End

Figure 5. The algorithm steps of motion reconstruction.

In this study, we used the position error of the human posture to calculate the reconstruction error between the reconstructed frames and the original frames [38]. Assuming m_1 is the original motion sequence, m_2 is the reconstruction motion sequence from the key-frames, the reconstruction error $E(m_1, m_2)$ is evaluated as [42]:

$$E(m_1, m_2) = \frac{1}{n} \sum_{i=1}^{n} D(p_1^i - p_2^i) \tag{12}$$

The distance of human posture is used to measure the position error of human posture:

$$D(p_1^i - p_2^i) = \sum_{k=1}^{m} \| p_{1,k}^i - p_{2,k}^i \|^2 \tag{13}$$

In this equation, m represents the total number of skeleton points, $p_{1,k}^i$ is the position of k point in i frame of the original motion sequence, and $p_{2,k}^i$ is the position of k point in i frame in the reconstruction sequence.

Extraction of Key-Frames on Clustering

A problem with the key-frames extraction on inter-frame pitch is that the compression rate of the key-frames with the same inter-frame threshold for different actions may vary considerably [40]. As the eight motions of Baduanjin are quite different, the key-frames extraction on the inter-frame pitch may cause some motions to be compressed too much, and some motions not compressed enough. Therefore, we also chose another way to extract key-frames on clustering. This method was used for key-frames with the pre-set compression rate [43].

1. K-means clustering algorithm

K-means clustering algorithm is an iterative partition clustering algorithm. In this key-frame extraction method, we used the K-means clustering algorithm to cluster the 3D coordinates ($[x, y, z]$) of the skeleton points in the original frame. Assuming that the total length of the original frames is N, i represents the i frame in N. p^i is the vectors of the 3D coordinate positions of all relevant skeleton points of the i frame in the original frames. Therefore, the vectors collection of the 3D coordinate data of every point of original frames is $(p^1, p^2, \ldots, p^i)$, $p^i \in R^N$. According to the K-means clustering algorithm, the data of skeleton points (R^N) in the frames is clustered into K ($K \le N$) clusters as follows [44]:

Step 1: Randomly select K cluster centroids from R^N are $u_1, u_2 \ldots u_K$;

Step 2: Repeat the following process to get convergence.

For the p^i corresponding to one frame, we calculated the distances from each cluster centroid ($u_j, j \in K$) and classified it into the class corresponding to the minimum distance [45]:

$$D = \operatorname{argmin} \sum_{i=1}^{N} \sum_{j=1}^{K} \|p^i - u_j\|^2 \tag{14}$$

In this equation, D represents the minimum distance between the cluster centroid and the centre of p^i, and when D is the smallest, p^i is classified into class j.

For each class j, the cluster centroid (u_j) of that class was recalculated:

$$u_j = \frac{\sum_{i=1}^{N} r_{ij} p^i}{\sum_{i=1}^{N} r_{ij}} \tag{15}$$

In this equation, r_{ij} indicates that when p^i is classified as j, it is 1; otherwise, it is 0.

2. Key-frames extraction

Using the above k-means clustering algorithm, we extracted K cluster centroids from the original frame. Each cluster is clustered from the 3D coordinates of the 17 points in the original frames. Therefore, one cluster centroid is constructed with 51 (17×3) vectors. Based on these cluster centroids, we extracted the key-frames by calculating the Euclidean distance between the cluster centroid of each point and the corresponding point coordinates in the original frames. The steps to extract key-frames are as follows:

Start

Input the 3D coordinate data of every point of the original frames:

$$\begin{aligned} \left(p^1, p^2 \ldots p^i\right), \quad & p^i \in R^N; \\ & p^i = (p^i_1, p^i_2 \ldots p^i_j),\ j = 17; \\ & p^i_j = [x^i_j, y^i_j, z^i_j] \end{aligned} \tag{16}$$

and the number of key-frames to be extracted is K;

Step 1: Using the k-means clustering algorithm to calculate cluster centroids of the K clusters are expressed as:

$$\begin{aligned} u_m &= (u_{m1}, u_{m2} \ldots u_{mj}),\ m \in (1,2,3 \ldots K),\ j = 17; \\ u_{mj} &= [x_{mj}, y_{mj}, z_{mj}] \end{aligned} \tag{17}$$

Step 2: Calculate the Euclidean distance of 3D coordinates between each point of the cluster and the corresponding point of the original frames:

$$
\begin{aligned}
C_m &= \min(u_m, p^i) \\
&= \sum_{j=1}^{17} \min(dis(u_{mj}, p^i_j)); \\
dis(u_{mj}, p^i_j) &= \|u_{mj} - p^i_j\|^2
\end{aligned}
\tag{18}
$$

$\min(dis(u_{mj}, p^i_j))$ means that after calculating the distances between m cluster and all original frames, the j point of pi which value of $dis(u_{mj}, p^i_j)$ is minimum is recorded as 1; otherwise, it is recorded as 0. i of p^i corresponding to the maximum value of C_m is a sequence of key-frames.

Step 3: Sequences of key-frames are arranged from small to large after extraction. If the first frame and the last frame in the original frames are not included in the key-frames, the first frame and the last frame must be added into key-frames.

End

In this key-frames extraction, the number of key-frames can be preset. The key-frames of the corresponding compression rate is obtained by presetting the compression rate as follows [42]:

$$
K = c\_rate * N
\tag{19}
$$

where K is the number of key-frames to be extracted, c_rate is the compression rate of the key-frame to be obtained, and N is the total number of original frames.

After extracting key-frames, we continued with the ways to motion reconstruction and evaluate reconstruction error as described above.

2.4.3. Evaluate Motion the Accuracy of Motions Data

In this study, we referred to previous studies [13,46] to evaluate the motion accuracy of student motions by assessing the differences between students' motions and teacher's motions. Due to the difference in speed between individual movements, different time series were considered when assessing the difference between two motions. We chose DTW, a well-established method, to account for different time series to evaluate the difference in the motions between teachers and students [47]. Since DTW compares the other methods, i.e., HMM and SAX, without a training stage, the taken time is shorter. First, the derived quaternions were normalized in unit length of a quaternion: $q = [w, x, y, z]$ can be described as: $\|q\| = 1$ and $w^2 + x^2 + y^2 + z^2 = 1$. Therefore, three components (x, y, z) out of the four components (w, x, y, z) of the quaternions can be used to represent the rotations of the skeleton points over a temporal domain. Then, we used DTW to evaluate the difference between two sequences of motions on the skeleton points. First, we assessed the difference between two motions on a single skeleton point. For example, there are two motion data on quaternions for a skeleton point from a teacher and a student, one from the teacher: $q_{tea}(t)$, one from a student: $q_{stu}(t)$. The length of the two sequences of quaternions are n and m:

$$
\begin{aligned}
q_{tea}(t) &= q_{tea}(1), q_{tea}(2), \ldots, q_{tea}(i), \ldots, q_{tea}(n) \\
q_{stu}(t) &= q_{stu}(1), q_{stu}(2), \ldots, q_{stu}(j), \ldots, q_{stu}(m)
\end{aligned}
\tag{20}
$$

The vector in the quaternion arrays consists of three components (x, y, z) of quaternions. A distance matrix $(n \times m)$ is constructed to align the quaternions of two sequences. The elements (i, j) in the matrix represent the Euclidean distance: $dis(qtea(i), qstu(j))$ between the two points $q_{tea}(i)$ and $q_{stu}(j)$:

$$
dis(q_{tea}(i), q_{stu}(j)) = \left|q_{tea}(i) - q_{stu}(j)\right|^2
\tag{21}
$$

In the distance matrix, many paths are from the upper-left corner to the lower-right corner of the distance matrix. We used Φk to represent any point on these paths: $\Phi k = (\Phi_{tea}(k), \Phi_{stu}(k))$ where:

$\Phi_{tea}(k)$: the value of k is $1, 2, \ldots, n$,

$\Phi_{stu}(k)$: the value of k is $1, 2, \ldots, m$,

Φk, the value of k is $1, 2, \ldots, T, (T = n \times m)$

We found a suitable path as the warping path, where the cumulative distance of path is the smallest of all paths [39]:

$$DTW(q_{tea}(t), q_{stu}(t)) = \min \sum_{k=1}^{T} dis(\Phi_{tea}(k), \Phi_{stu}(k)) \tag{22}$$

Then, the distance of $DTW(q_{tea}(t), q_{stu}(t))$ is obtained through dynamic programming as follows [47]:

$$
\begin{aligned}
&DTW(q_{tea}(t), q_{stu}(t)) = f(n, m); \\
&f(0,0) = 0; \\
&f(0,1) = f(1,0) = \infty; \\
&f(i,j) = dis(q_{tea}(i), q_{stu}(j)) + \min\{f(i-1,j), f(i,j-1), f(i-1,j-1)\}, (i = 1,2,\ldots,n; j = 1,2,\ldots,m)
\end{aligned}
\tag{23}
$$

To prevent the wrong matching by excessive time warping, the warping path was constrained near the diagonal of the matrix by setting the global warping window for DTW [48,49]. In this study, the global warping window is set as 10 percent of the entire window span: $0.1 \times \max(n, m)$. The cumulative distance of the warping path represents the difference of rotation between teacher and student on the skeleton points is shown in Equation (22). Then, the macro difference between students' motions and teacher's motions was evaluated by taking the average of the cumulative distances of all the skeleton points as follows:

$$D(m_{tea}, m_{stu}) = \frac{\sum\limits_{i=1}^{n} DTW(q_{tea}^{i}, q_{stu}^{i})}{n} \tag{24}$$

In this equation, m_{tea} represents the teacher motion sequence; m_{stu} represents the students' motion sequence, q^{i} is the vectors of the quaternion of i skeleton point in the two motion sequences, and the total number of skeleton points is n.

Finally, data of the differences were analysed using IBM SPSS Statistics 25.0 to assess if there were significant differences in the motion accuracy of the two groups of students (novice and senior students) on the whole and each point. We used the independent sample T-test on data with normal distribution and the Mann–Whitney U test on data with non-normal distribution.

3. Results

3.1. Demographic Characteristics of Participants

We recruited 21 participants for this study, including a martial arts teacher, nine undergraduate students who have not learned Baduanjin (novice students), and 11 undergraduate students who had completed the Baduanjin course (senior students). All participants gave their informed consent for inclusion before they participated in the study. The study was conducted in accordance with the Declaration of Helsinki, and the protocol was approved by the University of Malaya Research Ethics Committee (UM.TNC2/UMREC–558). The demographic characteristics of the students are shown in Table 2. For each mean duration of the eight motions shown in Table 3, we measured all participants three times with IMU, resulting in 63 motion data.

Table 2. Demographic characteristics of the students.

Group	Gender	Age (years) (mean ± SD)	Height (cm) (mean ± SD)	Weight (kg) (mean ± SD)
Novice Student	Male: 5	18.60 ± 0.55	169.40 ± 3.91	58.20 ± 4.60
	Female: 4	18.75 ± 0.96	161.25 ± 3.59	48.25 ± 2.06
Senior Student	Male: 4	20.25 ± 0.50	170.75 ± 4.11	60.50 ± 4.80
	Female: 7	20.00 ± 0.82	161.43 ± 3.84	48.57 ± 3.74

Table 3. Mean duration of Baduanjin.

Motion	Valid	Mean duration (s) (mean ± SD)
Motion 1	63	12.24 ± 2.49
Motion 2	63	21.52 ± 4.19
Motion 3	63	16.75 ± 4.14
Motion 4	63	14.79 ± 3.85
Motion 5	63	19.18 ± 5.25
Motion 6	63	16.95 ± 4.09
Motion 7	63	13.10 ± 3.93
Motion 8	63	1.51 ± 0.25

3.2. Differences in Motion Accuracy between Novice and Senior Students on Original Frames

Algorithms explained in the data analysis section were coded with Matlab R2018b. Independent sample T-tests and Mann–Whitney U tests were used to assess the differences in motion accuracy of novice and senior students.

Before assessing macro differences, we assessed the normality of original frames data using the Shapiro–Wilk test (see Table 4).

Table 4. Normality of data of groups using the Shapiro-Wilk test.

Motion	Group	Statistic	df	Sig.
1	Novice Student	0.788	27	0.000
	Senior Student	0.753	33	0.000
2	Novice Student	0.962	27	0.408
	Senior Student	0.973	33	0.575
3	Novice Student	0.956	27	0.299
	Senior Student	0.948	33	0.120
4	Novice Student	0.963	27	0.428
	Senior Student	0.979	33	0.758
5	Novice Student	0.963	27	0.428
	Senior Student	0.979	33	0.758
6	Novice Student	0.789	27	0.000
	Senior Student	0.852	33	0.000
7	Novice Student	0.881	27	0.005
	Senior Student	0.878	33	0.001
8	Novice Student	0.932	27	0.078
	Senior Student	0.890	33	0.003

From Table 4, we can see that the data of the groups on Motions 2, 3, and 4 were normally distributed ($p > 0.05$), whereas the others were not. Therefore, we assessed the differences in motion accuracy of Motions 2, 3, and 4 between novice and senior students using independent sample T-tests (see Table 5). The differences in the motion accuracy of other motions between novice and senior students were assessed using Mann–Whitney U tests (see Table 6).

Table 5. Differences in motion accuracy between novice and senior students on original frames (using the independent sample T-test).

Motion	Group	N [1]	Mean [2]	Std. Deviation	F	Sig.	t	Sig. [3]
2	Novice Student	27	640.76	74.38	2.289	0.136	4.275	0.000
	Senior Student	33	565.72	61.64			4.195	0.000
3	Novice Student	27	543.46	78.92	4.879	0.031	5.085	0.000
	Senior Student	33	455.75	54.30			4.903	0.000
4	Novice Student	27	536.45	41.44	0.061	0.806	5.805	0.000
	Senior Student	33	468.66	47.70			5.888	0.000

[1] Number of motions; [2] Mean of differences in motion between teacher and students; [3] 2-tailed.

Table 6. Differences in motion accuracy between Novice students and senior students on original frames (using the Mann-Whitney U test).

Motion	Group	N [1]	Mean Rank	Sum of Ranks	M-W U [2]	Wilcoxon W	Z	Asymp. Sig. [3]
1	Novice Student	27	38.52	1040.00	229.00	790.00	−3.217	0.001
	Senior Student	33	23.94	790.00				
5	Novice Student	27	41.96	1133.00	136.00	697.00	−4.599	0.000
	Senior Student	33	21.12	697.00				
6	Novice Student	27	35.93	970.00	299.00	860.00	−2.177	0.029
	Senior Student	33	26.06	860.00				
7	Novice Student	27	37.41	1010.00	259.00	820.00	−2.771	0.000
	Senior Student	33	24.85	820.00				
8	Novice Student	27	42.19	1139.00	130.00	691.00	4.688	0.000
	Senior Student	33	20.94	691.00				

[1] Number of motions; [2] Mann-Whitney U; [3] 2-tailed.

From Tables 5 and 6, we can see significant differences ($p < 0.05$ or $p < 0.01$) in motion accuracy of all eight motions between novice and senior students. The differences in motion accuracy between the teacher and senior students were lower than the differences in motion accuracy between the teacher and novice students.

We also evaluated the difference in motion accuracy on each skeleton point between novice and senior students (Figure 6).

From Figure 6, we found that out of the 17 points on eight motions of Baduanjin, there were significant differences in the motion accuracy between novice and senior students for some points. For example, in Motion 1, there were significant differences in motion accuracy between the two groups at the head and neck (points 8 and 9) and the right upper limb (points 10, 11, and 12).

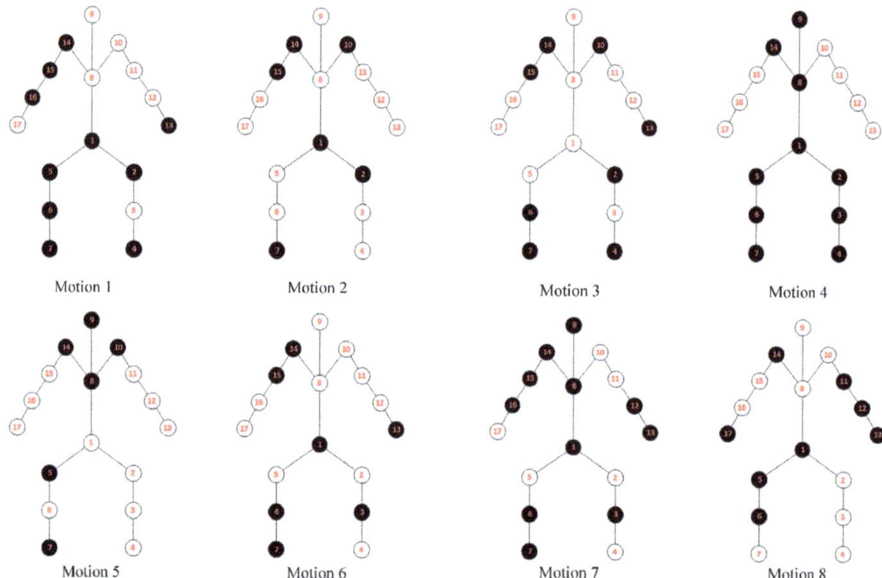

Note: ◯ means on this skeleton points, there is a significant difference in motion accuracy between novice and senior students

Figure 6. Differences in motion accuracy of points between novice and senior students on original frames.

3.3. Differences in Motion Accuracy between Novice and Senior Students on Key-Frames

3.3.1. Compression Rate and Reconstruction Error of Two Different Key-Frames Extraction Methods

Motion accuracy is assessed based on key-frames. In this study, we chose two methods to extract key-frames. In the key-frames extraction method on inter-frame pitch, we selected different thresholds (0.1, 0.5, 1.0, 1.5, 2.0) to extract key-frames and evaluated the compression rate and the reconstruction error of corresponding key-frames on different thresholds. The results are shown in Table 7.

Table 7. Compression rate and reconstruction error of corresponding key-frames on inter-frame pitch.

Threshold	Index	Motion							
		1	2	3	4	5	6	7	8
0.1	Rate [1]	60.45	80.57	55.84	62.00	86.84	76.58	58.83	67.30
	Error [2]	0.059	0.028	0.062	0.046	0.017	0.042	0.022	0.068
0.5	Rate	15.28	27.36	14.02	16.87	36.75	25.12	15.29	20.16
	Error	0.447	0.364	0.455	0.378	0.330	0.448	0.474	0.612
1.0	Rate	7.74	14.79	7.09	8.77	20.72	13.39	8.03	9.97
	Error	1.031	0.904	1.110	0.971	0.799	1.021	1.164	1.969
1.5	Rate	5.14	10.04	4.70	5.93	14.44	9.05	5.57	6.36
	Error	1.811	1.590	1.967	1.719	1.346	1.692	2.099	3.971
2.0	Rate	3.81	7.58	3.49	4.48	11.12	6.80	4.35	4.50
	Error	2.712	2.362	3.021	2.586	1.966	2.474	3.203	6.936

[1] Compression rate (%); [2] Reconstruction error.

Table 7 shows significant differences in the compression rates of the different motions extracted under the same threshold. We can see when the threshold value is set to 1 for obtaining key-frames using the inter-frame pitch, there was a difference in average compression rates ranging from 7.08% to 20.78% for the eight motions of Baduanjin. Moreover, when the threshold value increased, the number of key-frames decreased, which decreased the compression rate. However, the error of motion reconstruction also increased. Based on the data in Table 7, it can be seen that in the five preset values, the compression rate and reconstruction error of the extracted key-frames are relatively reasonable when the threshold is 1. In the other key-frames extraction method on clustering, we chose different compression rates (5, 10, 15, 20, 25) to extract key-frames and evaluate the reconstruction error on different key-frames. The results are shown in Table 8.

Table 8. Reconstruction error of corresponding key-frames on clustering.

Rate (%) [1]	Reconstruction Error							
	Motion 1	Motion 2	Motion 3	Motion 4	Motion 5	Motion 6	Motion 7	Motion 8
5	4.528	7.185	3.875	3.430	8.790	6.823	3.886	6.206
10	1.484	2.428	1.268	0.997	2.927	2.359	1.281	2.216
15	0.851	1.244	0.638	0.547	1.485	1.286	0.650	1.342
20	0.498	0.757	0.401	0.353	0.971	0.797	0.415	0.769
25	0.366	0.518	0.281	0.235	0.700	0.569	0.297	0.531

[1] Compression rate (%) of key-frames.

From Table 8, we can see that as the compression rate increases, the error of motion reconstruction decreases. When the compression rate increased from 5% to 15%, the reconstruction error dropped sharply. But when the compression ratio increased from 15% to 25%, the reconstruction error decrease tended to be smooth. It can be seen that, in the five preset values, the compression rate and reconstruction error of the extracted key frames were relatively reasonable when the preset compression rate is 15%.

3.3.2. Differences in Motion Accuracy on Key-Frames

The differences in motion accuracy on key-frames between novice and senior students are shown in Tables 9 and 10.

Table 9. Differences in motion accuracy on the key-frames on inter-frame pitch between novice and senior students.

Threshold	p Value							
	Motion 1	Motion 2	Motion 3	Motion 4	Motion 5	Motion 6	Motion 7	Motion 8
0.1	0.000	0.000	0.000	0.000	0.000	0.006	0.075 [1]	0.000
0.5	0.001	0.000	0.001	0.005	0.000	0.004	0.122 [1]	0.000
1.0	0.001	0.000	0.001	0.017	0.004	0.004	0.122 [1]	0.000
1.5	0.001	0.000	0.002	0.050 [1]	0.008	0.004	0.141 [1]	0.000
2.0	0.001	0.001	0.004	0.112 [1]	0.018	0.004	0.176 [1]	0.000

[1] $p \geq 0.05$.

Table 10. Differences in motion accuracy on the key-frames on clustering between novice and senior students.

Rate (%) [1]	p Value							
	Motion 1	Motion 2	Motion 3	Motion 4	Motion 5	Motion 6	Motion 7	Motion 8
5	0.000	0.000	0.000	0.000	0.000	0.001	0.003	0.000
10	0.000	0.000	0.000	0.000	0.000	0.024	0.003	0.000
15	0.000	0.000	0.000	0.000	0.000	0.020	0.004	0.000
20	0.000	0.000	0.000	0.000	0.000	0.031	0.004	0.000
25	0.000	0.000	0.000	0.000	0.000	0.024	0.004	0.000

[1] Compression rate (%) of key-frames.

From the results of the key-frames on clustering, the motion accuracy of the eight motions of novice and senior students were significantly different. This result is consistent with the result based on the original frames. However, on the key-frames of inter-frame pitch on five different thresholds, there was no significant difference in motion accuracy between the two groups in Motion 7.

The differences in motion accuracy of points between the two groups on key-frames were also evaluated. Figure 7 shows the results on the key-frames of inter-frame pitch when the setting threshold = 1.

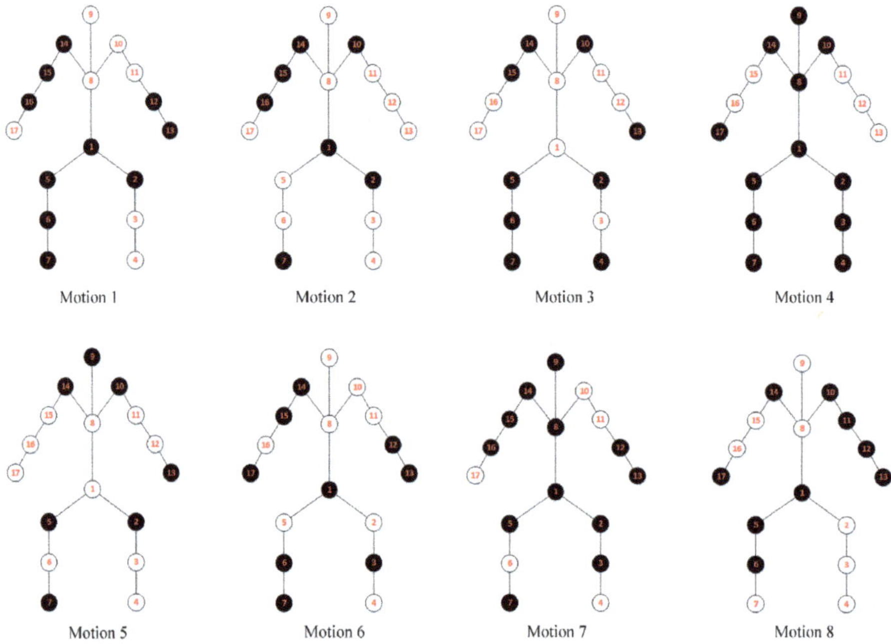

Motion 1   Motion 2   Motion 3   Motion 4

Motion 5   Motion 6   Motion 7   Motion 8

Note: ◯ means on this skeleton points, there is a significant difference in motion accuracy between novice and senior students

Figure 7. Differences in motion accuracy of points between novice and senior students on the key-frames of inter-frames pitch (Threshold = 1).

From Figures 6 and 7, we find that there was a difference between the results on the original frame and the key-frames on the inter-frame pitch. When there was a significant difference in motion accuracy between the two groups, we set the point to 1, otherwise, it was 0. Then, we evaluated the correlation between the results on the original frames and different key-frames on the Kendall correlation coefficient test (see Figures 8 and 9).

The results for key-frame extraction on inter-frame pitch show that when the threshold value was 0.1, the result of the differences in motion accuracy on the key-frames was highly correlated with the result based on the original frame (Kendall coefficient of points in each motion is higher than 0.7 except for Motion 7). However, when the threshold was 0.1, the compression rates of the key-frames were higher. As shown in Table 7, when the threshold was 0.1, the compression rate of each motion exceeded 50%. For key-frames extraction on clustering, there is a high correlation when the compression rate is 0.1. The Kendall coefficient of points in each motion is higher than 0.7 except for Motion 5, where the coefficient was 0.63.

We also tested the mean processing time for using DTW to calculate the distances between motions on original frames and key-frames (Table 11).

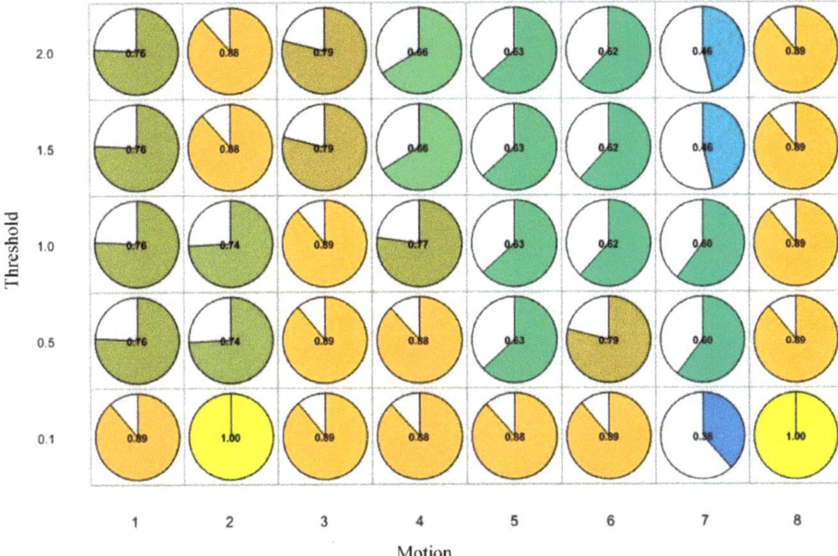

Figure 8. The Kendall coefficient of differences between skeleton points based on two difference methods (on the original frames and the key-frames on inter-frame pitch).

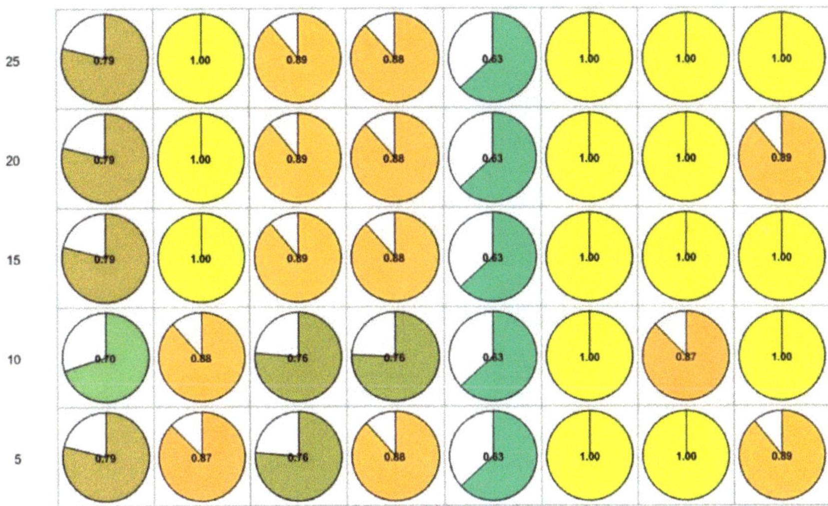

Figure 9. The Kendall coefficient of differences between skeleton points based on two difference methods (on the original frames and the key-frames on clustering).

Table 11. The mean processing time on the original frames and the key-frames.

Motions	The Mean Processing Time (s)		
	Original Frames	Key-Frames [1]	Key-Frames [2]
1	1.891	0.021	0.042
2	5.960	0.195	0.180
3	3.674	0.026	0.078
4	4.439	0.027	0.069
5	5.515	0.218	0.117
6	4.055	0.053	0.069
7	2.209	0.028	0.043
8	0.145	0.013	0.017

[1] Key-frames on inter-frames pitch (Threshold = 1); [2] Key-frames on clustering (compression rate = 15%).

From Table 11, the processing time on the key-frames is lower than original frames. Therefore, using key-frames can effectively decrease data processing time.

4. Discussion

When using mathematical methods, the macro differences between the motion data of novice students and the teacher were higher than the distances between the motion data of senior students and the teacher on eight motions of Baduanjin. Because the motion data of the experimental analysis are the rotation data of specific skeleton points measured by the IMU, if the teacher's motions were taken as the standard, the results show that the motions of senior students were closer to the standard motions. Therefore, IMU can effectively distinguish the differences in motion accuracy in Baduanjin between novice and senior students.

When using the original frames to evaluate the differences at 17 skeleton points in eight motions between novice and senior students, the results show the differences in motion accuracy between the two groups on skeleton points varied for the different motions. For Motion 1, the differences between the two groups were mainly concentrated on the head-spine segment and upper limbs, especially the right upper limb. The differences mean that the motion errors of novice students relative to senior students were mainly concentrated on these joints. The results are consistent with the common motion errors described in the official book: "When holding the palms up, the head is not raised enough, or the arms are not raised enough" [30]. However, for Motion 4, the common motion errors are described in the official book as: "Rotating head and arm are insufficient" [30]. The description shows that the main errors occur in the head-spine and bilateral upper limbs. However, significant differences of skeleton points were at bilateral upper limbs but not head-spine. This difference may be related to the small number of participates in this study.

In this study, we also used two methods to extract key-frames. The raw data can be effectively compressed to decrease the data storage space using extracting key-frames [40,41]. The repetitiveness of action exercises in the teaching process will generate an extremely large amount of raw data. From the results, both key-frames extraction methods can effectively compress the raw data. We also found that the data processing speed could be accelerated on key-frames. However, the compression rates of key-frames on different motions when using key-frames on inter-frame pitch were different. We found that the differences in skeleton points on the key-frames on inter-frame pitch were not consistent with the results on the original frames. However, there was high consistency between the results on the key-frames on clustering and the results on the original frames, especially when the compression rate was 15%. Therefore, we can use key-frames to replace the original frames to evaluate motion accuracy of Baduanjin in order to decrease data storage space and processing time.

However, the small number of participants in our study limits the application of the results. As the participants were from a university in China, the results might only be suitable for university students in China because different populations have variations in anatomical characteristics, physiological characteristics, and athletic ability.

Based on our results, IMU can effectively distinguish the difference in the motion accuracy of Baduanjin between novice and senior students. Therefore, in the following work, we can develop a system using IMU to evaluate the motion quality of students and provide feedback to teachers and students. Thus, it would be able to assist teachers in correcting errors in the motions of students immediately.

5. Conclusions

These initial results show that, based on the original frames, the IMU and the corresponding mathematical methods can effectively distinguish the motion accuracy of all eight motions of Baduanjin between novice and senior students. Furthermore, the IMU can identify the differences between the novice and senior students on the specific skeleton points of the eight motions of Baduanjin. The results regarding key-frames on clustering were highly correlated with the results of the original frames, which means, to a certain extent, that key-frames can replace the original frame to decrease the data storage space and processing time.

Author Contributions: Conceptualization, H.L, S.K., and H.J.Y.; Methodology, H.L, S.K., and H.J.Y.; Validation, H.L, S.K., and H.J.Y.; Data curation, H.L. and H.J.Y.; Writing—original draft preparation, H.L.; Writing—review and editing, S.K. and H.J.Y.; Supervision, S.K. and H.J.Y.; Funding acquisition, H.L. All authors have read and agreed to the published version of the manuscript.

Funding: This research was funded by Neijiang Normal University, Grant no: YLZY201912-1/-15, and the University of Malaya Geran Penyelidikan Fakulti, Grant no: GPF041A-2019.

Acknowledgments: We thank all participants and Neijiang Normal University for providing the facilities for the experiment.

Conflicts of Interest: The authors declare no conflict of interest.

References

1. The Guiding Outline of Physical Education in National Colleges and Universities. Available online: http://www.moe.gov.cn/s78/A10/moe_918/tnull_8465.html (accessed on 6 August 2020).
2. Several Opinions on Comprehensively Improving the Quality of Higher Education. Available online: http://www.moe.gov.cn/srcsite/A08/s7056/201203/t20120316_146673.html (accessed on 7 August 2020).
3. Healthy China 2030. Available online: http://www.gov.cn/zhengce/2016-10/25/content_5124174.htm (accessed on 3 May 2020).
4. Li, F.Y.; Chen, X.H.; Liu, Y.C.; Zhang, Z.; GUO, L.; Huang, Z.L. Research progress on the teaching status of fitness Qigong Ba Duan Jin. *China Med. Herald* **2018**, *15*, 63–66. [CrossRef]
5. Yan, Y.C.; Xia, Z.Q. Present situation and development countermeasures of traditional Wushu in colleges and universities in Hubei province. *J. Jishou Univ. (Soc. Sci. Ed.)* **2013**, *34*, 167–169. [CrossRef]
6. Yang, J.Y.; Qiu, P.X. Introspection of Wushu elective course teaching content reform in Zhejiang University of Technology. *J. Phys. Ed.* **2015**, *22*, 92–97. [CrossRef]
7. Lu, C.Y.; Song, L.; Shan, Z.S.; Wang, Z.Q. Present situation of physical education teachers and sports equipments after college expansion in Shandong province. *J. Tianjin Univ. Sport* **2006**, *1*, 26. [CrossRef]
8. Notice of Provincial Education Department on the Results of the Fifth Batch of Public Physical Education Courses in Colleges and Universities. Available online: https://wenku.baidu.com/view/bb4c0d8f876fb84ae45c3b3567ec102de2bddf83.html (accessed on 7 August 2020).
9. Zhan, Y.Y. Exploring a New System of Martial Arts Teaching Content in Common Universities in Shanghai. Master's Thesis, East China Normal University, Shanghai, China, 2015.
10. Xavier, R.-L.; Hakim, M.; Christian, L.; André, P. Validation of inertial measurement units with an optoelectronic system for whole-body motion analysis. *Med. Biol. Eng. Comput.* **2017**, *55*, 609–619. [CrossRef]

11. Pielka, M.; Janik, M.A.; Machnik, G.; Janik, P.; Polak, I.; Sobota, G.; Marszałek, W.; Wróbel, Z. A Rehabilitation System for Monitoring Torso Movements using an Inertial Sensor. In Proceedings of the Signal Processing: Algorithms, Architectures, Arrangements, and Applications (SPA), Poznan, Poland, 18–20 September 2019; pp. 158–163.

12. Yamaoka, K.; Uehara, M.; Shima, T.; Tamura, Y. Feedback of flying disc throw with Kinect and its evaluation. *Procedia Comput. Sci.* **2013**, *22*, 912–920. [CrossRef]

13. Chen, X.M.; Chen, Z.B.; Li, Y.; He, T.Y.; Hou, J.H.; Liu, S.; He, Y. ImmerTai: Immersive motion learning in VR environments. *J. Vis. Commun. Image Represent.* **2019**, *58*, 416–427. [CrossRef]

14. Elaoud, A.; Barhoumi, W.; Zagrouba, E.; Agrebi, B. Skeleton-based comparison of throwing motion for handball players. *J. Ambient Intell. Humaniz. Comput.* **2019**, *11*, 419–431. [CrossRef]

15. Thomsen, A.S.S.; Bach-Holm, D.; Kjærbo, H.; Højgaard-Olsen, K.; Subhi, Y.; Saleh, G.M.; Park, Y.S.; la Cour, M.; Konge, L. Operating room performance improves after proficiency-based virtual reality cataract surgery training. *Ophthalmology* **2017**, *124*, 524–531. [CrossRef]

16. Van der Kruk, E.; Reijne, M.M. Accuracy of human motion capture systems for sport applications; state-of-the-art review. *Eur. J. Sport Sci.* **2018**, *18*, 806–819. [CrossRef]

17. Corazza, S.; Mündermann, L.; Gambaretto, E.; Ferrigno, G.; Andriacchi, T.P. Markerless motion capture through visual hull, articulated icp and subject specific model generation. *Int. J. Comput. Vis.* **2010**, *87*, 156. [CrossRef]

18. Spörri, J.; Schiefermüller, C.; Müller, E. Collecting kinematic data on a ski track with optoelectronic stereophotogrammetry: A methodological study assessing the feasibility of bringing the biomechanics lab to the field. *PLoS ONE* **2016**, *11*. [CrossRef] [PubMed]

19. Panjkota, A.; Stancic, I.; Supuk, T. Outline of a Qualitative Analysis for the Human Motion in Case of Ergometer Rowing. In Proceedings of the 9th WSEAS International Conference on Simulation, Modelling and Optimization, Iwate Prefectural University, Iwate, Japan, 4–6 October 2009; pp. 182–186.

20. Schepers, H.M.; Veltink, P.H. Stochastic magnetic measurement model for relative position and orientation estimation. *Meas. Sci. Technol.* **2010**, *21*, 065801. [CrossRef]

21. Day, J.S.; Dumas, G.A.; Murdoch, D.J. Evaluation of a long-range transmitter for use with a magnetic tracking device in motion analysis. *J. Biomech.* **1998**, *31*, 957–961. [CrossRef]

22. Schuler, N.; Bey, M.; Shearn, J.; Butler, D. Evaluation of an electromagnetic position tracking device for measuring in vivo, dynamic joint kinematics. *J. Biomech.* **2005**, *38*, 2113–2117. [CrossRef] [PubMed]

23. Woodman, O.J. *An Introduction to Inertial Navigation*; Computer Laboratory, University of Cambridge: Cambridge, UK, 2007.

24. Poitras, I.; Bielmann, M.; Campeau-Lecours, A.; Mercier, C.; Bouyer, L.J.; Roy, J.-S. Validity of Wearable Sensors at the Shoulder Joint: Combining Wireless Electromyography Sensors and Inertial Measurement Units to Perform Physical Workplace Assessments. *Sensors* **2019**, *19*, 1885. [CrossRef]

25. McGinnis, R.S. Advancing Applications of IMUs in Sports Training and Biomechanics. Ph.D. Thesis, University of Michigan, Ann Arbor, MI, USA, 2013.

26. Advanced Wireless Motion Capture System: Perception Neuron PRO. Available online: https://www.noitom.com.cn/perception-neuron-pro.html (accessed on 5 August 2020).

27. Myers, C.S.; Rabiner, L.R. A comparative study of several dynamic time warping algorithms for speech recognition. *Bell Labs Tech. J.* **2013**, *60*, 1389–1409. [CrossRef]

28. Carmona, J.M.; Climent, J. A Performance Evaluation of HMM and DTW for Gesture Recognition. In *Proceedings of CIARP 2012: Progress in Pattern Recognition, Image Analysis, Computer Vision, and Applications*; Springer: Berlin/Heidelberg, Germany, 2012; pp. 236–243.

29. Badiane, M.; O'Reilly, M.; Cunningham, P. Kernel Methods for Time Series Classification and Regression. In Proceedings of the 26th AIAI Irish Conference on Artificial Intelligence and Cognitive Science, Trinity College Dublin, Dublin, Ireland, 6–7 December 2018.

30. Health Qigong Management Center of the State General Administration of Sports. *Health Qigong Baduanjin*; People's Sports Press: Beijing, China, 2018; Volume 3, pp. 215–217.

31. Perception Neuron 2.0. Available online: https://www.noitom.com.cn/perception-neuron-2-0.html (accessed on 20 August 2020).

32. Sers, R.; Forrester, S.E.; Moss, E.; Ward, S.; Zecca, M. Validity of the Perception Neuron inertial motion capture system for upper body motion analysis. *Measurement* **2019**, *149*, 107024. [CrossRef]

33. Dai, H.; Cai, B.; Song, J.; Zhang, D.Y. Skeletal Animation Based on BVH Motion Data. In Proceedings of the 2nd International Conference on Information Engineering and Computer Science, Wuhan, China, 25–26 December 2020; pp. 1–4.

34. Srivastava, R.; Sinha, P. Hand Movements and Gestures Characterization Using Quaternion Dynamic Time Warping Technique. *IEEE Sens. J.* **2016**, *16*, 1333–1341. [CrossRef]

35. Mukundan, R. Quaternions: From Classical Mechanics to Computer Graphics, and Beyond. In Proceedings of the 7th Asian Technology Conference in Mathematics, Melaka, Malaysia, 17–21 December 2002; pp. 97–105.

36. Kim, M.-H.; Chau, L.-P.; Siu, W.-C. Keyframe Selection for Motion Capture using Motion Activity Analysis. In Proceedings of the IEEE International Symposium on Circuits and Systems (ISCAS), Seoul, Korea, 20–23 May 2012; pp. 612–615.

37. Yan, C.; Qiang, W.; He, X.J. *Multimedia Analysis, Processing and Communications*; Springer: Berlin/Heidelberg, Germany, 2011; pp. 535–542.

38. Li, S.Y.; Hou, J.; Gan, L.Y. Extraction of motion key-frame based on inter-frame pitch. *Comput. Eng.* **2015**, *41*, 242–247. [CrossRef]

39. Jablonski, B. Quaternion Dynamic Time Warping. *IEEE Trans. Signal Process.* **2012**, *60*, 1174–1183. [CrossRef]

40. Yan, Y. An Adaptive Key Frame Extraction Method Based on Quaternion. Master's Thesis, Xiamen University, Xiamen, Fujian, China, 2011.

41. Liu, X.M.; Hao, A.M.; Zhao, D. Optimization-based key frame extraction for motion capture animation. *Vis. Comput.* **2013**, *29*, 85–95. [CrossRef]

42. Zhang, Y.; Cao, J. 3D Human Motion Key-Frames Extraction Based on Asynchronous Learning Factor PSO. In Proceedings of the Fifth International Conference on Instrumentation and Measurement, Computer, Communication and Control (IMCCC), Qinhuangdao, China, 18–20 September 2015; pp. 1617–1620.

43. Shi, X.B.; Liu, S.P.; Zhang, D.Y. Human action recognition method based on key frames. *J. Syst. Simul.* **2015**, *27*, 2401–2408. [CrossRef]

44. Sun, S.M.; Zhang, J.M.; Sun, C.M. Key Frame Extraction Based on Improved K-means Algorithm. *Comput. Eng.* **2012**, *38*, 169–172. [CrossRef]

45. Zhao, H.; Xuan, S.B. Optimization and Behavior Identification of Keyframes in Human Action Video. *J. Gr.* **2018**, *36*, 463–469. [CrossRef]

46. Alexiadis, D.S.; Daras, P. Quaternionic signal processing techniques for automatic evaluation of dance performances from MoCap data. *IEEE Trans. Multimed.* **2014**, *16*, 1391–1406. [CrossRef]

47. Keogh, E.; Ratanamahatana, C.A. Exact indexing of dynamic time warping. *Knowl. Inf. Syst.* **2005**, *7*, 358–386. [CrossRef]

48. Zhang, W.J.; Wang, J.J.; Zhang, X.; Zhang, K.; Ren, Y. A Novel Cardiac Arrhythmia Detection Method Relying on Improved DTW Method. In Proceedings of the IEEE 2nd Advanced Information Technology, Electronic and Automation Control Conference (IAEAC), Chongqing, China, 25–26 March 2017; pp. 862–867.

49. Chen, Q.; Hu, G.Y.; Gu, F.L.; Xiang, P. Learning Optimal Warping Window Size of DTW for Time Series Classification. In Proceedings of the 11th International Conference on Information Science, Signal Processing and their Applications (ISSPA), Montreal, QC, Canada, 2–5 July 2012; pp. 1272–1277.

Publisher's Note: MDPI stays neutral with regard to jurisdictional claims in published maps and institutional affiliations.

 *sensors*

Article

Biosensing and Actuation—Platforms Coupling Body Input-Output Modalities for Affective Technologies

Miquel Alfaras [1,2,*], William Primett [1,3], Muhammad Umair [4], Charles Windlin [5], Pavel Karpashevich [5], Niaz Chalabianloo [6], Dionne Bowie [4,7], Corina Sas [4], Pedro Sanches [5], Kristina Höök [5], Cem Ersoy [6] and Hugo Gamboa [3]

1   PLUX Wireless Biosignals, Avenida 5 de Outubro 70, 1050-059 Lisboa, Portugal; wprimett@plux.info
2   Departament d'Enginyeria i Ciència dels Computadors, RobInLab, Universitat Jaume I, Avinguda de Vicent Sos Baynat, s/n, 12071 Castelló, Spain
3   Departamento de Física, LIBPhys FCT—UNL Universidade NOVA de Lisboa, Largo da Torre, 2825-149 Caparica, Portugal; hgamboa@fct.unl.pt
4   Computing and Communications Department, InfoLab21, Lancaster University, Bailrigg, Lancaster LA1 4WA, UK; m.umair7@lancaster.ac.uk (M.U.); dionne.bowie@nhs.net (D.B.); c.sas@lancaster.ac.uk (C.S.)
5   Division of Media Technology and Interaction Design, School of Electrical Engineering and Computer Science, KTH Royal Institute of Technology, Brinellvägen 8, 114 28 Stockholm, Sweden; windlin@kth.se (C.W.); pavelka@kth.se (P.K.); sanches@kth.se (P.S.); khook@kth.se (K.H.)
6   Computer Engineering Department, Boğaziçi University, Rumeli Hisarı, 34470 Sarıyer/Istanbul, Turkey; niaz.chalabianloo@boun.edu.tr (N.C.); ersoy@boun.edu.tr (C.E.)
7   Research and Innovation Centre, Leeds Teaching Hospitals NHS Trust, Beckett St, Leeds LS9 7TF, UK
*   Correspondence: malfaras@plux.info

Received: 14 September 2020; Accepted: 16 October 2020; Published: 22 October 2020

Abstract: Research in the use of ubiquitous technologies, tracking systems and wearables within mental health domains is on the rise. In recent years, affective technologies have gained traction and garnered the interest of interdisciplinary fields as the research on such technologies matured. However, while the role of movement and bodily experience to affective experience is well-established, how to best address movement and engagement beyond measuring cues and signals in technology-driven interactions has been unclear. In a joint industry-academia effort, we aim to remodel how affective technologies can help address body and emotional self-awareness. We present an overview of biosignals that have become standard in low-cost physiological monitoring and show how these can be matched with methods and engagements used by interaction designers skilled in designing for *bodily engagement* and *aesthetic* experiences. Taking both strands of work together offers unprecedented design opportunities that inspire further research. Through *first-person soma design*, an approach that draws upon the designer's felt experience and puts the sentient body at the forefront, we outline a comprehensive work for the creation of novel interactions in the form of couplings that combine biosensing and body feedback modalities of relevance to affective health. These couplings lie within the creation of design toolkits that have the potential to render rich embodied interactions to the designer/user. As a result we introduce the concept of *"orchestration"*. By orchestration, we refer to the design of the overall interaction: coupling sensors to actuation of relevance to the affective experience; initiating and closing the interaction; habituating; helping improve on the users' body awareness and engagement with emotional experiences; soothing, calming, or energising, depending on the affective health condition and the intentions of the designer. Through the creation of a range of prototypes and couplings we elicited requirements on broader orchestration mechanisms. First-person soma design lets researchers look afresh at biosignals that, when experienced through the body, are called to reshape affective technologies with novel ways to interpret biodata, feel it, understand it and reflect upon our bodies.

Sensors **2020**, *20*, 5968

Keywords: human-computer interaction; affective technologies; interaction design; biosensing; actuation; somaesthetics; design toolkits

---

1. Introduction

The rise of tracking technologies has started to foster international collaborations that tackle the design of technologies for emotional awareness and regulation to support wellbeing and affective health. In fact, mental health research is trying to catch up with the affordances that ubiquitous technologies, wearable devices, and tracking systems offer in general, albeit not without challenges [1–3]. These can be addressed through interdisciplinary research bridging the gap between the fields of Human-Computer Interaction (HCI), Biosensing research, and Clinical psychology [4–6]. Projects such as AffecTech [6,7], have explored the development of digital platforms that position bodily affective awareness and engagements centrally, drawing on *somaesthetic design* [8]. Somaesthetic, or soma design for short, is grounded in somatic experiences, letting designers examine and improve on connections between sensation, feeling, emotion, subjective understanding and values. The soma design framework offers a coherent theoretical basis starting from the constitution and morphology of our human body and perception [8–10]. In particular, soma design emphasizes our ability to change and improve our aesthetic appreciation skills and perception. Other interaction design works within affecting computing exemplify ways to draw upon the body and its embodied metaphors [11,12]. This research builds on the growing HCI interest in affective technologies, whose ethical underpinnings could benefit from more consideration [13,14], addressing issues such as the pluralism of bodies, data privacy and ownership. The body has for a long time inspired emotion research across disciplines [15–18], as relevant connections exist between the body, emotion and movement, its interpretation, enacting and processing. Moreover, research studies point at the links between emotion and physical activity, for example, dance, exercise, movement, or paying attention to our body senses while immersed in nature [19–21]. Engagement with and through the body might therefore be a fruitful path to explore. There is room for an affective computing that does not look at the body as "an instrument or object for the mind, passively receiving sign and signals, but not actively being part of producing them"—as phrased by Höök when referring to dominant paradigms in commercial sports applications [22]. However, how to best address bodily movement and engagement beyond measuring cues and signals is unclear. Most studies in affective computing revolve around affect recognition from emotion detection and bodily data classification [23]. We take somaesthetic design—a design stance that draws upon the felt body and takes inspiration from experiencing it—and then combine it with the innovative integration of biosensors and actuators. The disruptive somaesthetic view, moves away from the idea of monitoring the body for the sake of bad habit reduction in pursuit of a healthy and long life [24]. Soma design, rather, lets us get attuned to our bodies and use sensations as a valuable resource instead of something to be improved to meet performance standards.

In this context, we present novel research on embodied interaction design couplings, that is, sensing-actuation combinations of aesthetically evocative body input-output modalities that render biodata shareable, body-centered, highly tangible or even able to be experienced collectively. The biosignals we address have become standard for physiological data tracking research, and are present in the low-cost BITalino biosensing platform [25]. Choosing them for the overview presented and further exploration is motivated by these two aspects, that is, standard and low-cost. The contribution of this paper is an approach to designing sensing-actuation orchestrations, that is, the ways in which body input-output systems and meanings are put in place, coupled, coordinated, customized, sequenced and exposed so that the underlying mechanisms can be better understood, challenged or extended—in other words, sketching, in hardware and software, tangible experiences that allow designers to design and improve overall orchestrated experiences addressing affective health. To make this design approach viable, we combine:

1. An overview description of biosignals used in ubiquitous low-cost personal sensing technologies that have gained strength in affective technology research and the possible actuation feedback elements that, when coupled, can lead to novel interactions embracing body awareness
2. The use of soma design [8], with an evaluative and explorative stance, to assess whether couplings are meaningful or evocative to address the question of how interaction design can further support interdisciplinary research of affective technologies

The individual components that take part in an interaction that integrates different body inputs and outputs must implement ways to communicate information, process and represent it, trigger events turning actions on/off and enabling interaction decisions. An orchestration of the protocols and interfaces involved could be beneficial for the design exploration or even for the introduction of use case scenarios, for example, closer to the actual psychotherapeutic practice [2,3,26]. In the discussion, we describe how these elements are shaping the future direction of our research, for example, extending interaction configuration tools with novel sensing-actuation couplings to better explore the design space of affective health technologies and their ethical underpinnings. Using technology for sensing and actuating upon our body, we can get access to bodily states from our physiology to then act in such a way that we help to alter or reassess our psychophysiological states. This construction process may be developed to extend our knowledge and expectations regarding the internal mechanics of our own body and serves as a bridge to design better informed affective health technologies. Moreover, not only does this approach aim to help having a better self-understanding but paves the way to put the body and its felt experience at the core of the design of such technologies.

The paper is organized as follows. In Section 2, we provide a brief overview of self-monitoring and affective technologies. Section 3 showcases a set of biosignals that we have had access to throughout our research and the information we can extract from their features in order to open a window to our internal psychophysiological processes. The features commonly available for each biosignal are listed to provide guidelines on what level of information is to be extracted. The actuation on the subject's body, addressed in Section 4, can be executed through a variety of mechanisms. We list actuation mechanisms that are available for interaction design using mainly consumer electronics. In Section 5, we present the design research approach that we have adopted, describing what the first-person perspective is and introducing soma design in this context. This design process has been applied to several explorations. The outcomes of our design explorations, coupling biosensing to actuation, are discussed (see Section 6). In this section, together with Section 7, we proceed by addressing coupling concepts and discussing the orchestration process. With the idea of orchestration, we highlight the role of technology-coordinated sequences and the possibilities brought by machine learning and advanced signal processing. We end by commenting on the ethical underpinnings of affective technology and somaesthetic design.

2. Body-Centric Affective Technologies

With the emergence of everyday personal sensing such as the sensing embedded in our permanently reachable phones, smart watches and fitness bracelets, HCI and ubiquitous computing scholars have highlighted the value of these technologies for innovative research. *Affective Computing* refers to computing that relates to, arises from, or deliberately influences emotions [27]. Technologies that we have seen permeate the *everyday* space with quantification, exercise tracking, and physical wellbeing, have also—perhaps in line with a more traditional affective computing view—made researchers dream of extended healthcare, diagnosis and monitoring applied as well to mental wellbeing [4,28–30]. As exemplified by Bardram and Matic [1], mental health research is catching up. In recent years, research on mobile and wearable technologies that track behavioral, psychological, and contextual signals has gained momentum in the field, albeit not without pending design challenges [31]. Following a research path toward ubiquitous technologies deployed in mental wellbeing domains may help to bring attention to such aspects as personalization, achieving forms of rapport or engagement not seen in traditional healthcare. The promise of affective computing is vast. In our view, we could argue that just as self-awareness plays a major role in the motivation of change in rehabilitation therapy, for example, in cardiac rehabilitation [32], psychotherapy could benefit from self-monitoring technologies revealing bodily dynamics. Awareness, for instance, may contribute both to a (re)assessment of emotions and behavioral change that are solid grounds of cognitive behavioral psychotherapy [33,34].

Emotion plays an integral role in design work, and design researchers are not exempt from its ups and downs [35–37]. As affective computing reaches maturity, alternative methods have emerged and reshaped traditional approaches to affect. In an effort to attend to emotions, rather than primarily recognizing them, researchers investigating what is known as the affect through interaction [22] prioritize making emotion available for reflection. In such line of thought, seeking emotion aside from context would not make sense. In this *"affect-through-interaction"* view, the role that emotion has had for a long time in artistic and design endeavors is acknowledged. This is exemplified by the analysis of Boehner et al. [38], later picked up by Howell et al. [39] to defy the role of personal sensing in design, in particular the role of biosensing. That is, by no means, to say that the progress that personal sensing has witnessed under the advent of affective computing should be diminished. Rather, dialogue with artificial intelligence research and attention to more cognitivist-oriented outcomes can strengthen the affect-interaction paradigm. From our standpoint, when designing technology-mediated experiences, we see the affect as a sociocultural, embodied, and interpretative construct. Hence, embarking on the challenge of creating use cases for novel technology that touch upon emotions, we start experiencing the body first (see Section 5). The examples and reflections laid down in this paper, the description of technologies we choose to design with, our AffecTech coupling results, and those we used as inspiration, convey directions in which we believe personal sensing, its mapping to actuation, and designing with the body are successfully integrated. Under the overarching lens of first-person design that provides strong foundations, paying respect to ethics, and "resisting the urge" [35] to engage users, we rediscover (and invite others to do so) technologies that are called upon to extend possibilities within affective interaction.

State of the Art

In the design space of affective interaction and physiological data, existing research has utilized visual and haptic technologies for affective feedback. *Affective Health* [40], for example, mapped skin conductance data measured from an electrodermal activity sensor (EDA) into a colorful spiral on a mobile phone screen. After using the mobile app for a month, users interpreted the skin conductance data as a tool to manage stress levels, track emotions, monitor personality, and even to change their behaviors. Khut [41] has been a pioneer in the area of designing heart rate based visual and sonic artworks for relaxation, both through a mobile app and large scale projections. HCI researchers have started to utilize alternative materials such as thermochromic ones to visually represent biosensing data. Howell et al. designed Ripple [42], a thermochromic-based shirt that changes colors responding to skin conductance. By using the garment over a two-day period, wearers were able to reflect on their emotions but they rarely questioned if the display was actually representing their feelings. In Reference [43], Umair et al. mapped skin conductance to haptic changes in addition to using visual thermochromic materials, that is, vibrations, heating, and squeezing effects. The feedback about the body properties measured is worn, felt or placed in contact with the body. The findings of these studies highlight that the material-driven qualities of such visual and haptic body interactions shape people's interpretation of how they identify, attribute, and regulate emotions in everyday life. Haptics have also been used with biosensors to regulate affect, which requires users to adapt their ongoing feelings. *EmotionCheck* [44] and *Doppel* [45] use vibrations simulating a slower heart rate sensation for the users and helped them decrease their anxiety. Recently, Miri et al. [46] used a vibration-based personalized slow-paced breathing pacer on the belly which delivered vibrations in a biphasic pattern for inhalation and exhalation and helped users in reducing anxiety during a stressor. With a research approach that explicitly sets out to design with the body—not as an object to be measured but "understanding the body as a site of creative thinking and imagination" [47]—, works at the intersection of biosensing, interaction design and affective technologies offer an opportunity to study how to support the design of interactions that make us connect with our bodies [9,48,49].

3. Sensing the Body

Biosignals are time representations of changes in energy produced in the body. These changes correspond to energy variations of different nature, such as electrical, chemical, mechanical, and thermal (as presented in Table 1). With the turn of the 21st century and the advent of the digital era, the advances in the field of electronic components that spurred the development of computing, instrumentation, and algorithms left their impact on medical and biosignal devices. Biosensing and electrophysiology technologies were greatly improved, ready for the study of body functions and health monitoring in the context of clinical research. As technologies grew, the miniaturization and reduction of costs contributed to the growth of biosensing monitoring technologies beyond clinical settings as well. Physiology signals and sources of tracking information are more available than ever, ranging from electromyography (EMG), electrocardiography (ECG), electroencephalography (EEG), electrodermal activity (EDA) to electrooculography (EOG) or eye movement tracking.

Table 1. Parameters and type of energy measured through body sensing. Adapted from Reference [50].

Energy	Changing Parameter	Measurement Examples
Mechanical	Position, force, torque, pressure	Muscle contractions, cardiac pressure, muscle movement
Electrical	Voltage, charge, current	EMG, ECG, EEG, EDA, EOG
Thermal	Temperature	Surface body temperature
Chemical	Concentrations, exchanged energy	pH, oxygen, hormonal concentrations

A direct consequence of such rapid expansion is the creation of the sports & health monitoring markets that fill up the mobile app stores and provide remarkable revenues in the ubiquitous computing paradigm that we live in. The democratization of the study of biosignals, however, comes with interesting possibilities such as a better understanding of the self and a richer, unprecedented way to interact with technologies that accompany us. This yields an opportunity to define alternative ways to live an affectively healthy life.

As the maturity of open access physiology databases [51] backs up the improvement of processing algorithms, low-cost hardware platforms help populate the open source space [52] where users embrace biosensing, share ideas and drive the future of biosignals applied in different areas. Furthermore, the biosignals that were once limited to hospitals and clinics, or in specialized research labs, addressed in classical texts of physiology, are nowadays accessible in virtually any context by means of wearable technologies. In the review of Heikenfeld et al. [53], an interesting account of the transition from lab tracking to wearables during the 20th century is offered along an in-depth overview of body sensing mechanisms not only restricted to electrophysiology. The field of affective computing has consistently found in biosignals a relevant source of information [54]. Besides, the fact that biosensing platforms have jumped off the clinic has contributed to embracing them alongside other technologies like movement tracking, traditionally linked to behavioral and psychophysiology labs.

We present a selection of studied biosignals (see Figure 1) that can be incorporated into the creation of new technologies for affective health. We focus on a subset of biosignals present in the BITalino revolution do-it-yourself (DIY) low-cost biosensing platform [25,55,56] that backed and inspired some of our research in affective technologies. These, although not an exhaustive list, are to some extent physiological signals that have become standard for physiology tracking research—slowly crossing disciplines and making their way into affective health tracking, interaction design, and other domains of interest. Moreover, with objectives that range from out-of-the-lab psychophysiology tracking [57–59] to new perspectives in interaction design [43,49,60] our work has often addressed biosignals through other available biosignal research platforms beyond BITalino, such as biosignalsplux [61], Empatica

E4 [62], Arduino accessories like the Grove GSR [63], or even commercial wearables such as the Samsung Gear S2 [64] among others.

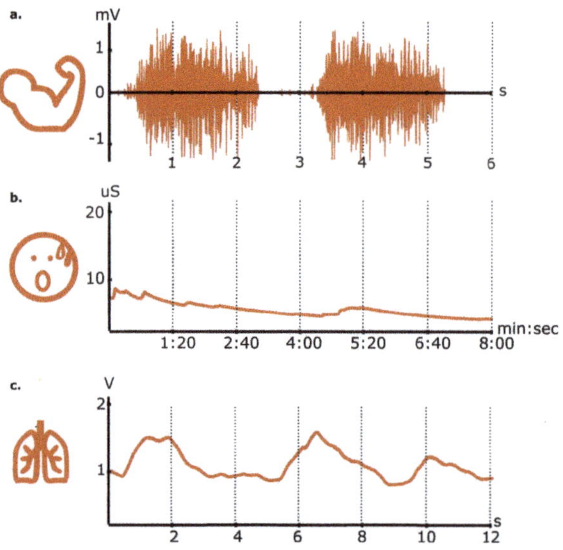

Figure 1. Visual representation of different biosignals: (**a**) Electromyography (EMG), (**b**) Electrodermal activity (EDA) and (**c**) Respiration signals. biosignals and icons obtained at PLUX S.A.

In this section we present a collection of these biosignals and offer a systematic but brief description on (1) How it works, summarizing the basic physiological principles that provide the biosignals energy observables; (2) What can be extracted from the collected biosignal; (3) Where the biosignal can typically be collected in the human body; (4) When, or how often, the signal should be sampled describing the concerns on the timing of the acquisition and in particular the typical sampling frequency of each biosignal; and (5) Limitations of the biosignal acquisition and processing with the challenges of noise or signal artifacts. All of them are examples of signals that we have addressed in our research. This non-exhaustive selection offers a good starting point for researchers interested to integrate biosignals in their design of technologies for wellbeing and mental health.

3.1. Surface Electromyography (sEMG)

How it works: The recording of the electrical activity produced by skeletal muscles receives the name of electromyography (EMG). Human muscles are made up of groups of muscle units that, when stimulated electrically by a neural signal, produce a contraction. The recording of the electrical activity of the muscles (voltage along time), traditionally relying on intrusive needle electrodes (intramuscular), is easily accessible nowadays by means of surface electrodes that capture the potentials of the fibers they lay upon. The result of this measurement is a complex surface electromyography signal (sEMG) that reveals data about movement and biomechanics of the contracted muscles (see Figure 1a).

What: Electromyography signals inform about the contraction of specific muscles and parts of the body. The EMG signal consists in the time representation of rapid voltage oscillations. Its amplitude range is approximately 5 mV. In terms of signal analysis, the EMG allows the assessment of several aspects such as muscle contraction duration, the specific timing at which movements or contractions are activated, the presence of muscular tension or fatigue, and the extent to which different fibers (area) are contracted. The analysis is conducted through noise filtering, together with feature extraction that yields contraction onset detection, the estimation of signal envelopes, and the computation of average frequencies. This lets subjects deepen their understanding of movement strategies, very relevant for embodied art and sports performance, improve muscle coordination, or even reveal existing movement patterns that they are unaware of.

Features: Onset instants; Max amplitude; Instant of maximum amplitude; Activation energy; Envelope.

Where: Having become the standard in EMG monitoring, bipolar surface electrodes consist of three electrodes. Two of them ($+/-$) must be placed close to each other, on the skin that lies on top of the muscle under study, along the fibers' direction, while the third one is placed in a bony area where no muscular activity is present. This allows the measurement of electrical potential differences with respect to a common reference, yielding a unique signal that represents the muscular activity of the area.

When/Frequency: Given the fast muscle-neural activation nature of EMG signals and the presence of different active muscles contributing to the same signal, muscle activity must be acquired at sampling rates no lower than 200 Hz frequencies. Working at 500 Hz is desirable, while a sampling rate of 1000 Hz guarantees the tracking of all the relevant events at a muscular level.

Limitations: Surface EMGs are intrinsically limited to the access to superficial muscles. This is compromised by the depth of the subcutaneous tissue at the site of the recording which depends on the weight of the subject, and cannot unequivocally discriminate between the discharges of adjacent muscles. Proper grounding (reference electrode attached to a bony inactive muscular region) is paramount to obtain reliable measurements. Motion artifacts and muscular crosstalk compromise the assessment of the muscle activity under study. In this context, interference from cardiovascular activity is not uncommon, particularly in areas such as chest and abdomen. The presence of power supplies and mains (powerline) in the vicinity poses the risk of 50 Hz–60 Hz interference.

3.2. Electrodermal Activity (EDA)

How it works: Electrodermal activity (EDA), also known as galvanic skin response (GSR), measures the electrical properties of the skin, linked to the activation of the autonomic nervous system (or more precisely the sympathetic nervous system). By applying a weak current upon two electrodes attached to the skin, it is possible to measure the variations of voltage that are present between the measuring points (see Figure 1b). When placed at specific locations on the skin, the measured electrical signals are affected by the sweat secreted by the glands that are found in the dermis.

What: Electrodermal activity signals inform about the activity of the sympathetic nervous system. Given its electrolyte composition, the sweat secreted by sweat glands has an impact on the electrical properties of the skin. This phenomenon, visibly monitored in voltage signals by means of electrical conductance (or impedance/resistance, conversely), facilitates the assessment of arousal effects. Arousal is the physiological response that stimuli such as emotional or cognitive stressors trigger. The measurement of electrodermal activity is usually decomposed in two major behaviors present and superposed in any skin response signal, that is, the skin conductance (tonic) level, with slowly varying dynamics, and the skin conductance (phasic) responses, that exhibit relatively faster dynamics. In terms of signal analysis, this decomposition is accompanied by the assessment of characteristics such as the rate of detected EDA events, detection of onsets, and the characteristic rise and recovery times.

Features: Onset instant; Skin Conductance Response (SCR) rise time; SCR 50% recovery time; Event rate; Skin Conductance Level (SCL).

Where: EDA measurements use two electrodes to monitor changes in electric potential between two locations on the skin. Electrodes must be placed a few centimeters apart for differences to be relevant. The nature of the measurement technique and the phenomenon itself, makes hand palm a suitable electrode location, for which either palm placement or finger phalanges, most subject to skin sweating, are optimal for the monitoring of electrodermal activity. Additionally, foot sole placement, also affected by sweating glands, is not uncommon in EDA measurements given that particular use cases or settings require access to hands for carrying out certain activities. For the alternative placements of the EDA sensors, such as forehead or wrist, the presence (or lack) of sweating glands remains a decisive factor in obtaining reliable measurements.

When/Frequency: Electrodermal activity is considered to be a slow physiological signal. Thus, sampling rate frequencies as low as 10 Hz allow a full representation of the skin conductance variations. Electrodermal activity peaks usually occur after few seconds from the exposure to a given stimulus (1–5 s).

Limitations: Electrodermal activity measurements use changes in electrical properties of the skin produced by sweating. Since sweating is not only triggered by arousal but also the human thermoregulation system, ambient heat and physical activity monitoring are aspects that limit the capabilities of EDA studies. In common practice, electrodermal sensors are usually prepared to obtain salient data from the most comprehensive userbase, providing relevant (measurable) changes regardless of the wide variety of sweating responses from subject to subject. However, it is not uncommon to find examples of subjects with either too high or too low skin conductance responses that complicate the measurements. Moreover, settings that involve an intense physical activity pose concerns on the electrode attachment and motion interference. The presence of power supplies and mains (power line) in the vicinity of the acquisition systems pose the risk of 50 Hz–60 Hz interference. With regard to feasibility, since traditional electrodermal activity studies rely on hands or feet electrode placement that compromises certain actions, attention needs to be given to the use case and activities that take place while monitoring, on a case by case basis.

3.3. Breathing Activity

How it works: Respiration (or breathing) sensors monitor the inhalation-exhalation cycles of breathing, that is, the process to facilitate the gas exchange that takes place in the lungs. In every breathing cycle, the air is moved into and out of the lungs. A breathing sensor uses either piezoelectric effects on bendable wearable bands or accessories (one of the most predominantly used technologies), respiratory inductance plethysmography on wired respiration bands around the thorax, microphonics on the nose/mouth airflow, plethysmographs (measuring air inflow) or radiofrequency, image and ultrasonic approaches. A review on breathing monitoring mechanisms is found in Reference [65]. For piezoelectric breathing sensors, thoracic or abdominal displacements (strain) produced in breathing cycles bend a contact surface that converts ressistive changes to continuous electrical signals (see Figure 1c).

What: A breathing signal informs about the respiration dynamics, that is, the dynamics of the process mediating gas exchange in the lungs, as well as supporting sound and speech production. The monitoring of the fundamental function of breathing brings in the assessment of breathing cycles and rates which in turn allows the study of apnoea-related problems (involving breathing interruptions), oxygen intake, metabolism of physical activity, and the effect of cognitive or emotional stressors in breathing. In terms of analysis, breathing cycles are studied using breathing rates, the maximum relative amplitude of the cycle, inhale-exhale volume estimation, inhale-exhale duration, and inspiration depth, that allow the characterization of several breathing patterns.

Features: Respiration rate; Inspiration duration; expiration duration; Inspiration-expiration ratio; Inspiration depth.

Where: A piezoelectric breathing sensor is usually located on the thoracic cavity or the belly, using a wearable elastic band. With adjustable strap and fastening mechanisms, the sensor can be placed slightly on one side where bending is most relevant, optimizing the use of the sensor range. These kinds of sensors, allow both the study of thoracic and abdominal breathing. With the development of conductive fabric, breathing sensors are making its way into the smart garment market in the form of T-shirts and underwear bands.

When/Frequency: Breathing is a relatively slow biosignal, with breathing rates often below 20 inhale/exhales per minute. A sampling rate frequency as low as 50 Hz is sufficient to capture the dynamics of respiration.

Limitations: While piezoelectric breathing sensors are prominent given the low cost and form factor advantages of wearable sensor platforms, deviations in placement have an effect in the relative range of the response signal. Movement artifacts, most relevant when physical activity is present, are a common source of problems. Respiration sensing techniques like the respiratory inductance plethysmography, compensate the highly localized piezoelectric approach with a sensor capturing the general displacement of the whole thoracic cavity, yielding a signal less prone to movement artifacts. The monitoring of breathing cycles is usually accurate, although the exploration of effects to be used as voluntary inputs in interactions, such as holding the breath, are not easily captured.

4. Actuation

The mechanisms to provide actuation in a form of feedback to the human take an important role in creating a complete interaction from sensing body properties to making the subject aware of them. Our research aims at linking biosensing to body actuation. Actuation is generally provided by mechanical elements that move and respond to input signals in order to either control or inform about a system. We stretch this definition to include feedback mechanisms such as screen-based visuals, although no mobile mechanical element is necessarily implied. In this section, we focus on actuation mechanisms that can be easily controlled and coupled to our body. We take a similar approach to the structure used to describe the biosignals, on explaining: how, what, where, when, and the actuator limitations and usage precautions for a selected list of actuators. The range of actuation mechanisms presented draws upon our research on affective technologies and interaction design, as well as inspirational works present interaction design research, but it should be seen as a non-exhaustive list of possibilities.

4.1. Screen-Based Visual Biofeedback

How it works: Screen-based visual biofeedback is the representation of body signals that inform about body changes happening along time. Its goal is to provide to the researcher a means to assess the dynamics of the aforementioned changes, helping to gain understanding and tracking the inner state of a given subject. Examples of this could be ECG feedback, respiration feedback, or movement tracking, usually employed in health metrics or sports performance research. Screen-based biosensing systems for feedback are standard practice in clinical settings and hospitals. Biofeedback use has for instance been adopted in psychotherapy, as research suggests that the technique provides a mechanism to self-regulate the emotions.

What: Screen-based visual biofeedback uses a 2D graphical interface and benefits from light, colors, strokes, and visual styles to represent a changing signal that evolves with time. Signal peaks and troughs appear in an axis showing the measurement magnitude in a given range, so that rapid and slow dynamics can be seen as the representation moves along the time axis when updated.

Where: Screen-based visual feedback takes place in a display, either a computer screen or a sensing platform display.

When: It is important that the represented signals are updated in real-time. Doing otherwise, although possible using delays or technology limitations, would compromise the ability of the actuation to convey the tracking meaning attributed to the practice of biofeedback. When sensing requirements pose concerns on the technical ability to render a smooth representation through time, approaches such as averaging or undersampled representations are used.

Limitations: Screen-based visual biofeedback connects easily with the mathematical properties that underlie the signals under study. However, signal processing procedures such as filtering, scaling or normalization are crucial in achieving a smooth and flowing representation. These come, of course, tightly dependent on the available computing capabilities. There are situations in which feedback users report finding difficulties or experiencing anxiety when engaging in the assessment of body rhythms. Moreover, visual information tends to remarkably capture the attention of the user, thus needing special care when used as an element of broader interaction (movement, performance, exercise) that could render a poorer experience quality or present a deviation from the aimed activity.

4.2. Sound Feedback

How it works: Sound feedback, when applied to biosignals, is the audio representation of body signals that uses sound properties to inform about body changes happening along time. Its goal is to exploit our sophisticated trained sense of hearing to convey meanings linked to body signal features, leading to the understanding and tracking of a given subject's biosignal dynamics.

What: Sound feedback uses the properties of sound, that is, volume, pitch or frequency (note), rhythm, harmony, timbre, and transients (attack, sustain, etc.) among others, to represent a signal (or its features) that changes over time. Its generation, often using speakers or headphones, is linked to properties of the signal. Alternative approaches draw upon several transducing paradigms, that is, different ways to convert electrical signals into sound (electromechanical as in the case of speakers, piezoelectric or others), often more limited such as buzzers or beepers made of basic vibrating elements that produce sound.

Where: Sound can be generated in speakers, devices that work converting electrical pulses to sound (air pressure) waves, allowing users to listen to the feedback without the need for additional equipment. Headphones, working by the same principle, can be used for the same purpose but only providing feedback to the person wearing them.

When: The human hearing range typically comprises frequencies between 20 Hz and 20,000 Hz. The oscillating frequency of the sound wave that is created is what gives it a particular tone (what we call a note). The different times at which sound waves are generated is what creates the meaning of rhythm and articulation.

Limitations: Audio generation and processing techniques are complex. Whilst high-level hardware and software tools can be exploited to make a complete system more accessible, there certainly remains a relevant learning-curve. The scenario in which audio feedback is deployed conditions a lot the effect achieved, given the fact that materials surrounding the sound generating system at use impose effects like reverberation, echoes, or absorption. Exposure to sound feedback for a prolonged period of time has some drawbacks. Sound volume can potentially harm our auditory system. Sound feedback that lack textural richness (e.g., a single sine-wave) has the risk of becoming unengaging for the user or potentially cause irritation.

4.3. Vibrotactile Actuation

How it works: Vibrotactile actuation uses motors to stimulate communication utilizing touch, and more precisely tactile vibrations. When linked to biosensors, it can use the properties of the so-called vibrations to convey features of the biosignal being tracked.

What: Vibrotactile actuation is a technology communication mechanism that uses touch vibrations to exploit the touching sense of humans. It is built upon motors, which can mostly be categorized under two types:

- Eccentric rotating mass vibration motor (ERM), with a small unbalanced mass on a DC motor that creates a centripetal force translated to vibrations when rotating.
- Linear resonant actuator (LRA), containing a small internal magnetic mass attached to a spring, which creates a displacement force in a single axis when driven by an AC signal, usually operating around a specific narrow frequency bandwidth that increases efficiency.

These motors usually take the form of small (few millimeters) enclosures with simple positive and negative ($+/-$) terminals to be driven, lowering the power and supporting (weight) requirements. The typical power supply needed for this kind of micromotors is of the order of 1–5 V.

Where: With weights below 1g, the small form factor of these motors makes them suitable for body explorations, often relying on patches, elastic bands, or holders. Typical uses include also vibrotactile-equipped wristbands or smartwatches. Besides the traditional game/remote controllers including vibrotactile feedback and actuating on the hands, the currently ubiquitous role of mobile phones has spread the use of vibration feedback and patterns for notification, alarms and other communication examples anywhere a phone can be placed or hold.

When: Small vibrotactile motors feature fast startup and breaking times and can actuate taking rotations up to 11,000 revolutions per minute (RPM), in the case of ERMs, and oscillations of the order of few hundreds of Hertz.

Limitations: Vibration comes often with undesired noises or sounds. While this is mitigated by rubber-made absorbing structures often integrated in the motors, use cases need to consider this aspect. While vibrotactile actuation offers the opportunity to explore a particular type of haptic feedback, the use of small motors limits the generated effects, in terms of amplitude, duration, and intensity perceived. To create vibration sequences, several motors are needed, provided integration software and hardware development efforts are carried out. The actuators often require extra drivers to widen the operating regime possibilities while maintaining electrical safety standards. As generally advised in the case of feedback modalities applied to the body, haptic feedback actuation has to go hand in hand with user experience studies, since prolonged exposure and certain placements can lead to discomfort.

4.4. Temperature Actuation

How it works: Temperature actuation uses heating and cooling elements to stimulate communication using heat passed by haptics, that is, through our sense of touch. When used with biosensors, it can use the properties of the heating/cooling dynamics of the material to convey characteristics of the biosignal being tracked.

What: Temperature actuation is a type of communication that is used as feedback drawing upon the human haptic (touch) sense. The properties of the temperature feedback depend on the materials that are used to convey the features of the information (e.g., biosignals, behavioral data, etc.) of interest. Most commonly used approaches rely on the conversion of an electrical current input into heat/cold outputs, mainly by using resistive elements that heat up when current flows through. Examples of these are nichrome wires, conductive threads, conductive fabrics, and thermoelectric coolers (Peltier elements).

Where: The nature of the heating elements determines where the temperature actuation can be placed. The flexibility of wires and fabric has led to many developments that extend wearable capabilities, producing smart garments that lie close to the skin. Implementation possibilities of these technologies comprise patches and configurations to be mounted in accessories (bags, caps, etc.), among others. In applying heat or cold, placement plays a key role given the different perceptual and comfort ranges that exist throughout the body skin.

When: Typically, heat is an actuation modality that acts slowly. Whereas thermoelectric coolers and nichrome could seem to behave otherwise, being able to be turned on in a fast manner thanks to conduction, there is usually an element that plays a dissipating (slow) role in heat dynamics either using convection or radiation. Heat transfer, hence, usually involves relatively slow dynamics.

Limitations: As it is the case with actuation having an effect upon the body, heat/cool feedback has to closely consider the user comfort and perceptual thresholds. Materials' properties (mainly heat conductivity) constrain the possibilities in terms of time. While options like actuation upon wide-areas or multi-actuator sequences emerge as interesting actuation paradigms, power requirements remain a challenge. Moreover, sensitivity to temperature varies widely from user to user and is affected by ambient temperature conditions.

4.5. Shape-Changing Actuation

How it works: Shape-changing actuation uses interfaces that exhibit changes in size, shape, or texture in order to, when linked to feedback, exploit the human visual and tactile perception to convey meanings and information content. By using shape changes that unfold over time, the actuation dynamics are brought forth letting the user be able to play with concepts such as time (increasingly/decreasingly fast, slow, abrupt) and volume or size to depict the desired information.

What: Shape-changing actuation interfaces use the change of physical form that, when linked to feedback, provide a certain output conveying meanings and properties of the signal that they are bound to. Shape-changing exploits a combination of the senses of sight and touch to convey meanings intrinsic in the dynamics of the information that they are linked to, such as rapid changes, stability, increase, decrease, and steady growth. Such interfaces, despite the parallelisms found in visual screen-based explorations done in visual computing, are emerging as an alternative, physical, and tangible way of interacting with technological devices [66]. Three of the most widely used examples of shape-changing actuation elements are:

- Shape-memory wire ("muscle wire", nitinol, flexinol): a unique type of wire, which can be deformed, stretched and bent at room temperature, able to restore its shape when heated (i.e., when exposed to the electrical current). The wire activates rapidly when the electrical current is applied or the wire is heated.
- Linear actuator: a mechanical device that converts electrical current into a linear movement along a given axis, as opposed to the circular motion of a conventional electric motor. When equipped with extra sensors (such as Hall effect sensors) they are able to provide precise information on their length(the absolute position of the moving element on the axis).
- Inflatable shapes: enclosed structures which can be inflated with fluids (typically gas: air, oxygen, nitrogen, helium, etc.), usually accompanied by pressure or volume control mechanisms.

Where: In order for the technology to best benefit from the focus on visual and tactile perception, shape-changing actuation implementations must remain under reach (sight or touch) to convey the meanings embedded in the changes of shape. This can entail direct contact with the body, with the potential to increase the felt shape meaning when in contact with a large body area, or where touch sensations are more developed, or within the field of vision of the user.

When: The wide range of shape-changing actuation possibilities comes with different actuation timings in it. While it is possible to work with shape-memory wires that are rapidly heated or compressed gas or pumps that quickly fill up a given inflatable, the time affordances of this kind of actuation do not generalize. Linear actuators, for instance, usually require a system of pistons and damping mechanisms that have an impact on the dynamics of the actuation while it unfolds over time. Moreover, it is often the case that the behavior of shape-changing interfaces is not symmetrical, for example, although an inflatable can be rapidly fed air by a pump or a reservoir, deflation valves have their own rules.

Limitations: Memory wires, although visually appealing, imply the utilization of high temperatures which challenges the use of haptic shape-changing feedback based on them. In turn, the strains achieved (or pulling forces) are generally weak, often leading to implementations that use several wires. It is very common that the developments using this kind of actuation include the application of protective heat layers. In the case of linear shape-changing actuators, movement is often accompanied by undesired noise and relatively slow dynamics. The actuators themselves, made of rigid moving elements, impose a certain rigidity to the overall actuation. Moreover, multiple units are often needed to create appealing effects. Shape-changing inflatables often present problems of fluid leaks, as well as different asymmetric behaviors for inflation and deflation. These can be tuned by further developments on valves and compartments but requires significant work. Besides, the type of pump poses specific fluid requirements and usually exhibits noise that interferes with the actuation designed.

5. Sensing and Actuation under the Soma Design Approach

Our affective technology research aims to create personal technologies that enable self-reflection [6]. With a focus on the body, design research is used as a way to enter introspectively to emotion self-reflection and potentially disrupt the way we relate to our mental well-being with technology-mediated interactions. Technology-mediated interactions, drawing upon ubiquitous computing capabilities (biosensing, wearables, monitoring applications, embodied actuation) could add novelty and be taken further to psychotherapy contexts. In this section, we introduce the design approach taken to accomplish meaningful biosensing-actuation couplings. This comprises the first-person design stance, the somaesthetic design ("design through the body") approach, and the path to orchestration (i.e., the mechanisms for the coordination and event recognition in technology-mediated body interactions and the connections and sequencing for evocative sensing-actuation experience design).

5.1. Designing from a First-Person Perspective

Often in user-centered design, designers conceive, test, and set requirements for the ultimate users that are placed at the center of the design efforts. In doing so (using a third-person perspective), users are relegated to a second line, in which from time to time, designers probe, test and interview the target users iteratively in order to modify and render the design outcome meaningful according to their needs. This approach, however, misses out the potential of stepping into the user's shoes. The first-person perspective [67], instead, constitutes a way to highlight the designers' user experiences, paying honest tribute to the potential end-users, and actively engaging in experiencing the design meanings and effects. The designer who follows the first-person perspective embarks in an iterative process of trying, testing, feeling, and evaluating the designed object or interaction. The design is tried by the designer herself/himself. This process provides meaningful insight into what the eventual user could get from the resulting design. When taking the first-person perspective, the designer is seen as the user, since eventually, anyone interacting with the technology shapes its meaning and how the technology behaves.

5.2. Somaesthetic Design

Somaesthetic design underscores the need to place importance on the aesthetic aspect of the felt bodily experiences, as a fundamental element of the design process. This is, for us, a great first step to attempt to create embodied technologies or interactions. The chosen approach, consequently, confers in our case a key role to the body in the design of personal technologies for affective health. Somaesthetics, introduced by philosopher R. Shusterman [10] is the result of the efforts of combining the body with aesthetics, with a strong emphasis on how the body plays a major role in how we feel, perceive and think the world. With somaesthetic design, an attempt is made to leverage the role of the feeling body when engaging in design experiences. This strategy, requiring certain training to grasp one's sensations, control the movement and perceive what we feel, offers a fresh approach to using the body as the main instrument to feel, assess, and appreciate the design affordances. The soma design manifesto [8] highlights, among other aspects, the need to engage slowly in the aesthetic appreciation of the technologies being designed, disrupting the habitual and inspiring users' drive to obtain interactions—biosignal-mediated in our case—that lead to novel ways to embrace technologies in our lives. Body practices to support the process of getting attuned to one's sensations are usually employed, as exemplified in Reference [9]. These often further support the notion of estrangement (or disrupting the habitual), that is, how designers can engage in actions, movements, or performances far from the habitual way of carrying them out. By doing so, the intricacies of a certain interaction become exposed, helping its analysis or reaching to novel possibilities to carry it out.

5.3. First-Person Biosensing

Having successfully been applied to design workshops that address the effects of actuation-based interaction for embodiment and self-reflection, the somaesthetic design offers a unique opportunity to address the challenging goal of bringing together biosensing and actuation in an evocative way, relevant to the user who would utilize personal technologies for emotional awareness and regulation. In taking a somaesthetic design approach, affective technology researchers find a path to make sensing workable, paving the way for discoveries that support the creation of new relevant technology-mediated experiences as targeted by the AffecTech project [6]. To explore how our internal physiological mechanisms can be revealed via a set of biosignals, a routine for experiencing sensing from a first-person perspective was followed, significantly inspired by previous haptic actuation explorations [48,68] and used in part in Reference [49]. Using the knowledge in biosignals acquisition and relevant information processing the designed routine aims to support the learning of newcomer students to the field of biosignals, potentially adding tangible interaction features that make the topic more accessible. This process was originally thought as having an expert guide on biosignals and a learner that would follow and report on the felt biosensing experiences during a 1-3 hour session depending on the number of different biosignals covered, but in practice moved to a more open exploration scheme based on switching roles where no expert/novice knowledge hierarchy is sought [49]. The first-person sensing experience goes beyond a theoretical explanation and the lab experience with pre-recorded signals or a recording session of biosignals. With this different approach, the students or researchers that want to be initiated follow an introspection process to discover in a deeper way how and what can change the internal mechanisms of the signals and the body. This process is not thought to take place in a formal lab experience or lecture teaching environment, but accompanied by a person experienced both in soma design and biosensing that would revisit the experience and rely on body practices to awaken the somaesthetic appreciation needed for the exercise (see for instance the work of Windlin et al. and Tsaknaki et al. [9,48]).

5.4. Orchestration

Orchestration mechanisms try to answer the overarching goal of achieving systems that facilitate the exploration of biosensor data and meaningful representation through actuation that addresses many more modalities than just visual feedback (see Section 4.1). To achieve this facilitation it is crucial to be able to combine and nicely coordinate the relationship between input (biosensors) and output (actuators). In this paper, the term orchestration defines the process of:

- Creating couplings, that is, combining biosensors and actuators in place
- Coordinating the technology-mediated body interactions
- Working on the sequence in which different modalities are addressed via the sensors and actuators. Deciding which one goes first
- Guiding users in understanding the captured biodata through their felt experience by means of the addressed modalities
- Providing the design exploration ground, showing capabilities, limitations and roles of the involved technologies that participate in these interactions
- Potentially laying out machine learning, feature extraction, smart event recognition, or signal processing tools that can be applied to render the interactions more intuitive or meaningful

5.5. A Soma Design Example: The Breathing Light, or How Light Actuation Inspires Design

Interaction design research work by Ståhl et al. such as the Breathing Light [69], preceding the cross-disciplinary efforts presented in this paper, proved inspirational to our research. The goal of the Breathing Light prototype is to help the users to find a safe place where they can take a break from daily routines, focus on inner body processes, and reflect. The prototype is built from fabric and string curtains creating a secluded space for the user's upper body. The Breathing Light system switches the users' attention to breathing, focusing on the experience of inhale/exhale cycles. Light has been chosen as a modality given its ability to subtly guide the attention of the participants inwards. Technology-wise, the Breathing Light is a lamp with a proximity sensor. The sensor measures the distance between the chest of the user and the lamp, which in practice becomes a breathing sensor under these conditions. The ambient light in the prototype is dimming in accordance with the breathing patterns: exhale with a dim-out and recovering when inhaling. The intensity of the ambient light is high enough to make it possible to follow the light pattern even with the eyes closed but is not high enough to distract the user. The participants reported that when they were lying under the Breathing Light module, they felt enclosed and taken care of. Limitations arose as it was a demanding task to set the timing, intensity, and warmth of the light. However, in turn, this interaction facilitated an intimate correspondence between the perception of the breathing and the light, which meant that the light was perceived as an extension of the body, providing a much richer experience of breathing.

6. Results: Designing Biosensing-Actuation Couplings

In the work of *Somadata* [49], soma design sessions and first-person accounts of the user/designer participants are highlighted to understand what constitutes a tangible or "felt encounter" with an otherwise disembodied design material, that is, biosignals. Our design approach is not that of a "solutionist" method that tries to quantitatively acquire data and formally evaluate how a coupling solves a given problem. This avoidance of a solution is a resource that has been leveraged in design fiction [70]. Our research does not try to tackle the recognition of certain given patterns in biodata to, for example, make users optimize behaviors (walking, running or fitness related activities) nor prevent anomalies (heart malfunction, fall detection, stress recognition). We use design topics, or challenges, at most—such as exploring synchrony between peers. We do not work with a given problem. Rather, a qualitative, explorative stance grounded on the body is at the basis of design discoveries that help us look at biosignals as design material to be shaped, changed and integrated in interaction design toolkits instead of taken as a given, unchallenged and immutable. Where we have failed with other coupling attempts, a selection of carefully crafted sensing-output combinations succeed in achieving what we call *soma data*, that is, biodata that is somatically experienced, leading to novel insight, collectively shareable and in line with a design context or goal. Soma data examples in Reference [49] include a mechanism for groups of two people to connect non-verbally through audio and synchronous movements, a way to share muscle activity insight (on the calf muscles) relevant to an activity of crossing a balancing pole and new way to understand EDA data thanks to a haptic heating effect—that we address in more detail. In this paper, we want to open the design space. Hence, we bring to discussion the underlying interaction mechanisms of this kind of experiences. Although prototypes could be evolved into final products and studied quantitatively, the research presented in this paper is, instead, driven by the inspiration gathered from works that use technology to connect experientially to our felt bodies [9,48,68]. We aim to incorporate biosensing in soma design toolkits, as a design material, and discern what is needed to support this design. When used in soma design workshops, first-person somaesthetic accounts of users exploring the couplings are taken to assess whether an input-output connection is meaningful with regard to body self-awareness. Orchestration decisions are integrated in the organization of wired, programmed effects and input-output mechanisms found in the couplings that successfully led us to what we consider interaction design discoveries [49].

However, our interaction design research seems to suggest that better orchestration mechanisms would render our technology-mediated interactions more aesthetic [9,48]. In this line of thought, we have conceptualized a workflow that integrates sensor devices and actuators into a shared network that is controlled through a server, responsible for data transmission between components. In this scenario, the designer has the ability to explore available devices either via a graphical user interface (GUI) or tangible user interface. Furthermore, the available devices can be connected with one another through the GUI to couple input signal streams (such as biosignals) to output signals (like the intensity of a haptic actuator) and allow one to create and fine-tune somatic associations. These associations are designed through defining triggers and responses to signals or even via training machine learning models that react to input signals and act on the output signals. The system should allow for a flexible use of devices, their signals and behaviors to uncover novel interactions (Figure 2). An early example of an interactive visual programming metaphor can be seen in Figure 6.

In this section, we describe *Scarfy*, one of our research outcome biosensing-actuation coupling prototypes (EDA to temperature), as well as current work in Breathing Synchrony and EMG couplings with audio. In our view, a potential orchestration platform should let the users choose what elements are present in every experience, such as sensing-actuation modalities (haptic, sound, light, heat, cold, airflow, shape-changing), count on visual programming interfaces (as used in the EMG-audio feedback experience), enable the possibility to run signal processing code snippets (e.g., breathing synchrony assessment and audio feedback), allow interactions that work more implicitly (movement monitoring, wireless/wearable devices), and set the time sequence structure and order in place, hence shifting the design focus away from the technology constraints and highlighting how experiences are enacted and elements are part of a whole.

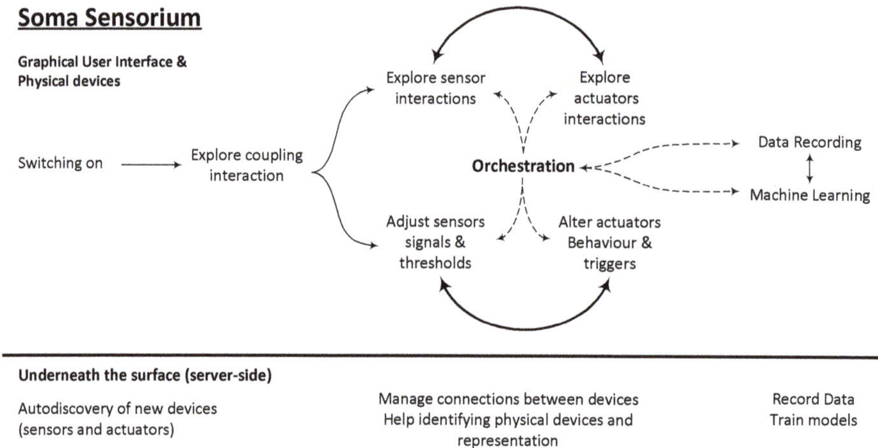

Figure 2. Mechanisms needed for the orchestration of couplings.

6.1. Scarfy: A Temperature Scarf to Make Electrodermal Activity Perceptible

Inspired by existing research and commercial work on haptic material actuators on the body [43, 71], we started exploring different materials and actuators to communicate biodata. We explored materials that are low-cost and safe to use for near body applications that take wearability and comfort into account. After trying out and working with different materials and actuators, that is, thermochromic, heat resistive materials, vibrotactile motors, shape memory alloys [43] and sensors, that is, electrodermal activity, heart rate and breathing, we started preparing a temperature-actuated scarf to promote interpersonal synchrony by linking skin conductance data. This coupling was aimed toward a soma-design session that explored the concept of interpersonal synchrony [49]. We chose the EDA signal, which has been often used to communicate increase and decrease of physiological arousal, to actuate heating and cooling. We used four 20x20 mm Peltier modules in series with a distance of 2.5 cm and enclosed them in a scarf. The resulting artifact can be easily worn on the neck and taken off as shown below (Figure 3).

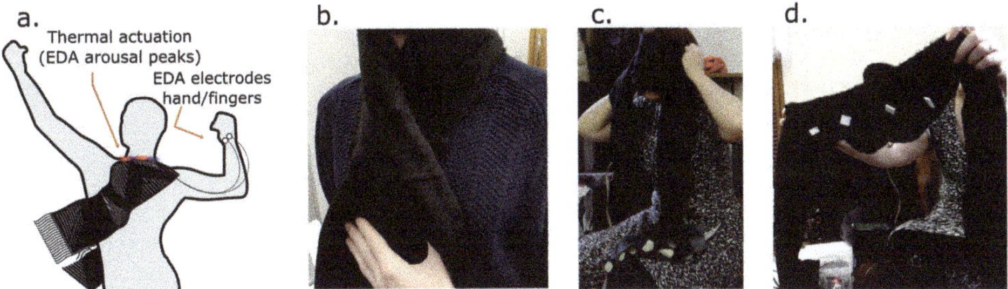

Figure 3. Scarfy: (**a**) EDA heat/cool temperature scarf coupling, (**b**) participants exploring actuation on the neck, (**c**) forehead and (**d**) showing the heating elements.

The Peltier modules are driven by Arduino boards with motor drivers. Their actuation is triggered by an EDA sensor. To mark the increase and decrease of changes in physiological arousal using temperature, we created four different patterns of heating and cooling as shown (see Figure 4). These patterns are *Appearing/disappearing heat/cool* actuating heat or cold in all the modules at once and then turning them off simultaneously. The second and third patterns as shown in Figure 4b,c are *Increasing heat/cool* meaning that heat or cool slightly turns up module by module, and *Decreasing heat/cool*, that is, gently reducing the thermal effect one by one. The fourth pattern is the *Moving heat/cool* pattern (Figure 4d) in which thermal actuation is alternated on the modules one by one following a spatial direction and keeping the temperature constant.

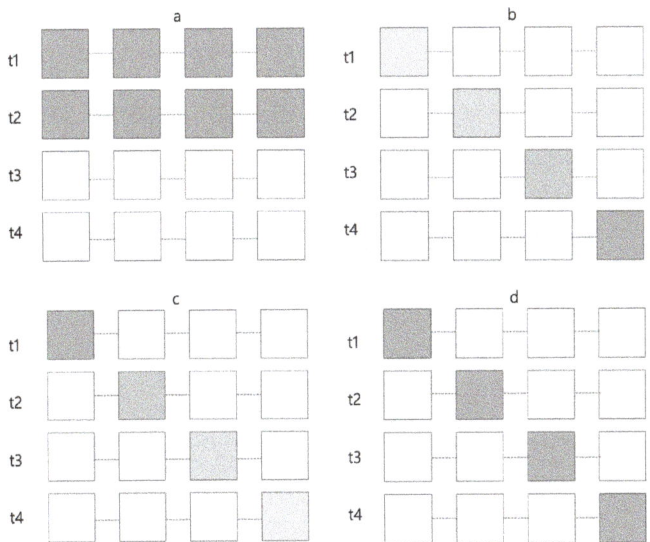

Figure 4. Scarfy EDA-temperature patterns with four Peltier module elements and how they change over time t_i: (**a**) Appearing/disappearing heat/cool, (**b**) Increasing heat/cool, (**c**) Decreasing heat/cool and (**d**) Moving heat/cool.

The purpose of this coupling and heating and cooling patterns was to communicate increasing and decreasing arousal in interpersonal settings. A set of designers' first-person accounts and insight on using the prototype in a design workshop focused on synchrony are found in Reference [49]. We wanted to explore how Scarfy can mediate synchrony between people and probe what heating and cooling patterns, intensity and duration would best support this quality. Besides we also wanted to explore feedback around technology *black-boxing* in Scarfy—that is, what design elements should be overtly exposed for customization—how it can be improved, and how it should be used in everyday life settings. Participants in design workshops approached Scarfy by wearing it around the neck, trying out different patterns and positions to feel the increase and decrease of heat and cold. While exploring different patterns and placement we felt that, although subjective, heat and cold have different scales, that is, the sensation of both heating and cooling feels different depending on the parts of the body it is applied to. While exploring the patterns on the body, we discussed how cold feels more pleasant than heat because of the placement of the modules inside a thick scarf fabric which is itself warm. Placing the scarf around the neck, we found that Peltier modules often do not touch the skin and need to be pressed in order to be felt. We were not limited to the neck area only in our exploration. We also explored several other areas such as the forehead (see Figure 3c), back, shoulder, and wrist. While exploring these other placements, we found that the considerable size of the scarf is harder to manage around these other areas. Therefore, we discussed that it would be better to place Peltier modules in smaller strap-on patches that can be placed and taken off easily. It would give us enough freedom to quickly explore patterns on different parts of the body. Finally, talking about arousal and the overall purpose, we discussed that the exploration should shape what meaning we assign to it, that is, whether you are trying to learn about yourself or you are trying to calm yourself down. In fact, ambiguity and the interpretability of electrodermal data have been a recent matter of study in human-computer interaction research [40], with some works questioning the user's meaning-making processes and challenges when new representations are appropriated, taken outside the lab [72,73]. For Scarfy, any researcher can explore several patterns and needs to figure out which one fits for his/her personal experience, that is, bodily awareness, calming yourself down, feeling your peers' arousal. Moreover, the interaction described in Reference [49], invites us to rethink how the aspects or features of the signal translated to actuation changes constrict the way biodata is perceived.

6.2. Breathing in Synchrony: From Physiological Synchrony to Audio Feedback

This coupling example, drawing upon the psychology concept of therapeutic alliance [74], takes respiration data from two users in the same physical space, where two BITalino devices stream data wirelessly to a host computer. In this example, two users participate in a timed breathing exercise together whilst their individual respiratory patterns are being measured with piezoelectric (PZT) bands placed around the diaphragm (see Figure 5). The data is aggregated on the host computer, executing a script that measures the collective breathing activity. From here, we apply shared biofeedback in the form of sound to stimulate synchrony awareness and physiological dialogue between users over time.

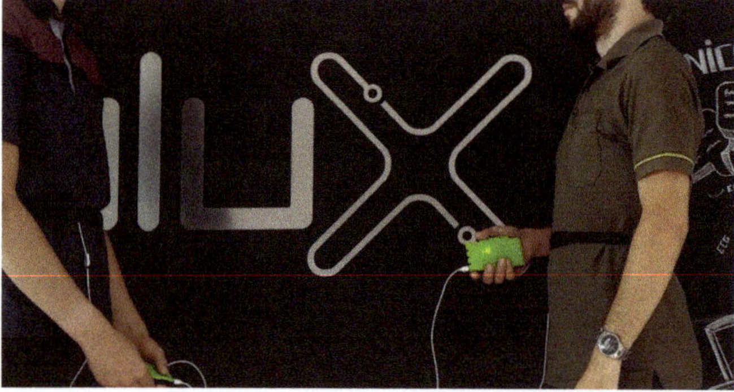

Figure 5. Breathing synchrony-audio experiment, based on the analysis of two BITalino piezoelectric abdominal respiration signals (image showing the two BITalino streaming simultaneously).

The exploration followed a stage of preliminary research on physiological synchrony features both in published research and drawing upon statistical measurements, potentially generalizable to signals other than breathing. We implemented the computation of linear regression coefficients, cosine similarity and correlations between filtered signal and derivatives. The process for mapping the user's activity audio output can be split into two main components. First, the device data is transmitted to a Python program, which is used to perform statistical analysis on the incoming signals, calculating a "magnitude of synchrony" using the features listed above. After a fifteen second warm-up period, the system accumulates a sufficient amount of data to determine mutual behavior, and the resulting values are encoded into Open Sound Control (OSC) messages that are continuously streamed to a local address, enabling the designer to map the data to appropriate parameters for sound feedback. With this generic protocol in place, we aim to embrace modularity, and advocate for the experimentation of sonic associations. In our tests, we used Cecilia's [75] built-in granular synthesis engine; this manipulates the playback of a pre-recorded soundscape divided into independent samples of 10 to 50 milliseconds [76].

6.3. Orchestrating an EMG-Audio Feedback Coupling

A depiction of an orchestration platform we achieved to create is that of the EMG-audio feedback coupling. Through a visual programming interface called PureData (Pd) [77,78], we connected a muscle activity signal with processing capabilities and a given audio pitch that changes properties according to the biosignal dynamics as muscles are contracted. The interface, shown in Figure 6, presents intuitive elements such as sliders and value boxes that facilitate the decision and modification of the coupling properties. This patch receives the EMG signal from a BITalino R-IoT device, a WiFi-enabled sensor platform, via Open Sound Control (OSC) [79,80] data packets. We mapped the EMG signal to sound in three steps: first, we took the absolute value of the signal—a full-wave rectification; second, we smoothed it with a low-pass filter to remove some of the oscillations; finally, the smoothed signal was mapped to the pitch of a sine wave generator. The higher the measured muscle contraction the higher the pitch the generator produces, and vice versa.

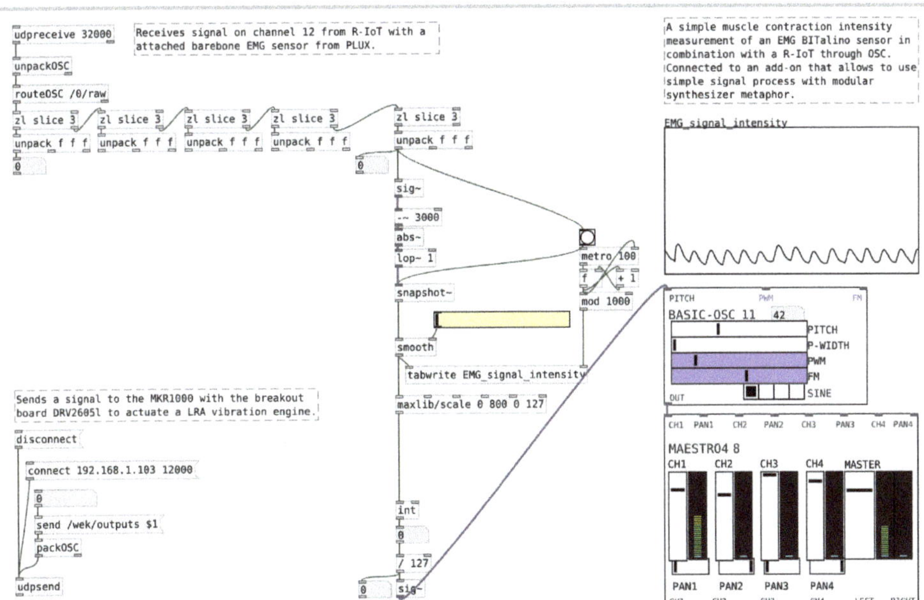

Figure 6. Orchestrating an EMG-audio feedback coupling in a PureData interface patch.

6.4. Understanding the Different Input/Output Tradeoffs

Through soma design sessions, we used first-person accounts of the designers or users of the created technology couplings to better understand the drawbacks and benefits that the input and output modalities pose. This subsection lists resulting remarks on the couplings and modalities studied (see Table 2). Scarfy, for example (Section 6.1), draws on previous studies [42,43] that propose novel EDA feedback. Although works such as Reference [41] are interesting in how they look differently at biodata to engage with, in our case we avoid visual feedback and design with the body to investigate effects that can be worn and felt. However, our results, letting us envision the platforms to support the design of couplings, lack the perspective of longitudinal studies highlighting users' data interpretation [40].

Table 2. Technology drawbacks and benefits.

	Drawbacks	Benefits
EDA temperature	• Limited EDA placement • High power supply needs • Circuitry-dependent orchestration • High temperature safety risks • Non-symmetrical effects (increase/decrease, heat dissipation)	• EDA data is shown tangibly (not only as peaks building up but also dissipating) • Low sampling rate, easily trackable with averages • Slow signal in line with the deliberate soma design stance • On-the-body effects (perceptible and physically grounding) • Easy to adjust, put on and take off in case of discomfort
Breathing synchrony input	• Advanced processing features that capture synchrony and multi-sensor behavior • Piezoelectric breathing sensor limitations (precise breathing rates but inaccurate breath holding detection)	• Multi-sensor (allowing multi-user and synchrony studies) • Highly controllable (useful for affective tracking but also interaction controls, for example, breathing amplitude and rate)
EMG input	• Placement for specific muscle tracking (trial and error needed) with high sampling rates (rapid muscle activity is precisely captured and multi-EMG muscle group and articulation monitoring possible)	• Low cost sensors with high sampling rates (rapid muscle activity is precisely captured and multi-EMG muscle group and articulation monitoring possible • Simple processing (signal energy and envelopes) to detect EMG bursts
Audio feedback	• Off-the-body actuation (needs a context or activity to relate to the physical body)	• Highly developed human hearing (high perception of pitch and rhythm changes) • Large consumer electronics audio possibilities (wireless speakers and headphones) • Many programming interfaces for audio. Music development area (many programming languages, libraries, platforms) • Existence of audio processing libraries in visual programming platforms → address orchestration platform GUIs

Our claim here is not that a soma design approach is the *best* or most efficient way to design with biosignals. Instead, we argue that soma design provides an interesting way to bridge between engineering- and interaction design perspectives and that this bridge in turn renders novel, creative, and relevant design concepts. In our work, it led to the creation of digitally-enabled experiences that succeeded in making us aware of sensations and reactions of our own bodies as well as those of our peers, at the same time as these explorations pinpointed technological challenges when sensors and actuators were used in ways they were not intended for. It helped us to move away from the predominant health optimization or fitness performance paradigm often present in physical and activity tracking devices that most biosensors are built for. In this sense, it provided a richer space for what biosensing might be used for.

7. Discussion

The coupling prototypes presented in this paper, in line with the reflections presented by Reference [49], led to what are arguably design discoveries in combining biosensing with body actuation. These are used to highlight the role of the body and ultimately make us, the users/designers, connect more intimately with it. Soma design is not a shortcut to circumvent the difficulties present when designing with biosensing, but a way to approach them differently. Interaction designers must face the same challenges that engineers or developers struggle with when evaluating what form factor, sampling rate or placement for a sensor is best for a given input or action of interest. Within our soma design exploration, though, issues such as noise in muscle tracking, electrode misplacement, or sensor undersampling that leads to no data variations are experienced through, for example, distorted sounds, excessive vibrations, or changeless temperature feedback, echoing what Fdili Alaoui writes about artists avoiding a problem-solving approach and turning technology resistance into creativity [81]. We believe that instances of couplings that have succeeded in bringing design insight should be integrated into a design or prototyping toolkit. Furthermore, our design approach offers the foundations to successfully integrate different sensing and actuation modalities in a way that is evocative to the body. With regard to Scarfy (Section 6.1), there is a direct link to the works that inspired an EDA coupling [42,43] to be more closely felt on the body emerge from an affective awareness goal. That is also the case of breathing, where Miri et al. [46] show haptic actuation examples with a clear affect intervention workflow. Our approach, instead, is that of supporting the design process. We do so by exploring what effects are possible and using the designer's own felt body to assess them. The change of focus is relevant. For instance, instead of using feedback as a breathing pacer [ibid.] or affect control mechanism, we delve into the experiential properties of the biosignal at hand and how they are shared or understood collectively. Although there is room for development, the paradigms that we used depict avenues in which we aim to widen the palette of interactions, refine orchestration mechanisms and connect to the underlying ethics of our way of designing bodily awareness or affective technologies.

7.1. Orchestration: The Soma Bits Toolkit

In previous work, we have brought forth the Soma Bits: a prototyping toolkit [48]. Acting as accessible "sociodigital materials", Soma Bits allow designers to pair digital technologies, with their whole body and senses, as part of an iterative design process. The Soma Bits have a form factor and materiality that allow actuators (heat, vibration, and shape-changing) and sensing (biosensors and pressure sensors) to be placed on and around the body (see Figure 7). They are comprised of a growing library of three-dimensional physical soft shapes, which are made of stretchable textile and memory foam. Each shape has at least one pocket, making it possible to insert different sensing or actuating components. By combining several actuators with shapes, one can orchestrate experiences, and explore the qualities of the sociodigital material directly on the body, by changing the parameters of the sensors and actuators and placing the shapes on different parts of the body.

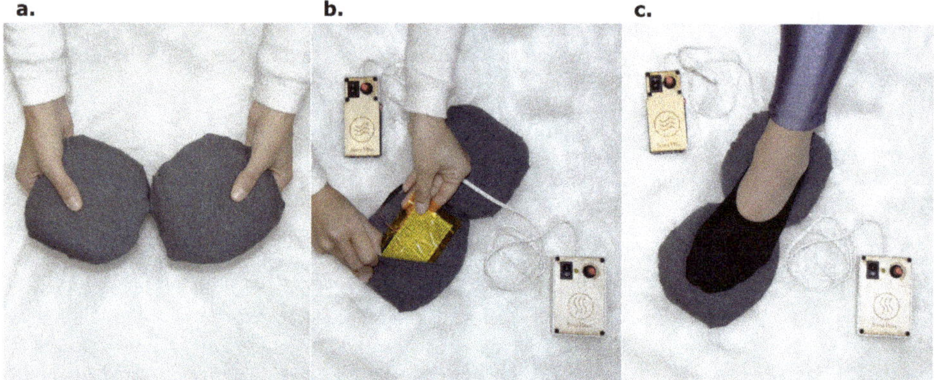

Figure 7. Elements of the Soma Bits design toolkit: (**a**) shapes, (**b**) temperature actuation, (**c**) vibration actuation.

The Soma Bits are easily (re-)configurable to enable quick and controllable creations of soma experiences which can be both parts of a first-person approach as well as shared with others. In the case of a first-person exploration, someone can, for example, experience and reflect on the properties of heat actuation on their foot and gain a bodily understanding of a heat modality, which can be later integrated into the design of interactive systems. In the case of exploring somaesthetic experiences shared with others, an example can be that one person feels on their spine the breathing patterns and rhythm of another person, translated into a shape-changing pattern that is experienced through the spine soma shape that is part of the Soma toolkit.

We have taken the first steps towards orchestrating collective behaviors of the Soma Bits by combining different Bits and allowing interaction designers to program complex behaviors (e.g., slowly shape-changing materials that heat up when a user presses on them). To achieve that, we aimed at a protocol and an interface for connecting Soma Bits together. In the middle, between sensing and actuation, we provide an "orchestration unit". This unit acts as a hub for the sensor and actuator network. The orchestration unit is controllable through OSC (Open Sound Control) [79,80]. This protocol enables the usage of common musical interfaces, such as controllers or sequencers, allowing end-users to program the bits without having to write code. We have also started using supervised learning algorithms to quickly help bootstrap interactions. These algorithms allow for mapping noisy sensor data to actuation, which in turn would allow for more complex behaviors to also be programmed into the bits without writing code. We tested the Soma toolkit focusing mainly on combinations of shapes and actuators during three workshops with interaction design researchers and students from several disciplines. Research purposes were explained to participants, who signed informed consent forms. The first was a one-day workshop at the Amsterdam University of Science that took place in October 2018, in which 30 master's students engaged in a soma design process having the Soma toolkit as the main medium to explore actuation and bodily experiences around the topic of empathy. The second workshop took place in December 2018 at the Mixed Reality Lab, at the University of Nottingham, UK, and was focused on the Soma Bit shapes addressing the topic of balance. First-person accounts of the designers involved in the study and an elaborate analysis of design outcomes of the workshop can be found at References [49,82]. Together, we explored for three days the Soma Bits in several design contexts, including VR applications and leg prosthetics for dancers. During this workshop, we introduced sensing, apart from actuation, through the BITalino prototyping platform [25,55,56]. We also initiated the design of couplings between sensing and actuation, for example by translating movement through acceleration, to sound. The third workshop deploying the Soma Bits was conducted in February 2019 in Milan (see examples in Reference [49]). In this workshop, researchers from several disciplines including psychology, engineering, and interaction design, experienced different prototypes that were brought to the workshop, in combination with the Soma Bits toolkit. The workshop lasted for a day, synchrony was the main topic underlying the prototype demonstrations, as well as the bodily and technological explorations. As a general reflection we observed that as soon as the Soma Bits toolkit was introduced to the design process, the workshop participants shifted their attention to experiencing the sensing-actuation technology through their bodies, rather than just on a conceptual or verbal level. On a broader level, the Soma Bits toolkit addresses the difficulty we experienced in past soma design processes—that of articulating sensations we want to evoke to others, and then maintaining these experiences in memory throughout a design process. Thus, the Soma Bits enable designers to know and experience what a design might feel like and to share that with others. The Soma Bits have become a living, growing library of shapes, sensors, and actuators and we continue using them in our design practices, as well as when engaging others in soma design processes.

7.2. Missing Bits: Shape-Changing Actuation

In the creation of a Soma Design toolkit, we aim for a wide range of tunable modalities for the user to explore and create the effects what communicates best for her/his soma. In this regard, we know that our prototyping efforts fall short on making shape-changing actuation available. HCI research has already shown some of the design potential behind linear actuators and inflatable shape-changing mechanisms that inspire us and guide our future research perspectives.

7.2.1. Linear Actuators

The linear actuator serves as a central piece of what we call the Soma Pixel system. Soma Pixel is a modular interchangeable sensor-actuator system (see Figure 8a), inspired by shape-changing projects carried out by MIT researchers from the Tangible Media and the Senseable groups:

- Project Materiable [83,84]
- Project Lift-Bit [85,86]

We aim at having a number of smart modules ("pixels"), capable of sensing the human body (pressure by weight, biosignals) and providing certain actuation. The shape-changing actuator (currently a linear one) serves as a skeleton for the "pixel". Sensors and other actuators are thought to be located in the upper part of the device. The modules can be easily rearranged in space, while "knowing" their relative position with respect to each other. At the moment, in the Soma Pixel setup, the linear actuator is coupled with a force sensor. Actuation happens when weight/force is being applied to the sensor. This is, nonetheless, accompanied with limitations, that is, the linear actuators cannot move (change length) fast and the motor inside of the linear actuator makes substantial noise when running.

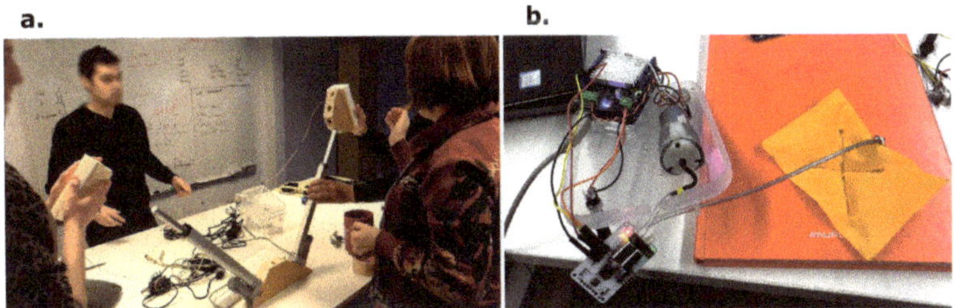

Figure 8. Shape change: (**a**) Prototyping with linear actuators (**b**) Inflatable shape.

7.2.2. Inflatable Shapes

In an ongoing research line, inflatable shapes are currently being used for constructing a "singing corset" prototype, as well as a part of a newer revision of the Soma Bits toolkit. The corset is built upon an Arduino-powered inflatable that will be integrated with the Soma Bits (see Figure 8b). However, the first iteration of inflatable shapes had difficulties with fast deflation. This was partly solved by adding another air pump, devoted specifically for exhausting air. Experimenting with the addition of separators inside an inflatable shape or splitting it into multiple inflatable sections may further improve exhaust performance. Another limitation is unavoidable minor air leakage, which will happen due to imperfections in manufacturing the actuator (sealing inflatable shape, valve timing).

7.3. Extending Orchestration: The Role of Biosignal Processing

A relevant part of our research has focused on the extension of biosignal feature extraction, processing, and analysis in real-time (to enrich real-time feedback possibilities within the lab and in more ecological settings). While initial sensing-actuation orchestration couplings have successfully made use of basic signal processing, further possibilities lie in improving the current algorithms and processing approaches. Any sort of biosignal acquisition, in particular in psychophysiology sensing, can be seen as the process of dealing with sequential and time-series data. Prior to the deployment of feature extraction and selection techniques, data must be preprocessed properly. ECG, EMG, EDA, IMU, respiration, and all the data collected from available sensors are in general sampled at different rates and present different properties that entail signal noise and instances of data that do not conform to (standard) expected representations. These can be seen as artifacts or "meaningless" data. Since the data obtained in real-world ambulatory settings is always noisy, presenting inconsistencies or missing values, preprocessing and cleaning is required (see examples in previous studies we conducted [57]). As it is the case for the features listed in Section 3, time domain and statistical features such as, in the case of designing with breathing and/or heart monitoring (ECG), the mean value of the rates, the mean value of the time between events, the standard deviation of those intervals and the root mean square of successive interval differences are accessible in the wild. Of particular interest are the less apparent frequency-based features which given the demanding spectral analysis computational requirements have just begun to appear in the nowadays more capable out-of-the lab devices. While certain features count on solid research support, pointing at the most scientifically validated ones, attention is given to the exploration of the potential mappings that can be built upon them. The use of multimodal data (namely, the use of several biosensors at the same time), extends the monitoring capabilities that can be brought to the couplings' design by widening the perspective and being able to keep track of body signals that are not the main focus of the interaction. This allows the designer to put in place validation mechanisms, gaining accuracy, and supporting the claims made with respect to the the targeted biosignal. If orchestration aims at the creation of meaningful technology-mediated interactions, the made connections need to rely on processing capabilities, event detection, and high-level feature extraction to overcome the limitations of too basic couplings that only put in place simple (signal) amplitude-to-intensity mappings. Moreover, the group somaesthetic design appreciation sessions where the felt experiences are brought together and debated (usually relying on inspirational body-centered exercises, reenactions, body sketching tools, and verbal communication) could be significantly enriched with the monitoring of signals acquired during the design experience. As design explorations have already started to show, there seems to be room for the creation of machine learning and processing capabilities that could render interactions with the technology more implicit (also embodied or intuitive).

7.4. Extending Orchestration: Machine Learning Capabilities and Personalization

Standard machine learning workflows are usually multi-step and involve the definition of problem-specific feature extraction methods, as well as in-depth expert knowledge of the problem at hand. Deep Learning (DL) [87] techniques, also part of our signal processing research efforts, shift the ML practitioner focus from that of feature extraction methods to feature learning. In particular, in end-to-end learning settings, DL models are directly fed with the raw input signal (i.e., without any form of pre-processing, or at best very mild), and use this to automatically extract (deep) features from it and make predictions based on them [23,88]. Its benefits apply as well to machine learning approaches aimed at improving pattern recognition capabilities of relevance in physiology or affective data [89]. In cases in which the input signal is high-dimensional or when timestamps are a relevant characteristic of the problem at hand (e.g., biosignals, multimodal), DL methodologies have been shown to outperform traditional pattern recognition techniques. While research for real-life DL deployments is still under progress, with algorithm computational resources being one of the main limitations nowadays, DL approaches emerge as mechanisms to overcome the difficulty of coming up with sensible hand-crafted features for a certain classification problem at hand. Provided advances towards DL in ambulatory scenarios are made, orchestration design (of sensing-to-actuation couplings) can potentially benefit from DL feature learning by lowering the technology expertise burden on the designer/experimenter side. Overall, this would allow experience design to focus more specifically on the affordances of the interaction rather than the processing mechanisms. To this end, work on Convolutional and Recurrent Neural Networks (CRNN) has shown to be able to deploy a Neural Network paradigm that obtains classification/recognition performance improvements by algorithm personalization, that is, avoiding specific calibrations to be done according to the user. Typically, personalized Machine Learning models are simply the idea of learning individual behaviors that train a model using the data collected only from the subject. These models are usually customized to the requirements of each individual, imposing requirements to consider the individual differences, yet still using data collected from the population. To some extent, new deep learning paradigms are circumventing these requirements.

In Section 3, we describe how embodied sensor technologies react differently in accordance with the unique biological characteristics of the body. Similarly, the perceived impact of a given actuation mechanism—as those described in Section 4—largely depends on the sensitivity to a given stimulus, as well as the natural bodily variations between different users. With this considered, we recognize the necessity to attune the system's parameters in order to produce mappings that facilitate meaningful interactions that are not overly obtrusive. While auto-calibration mechanisms have been implemented in the previous examples, which typically define and minimum/maximum parameter ranges, we foresee an extended benefit in adopting Interactive Machine Learning (IML) [90] frameworks as means to foster perspectives respecting body pluralism. Existing frameworks such as *Wekinator* [91] and *Teachable Machine* [92] facilitate the design of classification and regression-based mappings that are initiated and iteratively adjusted with example data provided by the user (e.g., Reference [93]). Furthermore, we would like to explore the use of Interactive Machine Learning to develop novel coupling relationships that go beyond linear mappings, as well as intuitive mappings between multimodal inputs and multi-dimensional outputs.

In our sound-based examples, visual programming environments heavily assisted the orchestration process. In both cases, the systems enabled users to visualize a continuous stream of mappable data in real-time, clearly exposing any unexpected behavior that may occur (for example, with the displacement of sensor electrodes). The node-based functionality of the frameworks allowed for a coherent representation of the dataflow and signal processing steps in order of execution, less abstract compared to a code-based script. During the process of developing the system, a user interface is generated in parallel on-the-fly as each node presents a GUI element that grants the designer access to parameters such as scaling and smoothing coefficients. In Section 6.2, a basic interface allowed users to test and compare a set of algorithms for sound mappings. This workflow can be beneficial for rapid experimentation with a variety of parameters and signal processing techniques that influence the interactive experience. It also presents a convenient solution for fine-tuning a complete system according to the user's experience.

7.5. Ethical Underpinnings

We have provided examples of research on sensing and actuation technologies, first-person somaesthetic approaches and experiments to foster the design and research of self-awareness (embodied, wearable) technologies with the potential to support self-reflection, emotion regulation, and affective health. In any design process, and particularly for technologies that may be used in affective or health contexts, it is important to consider the ethical implications of design, use, and research. Ethics simply concerns what is *good*, with a utilitarian perspective dealing with the greatest good for the greatest number of people. To assist in ethical decisions and practices, ethical frameworks outline key concepts such as beneficence and nonmaleficence, justice, responsibility, autonomy, privacy and confidentiality, and respect for the rights and dignity of others [13,14,94–97]. By reflecting on these ethical principles and standards, designers and researchers can ensure that their processes, technological developments, and research are done ethically and for the greater good.

The concept of the first-person somaesthetic design reflects qualities of good ethical practice through its focus on experiencing each aspect of the technology to facilitate its appropriate use, integration, and development. This extensive design process explores the benefits and potential adverse effects of the various sensors and actuators and uses this to shape orchestration and future design. Guiding users in interpreting the captured data throughout the felt experience and exposing or creating the meanings attached to personal sensing is a deliberate effort to make interactions intuitive but conspicuous. In Reference [67], designing with the body is seen as a route for designers to harmonize with their felt experiences, an alternative of particular relevance in light of the implicit interaction direction that personal tracking technologies seem to move into. The slow and deliberate soma design methods also reinforce the careful consideration of how technology may impact experience, and therefore the potential effects this may have on future users. Part of our research on orchestration, attempts to provide the design exploration ground that would show capabilities, limitations and roles of the involved technologies that participate in the designed interactions, in line with the first-person perspective. It is important to acknowledge that when stepping into the user's shoes, designers jump to the firing line. Potential effects for the user are experienced firsthand. Another ethical strength lies in the acknowledgement of the need for variety and customization of these experiences to suit the individual end-users and their specific needs, as avoiding the blindness for body differences is advocated early on in first-person work [67]. This is important for inclusive and diverse technologies which must consider not just individual differences across users, but also users with additional needs or impairments who may be disadvantaged or excluded from technologies which focus on only one modality, such as visual interfaces. The ethical somaesthetic design will involve consideration and incorporation of ethical principles and practices in the design process from conceptualization and throughout the design and development lifespan. Designers and researchers should be familiar with core ethical principles and should reflect on how these may shape their design practices, research, and technological developments. Research examples such as the work of Balaam et al. [35], point at ways to challenge the design practice, especially in emotion work, and avoid engaging participants by default but questioning and justifying their engagement. Autoethnographic and first-person design are seen as promising alternatives. Future research should explore how to incorporate user-centered design within the first-person design process. This will increase the validity of the premise that first-person experiences can be used to create devices to serve the diversity of human experiences. This is especially crucial for devices that may be used for emotion regulation or affective health, where different users may have different experiences, needs, and risks based on their unique histories and circumstances. Experiments such as the design explorations and couplings described in this paper can, therefore, be adapted to involve persons with varied lived experiences to explore how their experience of the soma toolkit or other technologies may differ from those of the designers. This is further connected to concepts of beneficence, non-maleficence, and justice which encompass issues related to benefits, risks, safety, fairness, and equal access for all. Technology offers opportunities to reduce barriers but to do so, design and development must consider how to deliver innovative and impactful technology while still being accessible and affordable. Designers also have a responsibility to consider the intended and unintended potential consequences of any new technology, and the need for appropriate design, guidance, and support to ensure safe use. While discussions of impact, effects, and outcomes are centered on end-results, it is crucial for these to be considered at the beginning of the design process and throughout to ensure the creation of good and safe technology. Also important to consider are issues of data security, privacy, and ownership of personal data, with the safe and secure handling of biosensor data an important design consideration with ethical and legal implications for its use and misuse. Finally, while somaesthetic couplings may offer opportunities for self-awareness and emotion regulation, designers must consider the balance between optimizing and pathologizing typical human experiences, as well as the potential stigma in encouraging the tracking and monitoring of affective health. As with all ethical issues, this must be considered throughout the design process. Like the somesthetic design, the ethical design must be a continuous process integrated throughout all aspects of the design experience and production.

8. Conclusions

Research on mobile and wearable technologies that track behavioral, psychological, and contextual signals has recently gained momentum in the field of mental health. At the same time, the rise of personal sensing has garnered the interest of HCI research. In this paper, we approached the design of sensing-actuation experiences intended for rich embodied interactions with relevance to affective health. To achieve this, we adopted first-person soma design to integrate biosignals that are commonly used in ubiquitous low-cost personal sensing together with actuation mechanisms studied in HCI. Our design exploration, giving special attention to the sentient body and acknowledging alternative ways to address affect within interaction, culminated with a set of coupling examples, in which we demonstrate data mapping strategies between various devices in the context of bodily and emotion awareness. The soma design approach applied to the creation of biosensing-actuation couplings for affective and self-awareness experiences is the main contribution of this paper. Through the couplings, we arrived at the concept of orchestration, defining the ways in which body input-output systems and meanings are put in place, the range of mappings and how they unfold. Soma design is a theoretically robust design approach that helps us sketch experiences to develop a (not necessarily dialogue-oriented) toolkit to facilitate creating affective technologies grounded on the body and enhanced by biosignals that are made available as design material. As a design toolkit, the examples created so far are instances of a wider collection of tools. The findings of our design explorations have unveiled a set of research directions (or requirements) to pursue in order to achieve broader orchestration mechanisms:

1. Further work on real-time machine learning tools to train biosignal-based input-output effects on the fly (in line with research in interactive machine learning [90–92]) and extend options to customize the features recognized and feedback received, according to the user preferences
2. Continue developing and testing programming interfaces that not only enable setting new sensing-actuation connections but leverage the user configuration and control while a coupling is being explored
3. Carry out more studies on multi-sensor experiences, as these point to the need for more advanced real-time signal processing and the rich multi-user interactions where outputs are the result of user collaborations or shared, bodily understanding of biodata

This insight aims to inspire developments in affective technologies and invites the joint work of engineering, interaction design, or even clinical disciplines that are traditionally disconnected from one another. Moreover, our discussion points at current limitations and paves the way for future research. We indicate sensing-actuation modalities that have been underexplored, then we consider the potential benefits of integrating refined machine learning algorithms and (developing) new orchestration interfaces to assist and democratize the crafting and customization process. As somatic perspectives are becoming more incorporated in areas of interaction design (research) and embraced with rigor, we foresee valuable intersections in other research domains.

Author Contributions: Authors H.G. and M.A. conceptualized and framed the paper; M.A., W.P. and N.C. curated the description of biosignals; M.U., W.P., C.W., P.K. and P.S. carried out the research on actuation mechanisms; C.W., P.K., P.S., M.U. and M.A. presented the results on first-person design combining the reported technologies; C.W., P.K., N.C., D.B. and M.A. carried out the discussion on further directions, orchestration and ethical underpinnings of the current multidisciplinary research; H.G., C.E., C.S. and K.H. reviewed and supervised the paper; All authors contributed to the original draft preparation and subsequent editing. All authors have read and agreed to the published version of the manuscript.

Funding: This work has been supported by Marie Skłodowska Curie Actions ITN AffecTech (ERC H2020 Project ID: 722022).

Acknowledgments: Authors would like to thank the whole AffecTech consortium for valuable scientific discussion and in particular researcher A.Patanè on the potential use of Deep Learning techniques. This paper would not have been possible without the useful criticism and input received from the design session's participants.

Conflicts of Interest: Authors M.A. and W.P are employed by PLUX Wireless Biosignals, a company manufacturing biosensensing devices. The rest of the authors declare no conflict of interest.

References

1. Bardram, J.E.; Matic, A. A Decade of Ubiquitous Computing Research in Mental Health. *IEEE Pervasive Comput.* **2020**, *19*, 62–72. [CrossRef]
2. Qu, C.; Sas, C. Exploring Memory Interventions in Depression through Lifelogging Lens. In Proceedings of the 32nd International BCS Human Computer Interaction Conference, Belfast, UK, 4–6 July 2018. [CrossRef]
3. Qu, C.; Sas, C.; Doherty, G. Exploring and Designing for Memory Impairments in Depression. In Proceedings of the 2019 CHI Conference on Human Factors in Computing Systems, Glasgow, UK, 4–9 May 2019. [CrossRef]
4. Abdullah, S.; Matthews, M.; Frank, E.; Doherty, G.; Gay, G.; Choudhury, T. Automatic detection of social rhythms in bipolar disorder. *J. Am. Med. Informatics Assoc.* **2016**, *23*, 538–543. [CrossRef] [PubMed]
5. TEAM ITN: Technology Enabled Mental Health. Available online: https://www.team-itn.eu (accessed on 25 July 2020).
6. AffecTech. Available online: https://www.affectech.org/ (accessed on 20 June 2020).
7. Repetto, C.; Sas, C.; Botella, C.; Riva, G. AFFECTECH: Personal Technologies for Affective Health. In Proceedings of the 15th European Congress of Psychology, Amsterdam, The Netherlands, 11–14 July 2017.
8. Höök, K. *Designing with the Body: A Somaesthetic Design Approach*; Design and Design Theory; MIT Press: Cambridge, MA, USA, 2018.
9. Tsaknaki, V.; Balaam, M.; Ståhl, A.; Sanches, P.; Windlin, C.; Karpashevich, P.; Höök, K. Teaching Soma Design. In Proceedings of the 2019 on Designing Interactive Systems Conference, San Diego, CA, USA, 23–28 June 2019. [CrossRef]
10. Shusterman, R. *Thinking through the Body*; Cambridge University Press: Cambridge, UK, 2012. [CrossRef]
11. Daudén Roquet, C.; Sas, C. Body Matters: Exploration of the Human Body as a Resource for the Design of Technologies for Meditation. In Proceedings of the 2020 ACM Designing Interactive Systems Conference, Eindhoven, The Netherlands, 6–10 July 2020; pp. 533–546. [CrossRef]
12. Sas, C.; Hartley, K.; Umair, M. ManneqKit Cards: A Kinesthetic Empathic Design Tool Communicating Depression Experiences. In Proceedings of the 2020 ACM Designing Interactive Systems Conference, Eindhoven, The Netherlands, 6–10 July 2020; pp. 1479–1493. [CrossRef]
13. Bowie, D.; Sunram-Lea, S.I.; Sas, C.; Iles-Smith, H. Evaluation of depression app store treatment descriptions and alignment with clinical guidance: Systematic search and content analysis. *JMIR Form. Res.* **2020**, accepted. [CrossRef]
14. Sanches, P.; Janson, A.; Karpashevich, P.; Nadal, C.; Qu, C.; Daudén Roquet, C.; Umair, M.; Windlin, C.; Doherty, G.; Höök, K.; et al. HCI and Affective Health: Taking Stock of a Decade of Studies and Charting Future Research Directions. In Proceedings of the 2019 CHI Conference on Human Factors in Computing Systems, Glasgow, UK, 4–9 May 2019. [CrossRef]
15. Niedenthal, P.M. Embodying Emotion. *Science* **2007**, *316*, 1002–1005. [CrossRef] [PubMed]
16. Höök, K. Knowing, Communication and Experiencing through Body and Emotion. *IEEE Trans. Learn. Technol.* **2008**, *1*, 248–259. [CrossRef]
17. Barrett, L.F.; Mesquita, B.; Ochsner, K.N.; Gross, J.J. The Experience of Emotion. *Annu. Rev. Psychol.* **2007**, *58*, 373–403. [CrossRef]
18. Nummenmaa, L.; Glerean, E.; Hari, R.; Hietanen, J.K. Bodily maps of emotions. *Proc. Natl. Acad. Sci. USA* **2013**, *111*, 646–651. [CrossRef]

19. Shafir, T.; Tsachor, R.P.; Welch, K.B. Emotion Regulation through Movement: Unique Sets of Movement Characteristics are Associated with and Enhance Basic Emotions. *Front. Psychol.* **2016**, *6*, 2030. [CrossRef]

20. Takayama, N.; Korpela, K.; Lee, J.; Morikawa, T.; Tsunetsugu, Y.; Park, B.J.; Li, Q.; Tyrväinen, L.; Miyazaki, Y.; Kagawa, T. Emotional, Restorative and Vitalizing Effects of Forest and Urban Environments at Four Sites in Japan. *Int. J. Environ. Res. Public Health* **2014**, *11*, 7207–7230. [CrossRef]

21. Kotera, Y.; Richardson, M.; Sheffield, D. Effects of Shinrin-Yoku (Forest Bathing) and Nature Therapy on Mental Health: A Systematic Review and Meta-analysis. *Int. J. Ment. Health Addict.* **2020**. [CrossRef]

22. Höök, K. Affective Computing: Affective Computing, Affective Interaction and Technology as Experience. In *The Encyclopedia of Human-Computer Interaction*, 2nd ed.; The Interaction- Design.org Foundation: Aarhus, Denmark, 2012; Chapter 12.

23. Bota, P.J.; Wang, C.; Fred, A.; da Silva, H.P. A Review, Current Challenges, and Future Possibilities on Emotion Recognition Using Machine Learning and Physiological Signals. *IEEE Access* **2019**, *7*, 140990–141020. [CrossRef]

24. Höök, K.; Ståhl, A.; Jonsson, M.; Mercurio, J.; Karlsson, A.; Johnson, E.C.B. Somaesthetic design. *Interactions* **2015**, *22*, 26–33. [CrossRef]

25. BITalino. 2013. Available online: http://www.bitalino.com (accessed on 20 May 2020).

26. Sas, C.; Coman, A. Designing personal grief rituals: An analysis of symbolic objects and actions. *Death Stud.* **2016**, *40*, 558–569. [CrossRef] [PubMed]

27. Picard, R.W. *Affective Computing*; MIT Press: Cambridge, MA, USA, 1997.

28. Cabibihan, J.J.; Javed, H.; Aldosari, M.; Frazier, T.; Elbashir, H. Sensing Technologies for Autism Spectrum Disorder Screening and Intervention. *Sensors* **2016**, *17*, 46. [CrossRef] [PubMed]

29. Matthews, M.; Doherty, G. In the Mood: Engaging Teenagers in Psychotherapy Using Mobile Phones. In Proceedings of the SIGCHI Conference on Human Factors in Computing Systems, Vancouver, BC, Canada, 7–12 May 2011; pp. 2947–2956. [CrossRef]

30. Matthews, M.; Voida, S.; Abdullah, S.; Doherty, G.; Choudhury, T.; Im, S.; Gay, G. In Situ Design for Mental Illness: Considering the Pathology of Bipolar Disorder in MHealth Design. In Proceedings of the 17th International Conference on Human-Computer Interaction with Mobile Devices and Services, Copenhagen, Denmark, 24–27 August 2015; pp. 86–97. [CrossRef]

31. Sas, C.; Höök, K.; Doherty, G.; Sanches, P.; Leufkens, T.; Westerink, J. Mental Wellbeing: Future Agenda Drawing from Design, HCI and Big Data. In *Companion Publication of the ACM Designing Interactive Systems Conference 2020, Eindhoven, The Netherlands, 6–10 July 2020*; Association for Computing Machinery: New York, NY, USA, 2020; pp. 425–428. [CrossRef]

32. Maitland, J.; Chalmers, M. Self-Monitoring, Self-Awareness, and Self-Determination in Cardiac Rehabilitation. In Proceedings of the SIGCHI Conference on Human Factors in Computing Systems, Atlanta, GA, USA, 10–15 April 2010; pp. 1213–1222. [CrossRef]

33. Morris, M.E.; Kathawala, Q.; Leen, T.K.; Gorenstein, E.E.; Guilak, F.; Labhard, M.; Deleeuw, W. Mobile Therapy: Case Study Evaluations of a Cell Phone Application for Emotional Self-Awareness. *J. Med Internet Res.* **2010**, *12*, e10. [CrossRef] [PubMed]

34. Miri, P.; Uusberg, A.; Culbertson, H.; Flory, R.; Uusberg, H.; Gross, J.J.; Marzullo, K.; Isbister, K. Emotion Regulation in the Wild: Introducing WEHAB System Architecture. In Proceedings of the the 2018 CHI Conference on Human Factors in Computing Systems, Montreal, QC, Canada, 21–26 April 2018. [CrossRef]

35. Balaam, M.; Comber, R.; Clarke, R.E.; Windlin, C.; Ståhl, A.; Höök, K.; Fitzpatrick, G. Emotion Work in Experience-Centered Design. In Proceedings of the 2019 CHI Conference on Human Factors in Computing Systems, Glasgow, UK, 4–9 May 2019; pp. 1–12. [CrossRef]

36. Sas, C.; Zhang, C. Do Emotions Matter in Creative Design? In Proceedings of the 8th ACM Conference on Designing Interactive Systems, Aarhus, Denmark, 16–20 August 2010; pp. 372–375. [CrossRef]

37. Sas, C.; Zhang, C. Investigating Emotions in Creative Design. In Proceedings of the 1st DESIRE Network Conference on Creativity and Innovation in Design, Desire Network, Aarhus, Denmark, 16–17 August 2010.

38. Boehner, K.; DePaula, R.; Dourish, P.; Sengers, P. How emotion is made and measured. *Int. J. Hum. Comput. Stud.* **2007**, *65*, 275–291. [CrossRef]

39. Howell, N.; Chuang, J.; De Kosnik, A.; Niemeyer, G.; Ryokai, K. Emotional Biosensing: Exploring Critical Alternatives. *Proc. ACM Hum. Comput. Interact.* **2018**, *2*. [CrossRef]

40. Sanches, P.; Höök, K.; Sas, C.; Ståhl, A. Ambiguity as a Resource to Inform Proto-Practices: The Case of Skin Conductance. *ACM Trans. Comput. Hum. Interact.* **2019**, *26*. [CrossRef]

41. Khut, G.P. Designing Biofeedback Artworks for Relaxation. In Proceedings of the 2016 CHI Conference Extended Abstracts on Human Factors in Computing Systems, San Jose, CA, USA, 7–12 May 2016; pp. 3859–3862. [CrossRef]

42. Howell, N.; Devendorf, L.; Vega Gálvez, T.A.; Tian, R.; Ryokai, K., Tensions of Data-Driven Reflection: A Case Study of Real-Time Emotional Biosensing. In Proceedings of the 2018 CHI Conference on Human Factors in Computing Systems, Montreal, QC, Canada, 21–26 April 2018.

43. Umair, M.; Sas, C.; Latif, M.H. Towards Affective Chronometry: Exploring Smart Materials and Actuators for Real-time Representations of Changes in Arousal. In Proceedings of the 2019 on Designing Interactive Systems Conference, San Diego, CA, USA, 23–28 June 2019. [CrossRef]

44. Costa, J.; Adams, A.T.; Jung, M.F.; Guimbretière, F.; Choudhury, T. EmotionCheck: Leveraging Bodily Signals and False Feedback to Regulate Our Emotions. In Proceedings of the 2016 ACM International Joint Conference on Pervasive and Ubiquitous Computing, Heidelberg, Germany, 12–16 September 2016. [CrossRef]

45. Azevedo, R.T.; Bennett, N.; Bilicki, A.; Hooper, J.; Markopoulou, F.; Tsakiris, M. The calming effect of a new wearable device during the anticipation of public speech. *Sci. Rep.* **2017**, *7*. [CrossRef]

46. Miri, P.; Jusuf, E.; Uusberg, A.; Margarit, H.; Flory, R.; Isbister, K.; Marzullo, K.; Gross, J.J. Evaluating a Personalizable, Inconspicuous Vibrotactile(PIV) Breathing Pacer for In-the-Moment Affect Regulation. In Proceedings of the 2020 CHI Conference on Human Factors in Computing Systems, Honolulu, HI, USA, 25–30 April 2020. [CrossRef]

47. Loke, L.; Schiphorst, T. The Somatic Turn in Human-Computer Interaction. *Interactions* **2018**, *25*, 54–5863. [CrossRef]

48. Windlin, C.; Ståhl, A.; Sanches, P.; Tsaknaki, V.; Karpashevich, P.; Balaam, M.; Höök, K. Soma Bits: Mediating technology to orchestrate bodily experiences. In Proceedings of the Research Through Design RTD Conference, Delft, The Netherlands, 19–22 May 2019. [CrossRef]

49. Alfaras, M.; Tsaknaki, V.; Sanches, P.; Windlin, C.; Umair, M.; Sas, C.; Höök, K. From Biodata to Somadata. In Proceedings of the 2020 CHI Conference on Human Factors in Computing Systems, Honolulu, HI, USA, 25–30 April 2020. [CrossRef]

50. Semmlow, J.L.; Griffel, B. *Biosignal and Medical Image Processing*, 3rd ed.; CRC Press: Boca Raton, FL, USA, 2014.

51. Goldberger, A.L.; Amaral, L.A.N.; Glass, L.; Hausdorff, J.M.; Ivanov, P.C.; Mark, R.G.; Mietus, J.E.; Moody, G.B.; Peng, C.K.; Stanley, H.E. PhysioBank, PhysioToolkit, and PhysioNet: Components of a New Research Resource for Complex Physiologic Signals. *Circulation* **2000**, *101*, 215–220. [CrossRef]

52. da Silva, H.P. Physiological Sensing Now Open to the World: New Resources Are Allowing Us to Learn, Experiment, and Create Imaginative Solutions for Biomedical Applications. *IEEE Pulse* **2018**, *9*, 9–11. [CrossRef] [PubMed]

53. Heikenfeld, J.; Jajack, A.; Rogers, J.; Gutruf, P.; Tian, L.; Pan, T.; Li, R.; Khine, M.; Kim, J.; Wang, J.; et al. Wearable sensors: modalities, challenges, and prospects. *Lab Chip* **2018**, *18*, 217–248. [CrossRef]

54. Giannakakis, G.; Grigoriadis, D.; Giannakaki, K.; Simantiraki, O.; Roniotis, A.; Tsiknakis, M. Review on psychological stress detection using biosignals. *IEEE Trans. Affect. Comput.* **2019**, 1. [CrossRef]

55. da Silva, H.P.; Fred, A.; Martins, R. Biosignals for Everyone. *IEEE Pervasive Comput.* **2014**, *13*, 64–71. [CrossRef]

56. Batista, D.; da Silva, H.P.; Fred, A.; Moreira, C.; Reis, M.; Ferreira, H.A. Benchmarking of the BITalino biomedical toolkit against an established gold standard. *Healthc. Technol. Lett.* **2019**, *6*, 32–36. [CrossRef] [PubMed]

57. Can, Y.S.; Chalabianloo, N.; Ekiz, D.; Ersoy, C. Continuous Stress Detection Using Wearable Sensors in Real Life: Algorithmic Programming Contest Case Study. *Sensors* **2019**, *19*, 1849. [CrossRef]

58. Can, Y.S.; Chalabianloo, N.; Ekiz, D.; Fernandez-Alvarez, J.; Repetto, C.; Riva, G.; Iles-Smith, H.; Ersoy, C. Real-Life Stress Level Monitoring using Smart Bands in the Light of Contextual Information. *IEEE Sens. J.* **2020**, *20*. [CrossRef]

59. Can, Y.S.; Chalabianloo, N.; Ekiz, D.; Fernandez-Alvarez, J.; Riva, G.; Ersoy, C. Personal Stress-Level Clustering and Decision-Level Smoothing to Enhance the Performance of Ambulatory Stress Detection With Smartwatches. *IEEE Access* **2020**, *8*, 38146–38163. [CrossRef]

60. Umair, M.; Latif, M.H.; Sas, C. Dynamic Displays at Wrist for Real Time Visualization of Affective Data. In Proceedings of the 2018 ACM Conference Companion Publication on Designing Interactive Systems, Hong Kong, China, 9–13 June 2018. [CrossRef]

61. PLUX S.A. biosignalsplux. Available online: http://www.biosignalsplux.com (accessed on 20 May 2020).

62. EMPATICA. E4 Wristband. Available online: https://e4.empatica.com/e4-wristband (accessed on 10 June 2020).

63. Seeed Studio. Arduino Grove GSR. 2014. Available online: https://wiki.seeedstudio.com/Grove-GSR_Sensor/ (accessed on 10 June 2020).

64. Samsung. Gear S2. 2015. Available online: https://www.samsung.com/wearables/gear-s2/features (accessed on 10 June 2020).

65. Massaroni, C.; Nicolò, A.; Presti, D.L.; Sacchetti, M.; Silvestri, S.; Schena, E. Contact-Based Methods for Measuring Respiratory Rate. *Sensors* **2019**, *19*, 908. [CrossRef]

66. Alexander, J.; Follmer, S.; Hornbaek, K.; Roudaut, A. Shape-Changing Interfaces (Dagstuhl Seminar 17082). *Dagstuhl Rep.* **2017**, *7*, 102–108. [CrossRef]

67. Höök, K.; Caramiaux, B.; Erkut, C.; Forlizzi, J.; Hajinejad, N.; Haller, M.; Hummels, C.; Isbister, K.; Jonsson, M.; Khut, G.; et al. Embracing First-Person Perspectives in Soma-Based Design. *Informatics* **2018**, *5*, 8. [CrossRef]

68. Akner-Koler, C.; Ranjbar, P. Integrating Sensitizing Labs in an Educational Design Process for Haptic Interaction. *Formakademisk Forskningstidsskrift Des. Des.* **2016**, *9*. [CrossRef]

69. Ståhl, A.; Jonsson, M.; Mercurio, J.; Karlsson, A.; Höök, K.; Johnson, E.C.B. The Soma Mat and Breathing Light. In Proceedings of the 2016 CHI Conference Extended Abstracts on Human Factors in Computing Systems—CHI EA '16, San Jose, CA, USA, 7–12 May 2016. [CrossRef]

70. Søndergaard, M.L.J.; Hansen, L.K. Intimate Futures: Staying with the Trouble of Digital Personal Assistants through Design Fiction. In Proceedings of the 2018 Designing Interactive Systems Conference, Hong Kong, China, 9–13 June 2018. [CrossRef]

71. EmbrLabs. Embr Wave: Personal Thermostat that You Wear as a Bracelet. Available online: https://embrlabs.com/embr-wave/ (accessed on 10 June 2020).

72. Umair, M.; Alfaras, M.; Gamboa, H.; Sas, C. Experiencing Discomfort: Designing for Affect from First-Person Perspective. In Proceedings of the 2019 ACM International Joint Conference on Pervasive and Ubiquitous Computing and Proceedings of the 2019 ACM International Symposium on Wearable Computers, London, UK, 9–13 September 2019; pp. 1093–1096. [CrossRef]

73. Umair, M.; Sas, C.; Alfaras, M. ThermoPixels: Toolkit for Personalizing Arousal-Based Interfaces through Hybrid Crafting. In Proceedings of the 2020 ACM Designing Interactive Systems Conference, Eindhoven, The Netherlands, 6–10 July 2020; pp. 1017–1032. [CrossRef]

74. Koole, S.L.; Tschacher, W. Synchrony in psychotherapy: A review and an integrative framework for the therapeutic alliance. *Front. Psychol.* **2016**, *7*, 1–17. [CrossRef]

75. Ajax Sound Studio. Cecilia. 2015. Available online: http://ajaxsoundstudio.com/software/cecilia/ (accessed on 9 September 2020).

76. Roads, C. Introduction to Granular Synthesis. *Comput. Music. J.* **1988**, *12*, 11. [CrossRef]

77. Puckette, M. Pure Data: another integrated computer music environment. In Proceedings of the Second Intercollege Computer Music Concerts, Tachikawa, Japan, 7 May 1997; pp. 37–41.

78. PureData. Available online: http://www.puredata.info (accessed on 9 September 2020).

79. Schmeder, A.; Freed, A.; Wessel, D. Best Practices for Open Sound Control. In Proceedings of the Linux Audio Conference, Utrecht, The Netherlands, 1–4 May 2010.

80. Introduction to OSC | opensoundcontrol.org. Available online: http://opensoundcontrol.org/introduction-osc (accessed on 9 September 2020).

81. Fdili Alaoui, S. Making an Interactive Dance Piece: Tensions in Integrating Technology in Art. In Proceedings of the 2019 on Designing Interactive Systems Conference, San Diego, CA, USA, 23–28 June 2019. [CrossRef]

82. Tennent, P.; Marshall, J.; Tsaknaki, V.; Windlin, C.; Höök, K.; Alfaras, M. Soma Design and Sensory Misalignment. In Proceedings of the 2020 CHI Conference on Human Factors in Computing Systems, Honolulu, HI, USA, 25–30 April 2020; pp. 1–12. [CrossRef]

83. Tangible Media. MIT Media Lab. Project Materiable. 2016. Available online: http://tangible.media.mit.edu/project/materiable (accessed on 3 July 2020).

84. Nakagaki, K.; Vink, L.; Counts, J.; Windham, D.; Leithinger, D.; Follmer, S.; Ishii, H. Materiable: Rendering Dynamic Material Properties in Response to Direct Physical Touch with Shape Changing Interfaces. In Proceedings of the 2016 CHI Conference on Human Factors in Computing Systems, San Jose, CA, USA, 7–12 May 2016. [CrossRef]

85. Carlo Ratti Associati. Project Lift-Bit. 2016. Available online: https://carloratti.com/project/lift-bit/ (accessed on 3 July 2020).

86. Paul, R. Lift-Bit Is the World's First Digitally-Transformable Sofa. 2016. Available online: https://www.6sqft.com/lift-bit-is-the-worlds-first-digitally-transformable-sofa/ (accessed on 3 July 2020).

87. Goodfellow, I.; Bengio, Y.; Courville, A. *Deep Learning*; The MIT Press: Cambridge, MA, USA, 2016.

88. Patanè, A.; Kwiatkowska, M. Calibrating the Classifier: Siamese Neural Network Architecture for End-to-End Arousal Recognition from ECG. In *Machine Learning, Optimization, and Data Science*; Springer International Publishing: Cham, Switzerland, 2019; pp. 1–13. [CrossRef]

89. Patanè, A.; Ghiasi, S.; Scilingo, E.P.; Kwiatkowska, M. Automated Recognition of Sleep Arousal Using Multimodal and Personalized Deep Ensembles of Neural Networks. In Proceedings of the 2018 Computing in Cardiology Conference (CinC), Maastricht, The Netherlands, 23–26 September 2018. [CrossRef]

90. Fails, J.A.; Olsen, D.R. Interactive Machine Learning. In Proceedings of the 8th International Conference on Intelligent User Interfaces, Miami Beach, FL, USA, 12–15 January 2003. [CrossRef]

91. Fiebrink, R.A. Real-Time Human Interaction with Supervised Learning Algorithms for Music Composition and Performance. Ph.D. Thesis, Princeton University, Princeton, NJ, USA, 2011. [CrossRef]

92. Carney, M.; Webster, B.; Alvarado, I.; Phillips, K.; Howell, N.; Griffith, J.; Jongejan, J.; Pitaru, A.; Chen, A. Teachable Machine: Approachable Web-Based Tool for Exploring Machine Learning Classification. In Proceedings of the 2020 CHI Conference on Human Factors in Computing Systems, Honolulu, HI, USA, 25–30 April 2020. [CrossRef]

93. Gillies, M. Understanding the Role of Interactive Machine Learning in Movement Interaction Design. *ACM Trans. Comput. Hum. Interact.* **2019**, *26*. [CrossRef]

94. American Psychological Association. Ethical Principles of Psychologists and Code of Conduct. 2016. Available online: http://www.apa.org/ethics/code/ethics-code-2017.pdf (accessed on 20 June 2020).

95. Bowie, D.; Sunram-Lea, S.I.; Iles-Smith, H. A Systemic Ethical Framework for Mobile Mental Health: From Design to Implementation. NIHR MindTech MIC National Symposium 2018: Improving Lives with Digital Mental Healthcare. 2018. Available online: https://eprints.lancs.ac.uk/id/eprint/131847/1/A_systemic_ethical_framework_for_mobile_mental_health_From_design_to_implementation.pdf (accessed on 9 September 2020).

96. British Psychological Society. Codes of Ethics and Conduct. 2018. Available online: https://www.bps.org.uk/sites/bps.org.uk/files/Policy/Policy-Files/BPSCodeofEthicsandConduct(UpdatedJuly2018).pdf (accessed on 20 June 2020)

97. Qu, C.; Sas, C.; Roquet, C.D.; Doherty, G. Functionality of top-rated mobile apps for depression: Systematic search and evaluation. *JMIR Mental Health* **2020**, *7*. [CrossRef]

Publisher's Note: MDPI stays neutral with regard to jurisdictional claims in published maps and institutional affiliations.

MDPI

St. Alban-Anlage 66

4052 Basel

Switzerland

Tel. +41 61 683 77 34

Fax +41 61 302 89 18

www.mdpi.com

Sensors Editorial Office

E-mail: sensors@mdpi.com

www.mdpi.com/journal/sensors